A History of Modern C̶o̶m̶p̶u̶t̶i̶n̶g̶

A History of Modern Computing

Paul E. Ceruzzi

The MIT Press
Cambridge, Massachusetts
London, England

First MIT Press paperback edition, 2000

© *1998 Massachusetts Institute of Technology*

This book was set in New Baskerville by Techset Composition Ltd., Salisbury, UK, and was printed and bound in the United States of America.

Library of Congress Cataloging-in-Publication Data

Ceruzzi, Paul E.
     A history of modern computing / Paul E. Ceruzzi.
          p.   cm. — (History of computing)
     Includes bibliographical references and index.
     ISBN 0-262-03255-4 (hardcover : alk. paper), 0-262-53169-0 (pb)
     1. Computers—History. 2. Electronic data processing—History.
     I. Title. II. Series.
     QA76.17.C47   1998   98-22856
     004′ .09′045—dc21   CIP

Dedication

I wrote this book in an office at the Smithsonian Institution's National Air and Space Museum, one of the busiest public spaces in the world. On a typical summer day there may be upwards of 50,000 visitors to the museum—the population of a small city. These visitors—with their desire to know something of modern technology—were a great inspiration to me. Their presence was a constant reminder that technology is not just about machines but about people: the people who design and build machines and, more importantly, the people whose lives are profoundly affected by them. It is to these visitors that I respectfully dedicate this book.

Contents

Preface

A few years ago I met with a group of historians who were planning to write a comprehensive history of computing from its origins to recent years. Led by William Aspray, we surveyed the field and each agreed to write about the topics we knew best. After we had each had a chance to start work, we met again and compared what we had done. We found that we had succeeded pretty well in telling the story of computing before 1945; but after 1945—at precisely the moment when electronic digital computing emerged—we had failed. After several more attempts, we gave up. Computing after 1945 was too complex, too daunting, and too little buttressed by a theoretical framework to allow for solid historical assessment (in some form other than simply listing one machine after another, which we refused to do). We decided to publish the work we had done on events before 1945 and to postpone the rest of the project.[1]

I believe that the time is now ripe to write the rest of that book, or at least a major part of it. What follows is a history of computing from the completion of the ENIAC in 1945 to the networks of personal computers at work and home in the 1990s. The focus is on the United States and on computing systems that were commercially sold and installed in large quantities. That leaves out some parts of the story, but I think it covers enough to offer readers a sense of what has happened to computing in the past fifty years.

The pace of innovation—one of the reasons for the failure of the earlier attempt to write this history—has not slowed. But a half-century of innovation has revealed several patterns on which a structure can be built. These patterns seem to have held fast through successive waves of technology, although I must also admit that the next few years may very well transform computing so much as to render some of this narrative obsolete. Future generations may look on the transition from large mainframes to desktop personal computers (a transition that I examine at length in this book) as a sort of nonevent, much as we now look on the development of key-driven, as opposed to cranked, adding machines in the 1880s. "What Is Past Is Prologue"—the inscription that greets visitors to the U.S. National Archives in Washington—applies to the history of computing with a vengeance. Perhaps computing advances will slow to a crawl while the world assimilates the advances of recent years. I do not think so, but that has happened with other technologies.

Regardless of what the coming decades bring, however, I believe that the story of computing from the end of World War II to the mid-1990s will come to be seen as beginning one of the great transformations of American life, and I also believe that now is a good time to start telling that story.

Acknowledgments

Many institutions and people assisted me in the writing of this book.

I received help in locating manuscripts and papers from the National Archives, the Library of Congress Manuscript Division, the Eisenhower Library, the Hagley Museum and Library, the Charles Babbage Institute, The Computer Museum, the San Jose, California, Public Library, the Stanford University Archives, the National Museum of American History Archives, Digital Equipment Corporation Archives, and the French Library.

I received helpful comments on drafts of chapters or the whole manuscript from Tim Bergin, Robert Smith, Bruce Seely, Michael Neufeld, Allan Needell, Jon Eklund, Steve Lubar, anonymous referees at *Technology and Culture* and at *History and Technology*, members of the National Air and Space Museum's Contemporary History Seminar, Bill Aspray, Martin Campbell-Kelly, J. A. N. Lee, Eric Weiss, Gordon and Gwen Bell, Herb Grosch, Mike Williams, Paul Forman, Oscar Blumtritt, John Wharton, and G. P. Zachary.

I was assisted in my photo research by Harold Motin, Toni Thomas, and the staff of the National Air and Space Museum's Photographic Services Department.

I also wish to acknowledge the following persons for their support: Diane Wendt, Peggy Kidwell, Connie Carter, Michael Mahoney, Alice Jones, and Jamie Parker Pearson. I wish also to thank Larry Cohen, Sandra Minkkinen, and Chryseis O. Fox, of the MIT Press, for their hard work and dedication to this project.

Financial Support came in part from the Smithsonian Institution's Regents' Publication Fund.

Introduction: Defining "Computer"

Computers were invented to "compute": to solve "complex mathematical problems," as the dictionary still defines that word.[1] They still do that, but that is not why we are living in an "Information Age." That reflects *other* things that computers do: store and retrieve data, manage networks of communications, process text, generate and manipulate images and sounds, fly air and space craft, and so on. Deep inside a computer are circuits that do those things by transforming them into a mathematical language. But most of us never see the equations, and few of us would understand them if we did. Most of us, nevertheless, participate in this digital culture, whether by using an ATM card, composing and printing an office newsletter, calling a mail-order house on a toll-free number and ordering some clothes for next-day delivery, or shopping at a mega-mall where the inventory is replenished "just-in-time." For these and many other applications, we can use all the power of this invention without ever seeing an equation. As far as the public face is concerned, "computing" is the least important thing that computers do.

But it was to solve equations that the electronic digital computer was invented. The word "computer" originally meant a *person* who solved equations; it was only around 1945 that the name was carried over to machinery.[2]

That an invention should find a place in society unforeseen by its inventors is not surprising.[3] The story of the computer illustrates that. It is not that the computer ended up not being used for calculation—it *is* used for calculation by most practicing scientists and engineers today. That much, at least, the computer's inventors predicted. But people found ways to get the invention to do a lot more. How they did that, transforming the mathematical engines of the 1940s to the networked information appliance of the 1990s, is the subject of this book.

Figure 0.1
Human "computers" at work at North American Aviation, Los Angeles, in the early 1950s. The two women in the lower center of the photo are using Friden calculators, and the man at the lower left is looking up a number on a slide rule. The rear half of the room is set up for drafting. Absent from this room are any punched-card machines, although aircraft engineers did use them for some applications, as described in the text. (*Source:* NASM.)

The Computer Revolution and the History of Technology

In the early 1980s, when I had just taken my first job as a member of the history department of a state university, I mentioned to one of my colleagues that I was studying the history of computing. "Why computing?" he replied. "Why not study the history of washing machines?" I thought he was joking, maybe making fun of the greenhorn just arrived in the faculty lounge. But he was serious. After all, he had a washing machine at home, it was a complex piece of technology, and its effect on his daily life was profound. Surely it had a history. But computers? Those were exotic things he had heard of but experienced only indirectly.

In the 1990s that question would not be asked, because few would argue that computers are not important. We live in an age transformed by computing.[4] This is also the reason why we need to understand its origins. But terms like "Information Age" or "Computer Revolution" are not ones I like. They mislead as much as inform. Technological revolutions certainly do occur, though not often. The story of how a new technology finds its place in a society is always more subtle and complex than implied by the phrase "X Revolution," or "X Age," where "X" stands for jet aircraft, nuclear energy, automobiles, computers, information, space, the Internet, microelectronics, and so on.[5] The daily press tends to overstate the possible effects of each new model of a chip, each new piece of software, each new advance in networking, each alliance between computing and entertainment firms: surely they will change our lives for the better. A few weeks later the subject of these glowing reports is forgotten, replaced by some new development that, we are assured, is the *real* turning point.[6]

Yet who would deny that computing technology has been anything short of revolutionary? A simple measure of the computing abilities of modern machines reveals a rate of advance not matched by other technologies, ancient or modern. The number of computers installed in homes and offices in the United States shows a similar rate of growth, and it is not slowing down. Modern commercial air travel, tax collection, medical administration and research, military planning and operations—these and a host of other activities bear the stamp of computer support, without which they would either look quite different or not be performed at all. The history of computing commands—as it probably should—more attention from the public than the history of the washing machine. The colleague who in 1981 dismissed the study of computing no longer prepares his papers on a manual typewriter, I suspect. Historians are among the most fanatic in embracing the latest advances in computer-based aids to scholarship.[7]

Is the electronic computer only one of many large-scale, high-technology systems that have shaped the twentieth century? To what extent is it unique as an information-processing machine? To what extent is computing after 1945 different from the information-handling activities of an earlier age? The popular literature tends to stress computing's uniqueness, hand in hand with breathless accounts of its revolutionary impacts. Some writers cast this revolution as a takeover by a "clean" technology, with none of the pollution or other side effects of the technologies of the Iron Age.[8] If the computer is revolutionizing our lives, who is on the losing side; who are the loyalists that computing must banish from this new world? Or is computing like the ruling party of Mexico: a permanent, benign, institutionalized "revolution"? The narrative that follows will, I hope, provide enough historical data to answer these questions, at least tentatively.[9]

Current studies of computing give conflicting answers to these questions. Some show the many connections between modern computing and the information-handling machinery and social environments that preceded it.[10] Some make passing references to computing as one of many technologies that owe their origins to World War II research. Many stress the distinction between computing and other products of wartime weapons laboratories; few examine what they have in common.[11] Still others make little attempt to discover any connection at all.[12]

In writing about the emergence of electrical power systems in the United States and Europe, Thomas Parke Hughes introduced the notion of technological systems, into which specific pieces of machinery must fit.[13] His work is too rich and complex to be summarized here, but a few aspects are particularly relevant to the history of computing. One is that "inventors" include people who innovate in social, political, and economic, as well as in technical, arenas. Sometimes the inventor of a piece of hardware is also the pioneer in these other arenas, and sometimes not. Again and again in the history of computing, especially in discussing the rise of Silicon Valley, we shall encounter an entrepreneur with a complex relationship to a technical innovator. This narrative will also draw on another of Hughes's insights: that technology advances along a broad front, not along a linear path, in spite of terms like "milestone" that are often used to describe it.

The history of computing presents problems under this systems approach, however. One definition of a modern computer is that it *is* a system: an arrangement of hardware and software in hierarchical layers. Those who work with the system at one level do not see or care about what is happening at other levels. The highest levels are made up of "software"—by definition things that have no tangible form but are best described as methods of organization. Therefore, it might be argued, one need not make any special effort to apply the systems approach to the history of computing, since systems will naturally appear everywhere. This is another example of computing's uniqueness. Nevertheless, the systems approach will be applied in this narrative, because it helps us get away from the view of computing solely as a product of inventors working in a purely technical arena.

Another approach to the history of technology is known as "social construction." Like the systems approach, it is too rich a subject to be summarized here.[14] Briefly a social constructionist approach to the history of computing would emphasize that there is no "best" way to design computing systems or to integrate them into social networks. What emerges as a stable configuration—say, the current use of desktop systems and their software—is as much the result of social and political negotiation among a variety of groups (including engineers) as it is the natural emergence of the most efficient or technically best design. A few historians of computing have adopted this approach,[15] but most have not, preferring to describe computing's history as a series of technical problems met by engineering solutions that in hindsight seem natural and obvious.

However, a body of historical literature that has grown around the more recent history of computing does adopt a social constructionist approach, if only informally. The emergence of personal computing has been the subject of popular books and articles by writers who are either unfamiliar with academic debates about social construction or who know of it but avoid presenting the theory to a lay audience. Their stories of the personal computer emphasize the idealistic aspirations of young people, mainly centered in the San Francisco Bay area and imbued with the values of the Berkeley Free Speech Movement of the late 1960s. For these writers, the personal computer came not so much from the engineer's workbench as from sessions of the Homebrew Computer Club between 1975 and 1977.[16] These histories tend to ignore advances in fields such as solid state electronics, where technical matters, along with a different set of social forces, played a significant role. They also do little to incorporate the role of the U.S. Defense Department and NASA (two of the largest employers in Silicon Valley) in shaping the technology. These federal agencies represent social and political, not engineering, drivers. I shall draw on Hughes's concepts of social construction and his systems approach throughout the following narrative; and we will find abundant evidence of social forces at work, not only during the era of personal computing but before and after it as well.

Themes

The narrative that follows is chronological, beginning with the first attempts to commercialize the electronic computer in the late 1940s and ending in the mid–1990s, as networked personal workstations became common. I have identified several major turning points, and these get the closest scrutiny. They include the computer's transformation in the late 1940s from a specialized instrument for science to a commercial product, the emergence of small systems in the late 1960s, the advent of personal computing in the 1970s, and the spread of networking after 1985. I have also identified several common threads that have persisted throughout these changes.

The first thread has to do with the internal design of the computer itself: the way that electronic circuits are arranged to produce a machine that operates effectively and reliably. Despite the changes in implementation from vacuum tubes to integrated circuits, the flow of information within a computer, at one level at least, has not changed. This design is known as the "von Neumann Architecture," after John von Neumann (1903–1957), who articulated it in a series of reports written in 1945 and 1946.[17] Its persistence over successive waves of changes in underlying hardware and software provides the historian with at least one path into the dense forest of recent history. How successive generations of machines departed from the concepts of 1945, while retaining their essence, also forms a major portion of the story.

Many histories of computing speak of three "generations," based on whether a computer used vacuum tubes, transistors, or integrated circuits. In fact, the third of these generations has lasted longer than

Figure 0.2
Computing with machines at the same company ten years later. A pair of IBM 7090 computers assist in the design and testing of the rocket engines that will later take men to the Moon and back. The most visible objects in this scene are the magnetic tape drives, suggesting that storage and retrieval of information are as much a part of "computing" as is arithmetic. Obviously fewer people are visible in this room than in the previous photo, but it is worth noting that of the four men visible here, two are employees of IBM, not North American Aviation. (*Source*: Robert Kelly, *Life* Magazine, © Time Inc., 1962.)

the first two combined, and nothing seems to be on the horizon that will seriously challenge the silicon chip. Silicon integrated circuits, encased in rectangular black plastic packages, are soldered onto circuit boards, which in turn are plugged into a set of wires called a bus: this physical structure has been a standard since the 1970s. Its capabilities have progressed at a rapid pace, however, with a doubling of the chip's data storage capacity roughly every eighteen months. Some engineers argue that this pace of innovation in basic silicon technology is the true driving force of history, that it causes new phases of computing to appear like ripe fruit dropping from a tree. This view is at odds with what historians of technology argue, but the degree to which it is accepted and even promoted by engineers makes it a compelling argument that cannot be dismissed without a closer look.

Computing in the United States developed after 1945 in a climate of prosperity and a strong consumer market. It was also during the Cold War with the Soviet Union. How the evolution of computing fits into that climate is another theme of this story. The ENIAC itself, the machine that began this era, was built to meet a military need; it was followed by other military projects and weapons systems that had a significant impact on computing: Project Whirlwind, the Minuteman ballistic missile, the Advanced Research Projects Agency's ARPANET, and others. At the same time, the corporation that dominated computing, IBM, built its wealth and power by concentrating on a commercial rather than a military market, although it too derived substantial revenues from military contracts. In the 1970s, as the military was subsidizing computer development, another arm of the U.S. government, the Justice Department, was attempting to break up IBM, charging that it had become too big and powerful.

The military's role in the advancement of solid state electronics is well known, but a closer look shows that role to be complex and not always beneficial.[18] The term "military" is misleading: there is no single military entity but rather a group of services and bureaus that are often at odds with one another over roles, missions, and funding. Because the military bureaucracy is large and cumbersome, individual "product champions" who can cut through red tape are crucial. Hyman Rickover's role in developing nuclear-powered submarines for the Navy is a well-known example. Military support also took different forms. At times it emphasized basic research, at others specific products. And that relationship changed over the decades, against a backdrop of, first, the nascent Cold War, then the Korean conflict, the Space Race, the Viet

Nam War, and so on. From 1945 onward there have always been people who saw that computing technology could serve military needs, and that it was therefore appropriate to channel military funds to advance it. Not everyone shared that view, as the following chapters will show; but military support has been a constant factor.

The breakup of the Soviet Union and the end of the Cold War have brought into focus some aspects of that conflict that had been hidden or suppressed. One was the unusually active role played by scientists and university researchers in supporting the effort.[19] Another was the unique role that information, and by implication, information-handling machines, played. Information or "intelligence" has been a crucial part of all warfare, but as a means to an end. In the Cold War it became an end in itself. This was a war of code-breaking, spy satellites, simulations, and "war games." Both science-based weapons development and the role of simulation provided a strong incentive for the U.S. Defense Department to invest heavily in digital computing, as a customer and, more importantly, as a source of funds for basic research.[20]

The role of IBM, which dominated the computer industry from about 1952 through the 1980s, is another recurring theme of this narrative. Its rise and its unexpected stumble after 1990 have been the subject of many books. IBM itself sponsored a series of excellent corporate histories that reveal a great deal about how it operated.[21] One issue is how IBM, a large and highly structured organization with its own first-class research facilities, fared against start-up companies led by entrepreneurs such as William Gates III, William Norris, Ken Olsen, or Max Palevsky. These people were able to surmount the barriers to entry into the computer business that IBM erected, while a host of others tried and failed. What, besides luck, made the difference? How did IBM continue to dominate in an environment of constant and often disruptive technological innovation? Unlike the start-up companies, IBM had an existing customer base to worry about, which prevented it from ever starting with a clean slate. Its engineering and sales force had to retain continuity with the punched-card business that had been its prewar mainstay, even as its own research laboratories were developing new technologies that would render punched cards obsolete. Likewise the start-up companies, once they became established and successful, faced the same problem of dealing with new technology. We may thus compare IBM's strategy with the strategies of its new competitors, including Digital Equipment Corporation, Wang Labs, Control Data, and Microsoft.

Another theme concerns a term, unknown in the late 1940s, that dominates computing in the 1990s: software.[22] Chapter 3 chronicles the early development of software, but the topic crops up occasionally before that, and in the chapters that follow it appears with more frequency. In the 1950s computer companies supplied system software as part of the price of a computer, and customers developed their own applications programs. More than one purchaser of an early computing system winced at the army of systems analysts, programmers, and software specialists that had to be hired into the company to manage a machine that was supposed to eliminate clerical workers. It was not until 1990 that commercial software came to the fore of computing, as hardware prices dropped and computer systems became more reliable, compact, and standardized.

The literature on the history of computing recognizes the importance of software, but this literature is curiously divided into two camps, neither of which seems to recognize its dependence on the other. In one camp we find a glut of books and magazine articles about personal computer software companies, especially Microsoft, and the fortunes made in selling the DOS and Windows operating systems for PCs. Some chronicle the history of UNIX, an influential operating system that has also had success in the commercial marketplace. These accounts lack balance. Readers are naturally interested in the enormous sums of money changing hands, but what does this software do, and why do the operating systems look the way they do? Moreover, few of these chronicles connect these systems to the software developed in the first two decades of computing, as if they had nothing to do with each other. In fact, there are strong connections.

Another camp adheres to higher standards of scholarship and objectivity, and gives appropriate emphasis to computing before the advent of the PC. But this body of literature has concentrated its efforts on programming languages. In a sense, this approach mirrors activity in computing itself in the early days, when it was not hard to find people working on new and improved programming languages, but was hard to find people who worried about integrating these languages into systems that got work done efficiently and made good use of a customer's time.[23] We now know a great deal about the early development of FORTRAN, BASIC, COBOL, and a host of other more obscure languages, yet we know little of the systems those languages were a part of.

A final theme is the place of information in a democratic society. Computers share many values associated with the printing press, the freedom of which is guaranteed by the First Amendment to the U.S. Constitution. But computers are also agents of control.[24] Are the two attributes at odds with each other? The first customers for commercial computers were military or government agencies, who hoped these machines could manage the information that was paralyzing their operations; at the same time, the popular press was touting "automation" as the agent of a new era of leisure and affluence for American workers. Project Whirlwind led, on the one hand, to SAGE, a centralized command-and-control system whose structure mirrored the command structure of the Air Force, which funded it; on the other, it led to the Digital Equipment Corporation, a company founded with the goal of making computers cheaper and more accessible to more people. The notion of personal computers as liberating technology will be discussed in detail in chapter 7; the question we shall ask is whether those ideals were perverted as PCs found their way into corporate offices in the 1980s. We shall also see that these same issues have surfaced once again as the Internet has exploded into a mass market.

Computer software is very much a part of this narrative, but one facet of software development is excluded—Artificial Intelligence (AI). AI explores the question of whether computers can perform tasks that, if done by human beings, would be universally regarded as indicating intelligence. Machine intelligence was first believed to be a problem of hardware, but for most of this history AI research has dealt with it by writing programs that run on the same stored-program digital computers (perhaps with some enhancements) that are made and sold for other applications. Artificial Intelligence spans a wide range—from fairly prosaic applications in daily commercial use to philosophical questions about the nature of humanity. What defines AI research is constantly changing: cheap pocket chess-playing machines are not AI, for example, but advanced chess playing by computer still is. To paraphrase Alan Turing, the history of AI is perhaps better written by a computer than by a person.

This book focuses on the history of computing as it unfolded in the United States. Western Europe, especially England, was also a site where pioneering electronic computing machines were built, first for military and then for commercial customers. By December 1943, when construction of the ENIAC had only begun, the British already had at least one electronic "Colossus" in operation. The British catering firm J. Lyons & Company had installed and was using a commercial computer, the LEO, well before the American UNIVACs found their first customers.[25] In

Germany, Konrad Zuse was also taking steps to commercialize inventions he had created for the German military during World War II. By the late 1950s, though, whatever lead the Europeans had was lost to American companies. The economist Kenneth Flamm suggests one reason for this: "European governments provided only limited funds to support the development of both electronic component and computer technology in the 1950s and were reluctant to purchase new and untried technology for use in their military and other systems."[26] There was little of the easy flow of information—and more important, people—between military and commercial computing in Europe. The following narrative will occasionally address European contributions, but for reasons of space will not chronicle the unfolding of the computer industry there.

This narrative will also touch only lightly on the history of computing in Japan. That story is different: Japan had a late start in computing, never producing vacuum tube computers at all. Japanese firms made remarkable advances in integrated circuit production, however, and had established a solid place in portions of the industry by the 1980s. The announcement in the early 1980s of a Japanese "Fifth Generation" program, intended to leapfrog over U.S. software expertise, created a lot of anxiety in the United States, but the United States retained its leadership in software into the 1990s.[27] How Japanese firms gained a foothold is discussed briefly in chapter 5.

The end of the Cold War, and with it the opening of Soviet archives, may help us better understand the development of computing in the U.S.S.R. Throughout this era the Soviets remained well behind the United States in computing.[28] So far the reasons that have been given tend to be post hoc: because it was so, therefore it had to be so. But what of Soviet achievements in pure mathematics and physics, as well as in developing ballistic missiles, nuclear weapons, space exploration, and supersonic aircraft? One might expect that the Soviet military would have supported computing for the same reasons the U.S. Air Force supported Whirlwind and SAGE. We know that Soviet scientists began work on advanced digital computers as soon as the ENIAC was publicized. Yet when they needed advanced machines, the Soviets turned to their East European satellites (especially Hungary and Czechoslovakia), or else they reverse-engineered U.S. computers such as the IBM System/360 and the VAX. Building copies of these computers gave them access to vast quantities of software, which they could acquire by a purchase on the open market, or by espionage, but it also meant that they remained one or two hardware generations behind the United States.

Perhaps it was the perception that computers, being instruments that facilitate the free exchange of information, are antithetical to a totalitarian state. But U.S. computing from 1945 through the 1970s was dominated by large, centralized systems under tight controls, and these were not at odds with the Soviet political system. Such computers would have been perfect tools to model the command economy of Marxism-Leninism. Soviet planners would not have been alone. Throughout this era some Americans embraced computers for their potential to perform centralized economic modeling for the United States—with constitutional rights guaranteed, of course.[29] Perhaps the reason was the other side of the Western European coin: plenty of military support, but no transfer to a market-driven computer industry. Americans may have found that military support was "just right": enough to support innovation but not so focused on specific weapons systems as to choke off creativity. More research on the history of Soviet computing needs to be done.

Most of us know that computers are somehow different from washing machines in the ways they are affecting modern life. This book concludes with some observations about why that might be so. Throughout the narrative I question whether the computer is itself the impersonal agent of change, or even whether it is an autonomous force that people can do little to affect, much less resist. In my conclusion I revisit that question. I do not have an answer. My hope is that the chronicle presented in these chapters will enlighten those of us, lay and professional, who continue to ask.

1

The Advent of Commercial Computing, 1945–1956

"[Y]ou . . . fellows ought to go back and change your program entirely, stop this . . . foolishness with Eckert and Mauchly." That was the opinion of Howard Aiken, Harvard mathematician and builder of the Mark I calculator, expressed to Edward Cannon of the U.S. National Bureau of Standards in 1948. Aiken made that remark as a member of a National Research Council committee that had just recommended that the Bureau of Standards not support J. Presper Eckert and John Mauchly's proposal to make and sell electronic computers (figure 1.1). In Aiken's view, a commercial market would never develop; in the United States there was a need for perhaps for five or six such machines, but no more.[1]

Howard Aiken was wrong. There turned out to be a market for millions of electronic digital computers by the 1990s, many of them personal devices that fit easily into a briefcase. That would not have happened were it not for advances in solid state physics, which provided a way of putting the circuits of a computer on a few chips of silicon. Nevertheless, the nearly ubiquitous computers of the 1990s are direct descendants of what Eckert and Mauchly hoped to commercialize in the late 1940s.

The Eckert-Mauchly Computer Corporation did not remain an independent entity for long; it was absorbed by Remington Rand and became a division of that business-machine company. Eckert and Mauchly's computer, the UNIVAC, was a technical masterpiece but was eclipsed in the market by computers made by Remington-Rand's competitor, IBM. So one could say that they were indeed foolish in their underestimation of the difficulties of commercializing their invention. What was not foolish was their vision, not only of how to design and build a computer but also of how a society might benefit from large numbers of them.

Figure 1.1
Staff of the Eckert-Mauchly Computer Corporation, ca. 1948, in Philadelphia.
Eckert is at the lower left; Mauchly at the lower right. The apparatus behind
them is a portion of the BINAC, which the company was building for the
Northrop Aircraft Company. *Back row, left to right*: Albert Auerbach, Jean Bartik,
Marvin Jacoby, John Sims, Louis Wilson, Robert Shaw, Gerald Smoliar. *Front row*:
J. Presper Eckert, Frazier Welsh, James Wiener, Bradford Sheppard, John
Mauchly. (*Source*: Unisys Corporation.)

Computing after 1945 is a story of people who at critical moments
redefined the nature of the technology itself. In doing so they opened
up computing to new markets, new applications, and a new place in the
social order. Eckert and Mauchly were the first of many who effected
such a transformation. They took an expensive and fragile scientific
instrument, similar to a cyclotron, and turned it into a product that
could be manufactured and sold, if only in small quantities.[2] In the mid-
1950s the IBM Corporation developed a line of products that met the
information-handling needs of American businesses. A decade later,
alumni from MIT's Project Whirlwind turned the computer into a device
that one interacted with, a tool with which to augment one's intellectual
efforts. In the mid-1970s, a group of hobbyists and enthusiasts transformed it into a personal appliance. Around 1980, it was transformed

from a piece of specialized hardware to a standardized consumer product defined by its now-commercialized software. In the 1990s it is going through another transformation, turning into an agent of a worldwide nexus, a communications medium. The "computer age"— really a series of "computer ages"—was not just invented; it was willed into existence by people who wanted it to happen. This process of reinvention and redefinition is still going on.

The UNIVAC in Context

Eckert and Mauchly brought on the first of these transformations in 1951 with a computer they called "UNIVAC." The acronym came from "Universal Automatic Computer," a name that they chose carefully. "Universal" implied that it could solve problems encountered by scientists, engineers, and businesses. "Automatic" implied that it could solve complex problems without requiring constant human intervention or judgment, as existing techniques required. Before discussing its creation, one needs to understand how computing work was being done in different areas and why a single machine, a UNIVAC, could serve them equally well. One must also understand how existing calculating machines, the results of decades of refinement and use, were deficient. It was that deficiency that made room for the UNIVAC, which broke with past practices in many ways.

Punched Cards

During the Second World War, Eckert and Mauchly designed and built the ENIAC at the University of Pennsylvania's Moore School of Electrical Engineering. The ENIAC was an electronic calculator that inaugurated the era of digital computing in the United States. Its purpose was to calculate firing tables for the U.S. Army, a task that involved the repetitive solution of complex mathematical expressions. It was while working on this device that they conceived of something that had a more universal appeal.

The flow of information through the UNIVAC reflected Eckert and Mauchly's background in physics and engineering. That is, the flow of instructions and data in the UNIVAC mirrored the way humans using mechanical calculators, books of tables, and pencil and paper performed scientific calculations.[3] Although the vacuum tube circuits might have appeared novel, a scientist or engineer would not have found anything unusual in the way a UNIVAC attacked a problem.

However, those engaged in business calculations, customers Eckert and Mauchly also wanted their machine to serve, would have found the UNIVAC's method of processing unusual.[4] In the late nineteenth century, many businesses adopted a practice that organized work using a punched card machine; typically an ensemble of three to six different punched-card devices would comprise an installation.[5] To replace these machines with a computer, the business had also to adopt the UNIVAC's way of processing information. Punched-card machines are often called "unit record equipment." With them, all relevant information about a particular entity (e.g., a sales transaction) is encoded on a single card that can serve multiple uses by being run through different pieces of equipment; for example, to count, sort, tabulate, or print on a particular set of columns.[6] Historical accounts of punched-card machinery have described in great detail the functioning of the individual machines. More relevant is the "architecture" of the entire room—including the people in it—that comprised a punched-card installation, since it was that room, not the individual machines, that the electronic computer eventually replaced.

In a typical punched-card installation, the same operation was performed on all the records in a file as a deck of cards went through a tabulator or other machine (figure 1.2). The UNIVAC and its successors could operate that way, but they could also perform a long sequence of operations on a single datum before fetching the next record from memory. In punched-card terms, that would require carrying a "deck" of a single card around the room—hardly an economical use of the machinery or the people. Processing information gathered into a deck of cards was entrenched into business practices by the mid-1930s, and reinforced by the deep penetration of the punched-card equipment salesmen into the accounting offices of their customers.[7]

By the 1930s a few scientists, in particular astronomers, began using punched-card equipment for scientific problems. They found that it made sense to perform sequences of operations on each datum, since often the next operation depended on the results of the previous one. One such person was Wallace Eckert (no relation to J. Presper Eckert), who with the aid of IBM established the Thomas J. Watson Computing Bureau at Columbia University in New York in 1934. In 1940 he summarized his work in an influential book, *Punched Card Methods in Scientific Computation*. In it, he explained that punched-card machines "are all designed for computation where each operation is done on

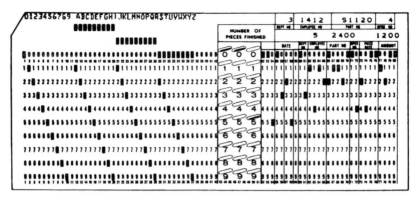

**WHAT THE
PUNCHED HOLE WILL DO**

1 It will add itself to something else.
2 It will subtract itself from something else.
3 It will multiply itself by something else.
4 It will divide itself by something else.
5 It will list itself.
6 It will reproduce itself.
7 It will classify itself.
8 It will select itself.
9 It will print itself on an IBM card.
10 It will produce an automatic balance forward.
11 It will file itself.
12 It will post itself.
13 It will reproduce and print itself on the end
of a card.
14 It will be punched from a pencil mark on the
card.
15 It will cause a total to be printed.
16 It will compare itself to something else.
17 It will cause a form to feed to a predeter-
mined position, or to be ejected automatically,
or to space one position to another.

The IBM card has 80 vertical columns. Each column accommo-
dates a hole (or holes) represnting a single number or letter.
Information to be included in the card is determined by requirements of the final reports and documents.

Data to be processed by IBM machines must be punched in the
card according to a standard arrangement. Consequently, columns of the card are grouped and reserved for the recording of
each fact about a business transaction.

An IBM card – once punched and verified – is a permanent
record. It can be read by machines to do transcribing and
other processing at high speed.

Figure 1.2
IBM punched card. From IBM Corporation, "IBM Data Processing Functions,"
Brochure 224-8208-5, ca. 1963. (*Source*: IBM Corporation.)

many cards before the next operation is begun."[8] He emphasized how one could use existing equipment to do scientific work, but he stated that it was not worth the "expense and delay involved" in building specialized machines to solve scientific problems.[9] A decade later, that was precisely what J. Presper Eckert and John Mauchly were proposing to do—go to great expense and effort to create a "universal" machine that could handle both business and scientific problems.

Ironically, Wallace Eckert was among the first to venture away from traditional punched-card practices and toward one more like the digital computers that would later appear. Despite his recommendation against building specialized equipment, he did have a device called a control switch designed at his laboratory. He installed this switch between the multiplier, tabulator, and summary punch. Its function was to allow short sequences of operations (up to 12) to be performed on a single card before the next card was read.[10] Following his advice, IBM built and installed two specially built punched-card machines at the U.S. Army's Ballistic Research Laboratory at Aberdeen, Maryland. IBM called these machines the "Aberdeen Relay Calculators"; they were later known as the PSRC, for "Pluggable Sequence Relay Calculator."[11]

In late 1945, three more were built for other military labs, and these were even more complex. During the time one of these machines read a card, it could execute a sequence of up to forty-eight steps. More complex sequences-within-sequences were also possible.[12] One computer scientist later noted that this method of programming demanded "the kind of detailed design of parallel subsequencing that one sees nowadays at the microprogramming level of some computers."[13] When properly programmed, the machines were faster than any other nonelectronic calculator. Even after the ENIAC was completed and installed and moved from Philadelphia to Aberdeen, the Ballistic Research Lab had additional Relay Calculators built. They were still in use in 1952, by which time the BRL not only had the ENIAC but also the EDVAC, the ORDVAC (both electronic computers), an IBM Card Programmed Calculator (described next), and the Bell Labs Model V, a very large programmable relay calculator.[14]

The Card-Programmed Calculator

The Aberdeen Relay Calculators never became a commercial product, but they reveal an attempt to adapt existing equipment to post–World War II needs, rather than take a revolutionary approach, such as the

UNIVAC. There were also other punched-card devices that represented genuine commercial alternatives to Eckert and Mauchly's proposed invention. In 1935 IBM introduced a multiplying punch (the Model 601); these soon became popular for scientific or statistical work. In 1946 IBM introduced an improved model, the 603, the first commercial IBM product to use vacuum tubes for calculating. Two years later IBM replaced it with the 604, which not only used tubes but also incorporated the sequencing capability pioneered by the Aberdeen machines. Besides the usual plugboard control common to other punched-card equipment, it could execute up to 60 steps for each reading of a card and setting of the plugboard.[15] The 604 and its successor, the IBM 605, became the mainstays of scientific computing at many installations until reliable commercial computers became available in the mid 1950s. It was one of IBM's most successful products during that era: over 5,000 were built between 1948 and 1958.[16]

One of IBM's biggest engineering customers, Northrop Aircraft of Hawthorne, California, connected a 603 multiplying punch to one of their tabulating machines. That allowed Northrop's users to print the results of a calculation on paper instead of punching them on cards. With a slight further modification and the addition of a small box that stored numbers in banks of relays, the machine could use punched cards run through the tabulator to control the sequences carried out by the multiplier.[17]

Logically, the arrangement was no different from an ordinary punched card installation, except that a set of cables and control boxes replaced the person whose job had been to carry decks of cards from one machine to the next. One of the Northrop engineers recalled years later that they rigged up the arrangement because they were running a problem whose next step depended on the results of the previous step. What this meant was that the normal decks of cards that ran through a machine were reduced to "a batch of one [card], which was awkward."[18] In other words, with cables connecting the machines, the installation became one that executed instructions sequentially and was programmable in a more flexible way than plugging cables.

IBM later marketed a version of this ensemble as the Card-Programmed Calculator (CPC).[19] Perhaps several hundred in all were installed between 1948 and the mid 1950s—far fewer than the thousands of tabulators, punches, and other equipment installed in the traditional way. But even that was many times greater than the number of electronic computer installations worldwide until about 1954. For engineering-oriented companies like Northrop, the CPC filled a pressing need that could not wait for the problems associated with marketing stored-program computers to be resolved.[20]

The Aberdeen calculators and the 604 were transitional machines, between calculators, tabulators, and genuine computers like the UNIVAC. The CPC carried the punched-card approach too far to be of value to computer designers. By the time of its introduction, it was already clear that the design used by the UNIVAC, in which both the instructions and the data were stored in an internal memory device, was superior. The Card-Programmed Calculator's combination of program cards, plugboards, and interconnecting cables was like the epicycles of a late iteration of Ptolemaic cosmology, while the Copernican system was already gaining acceptance.[21] Customers needing to solve difficult engineering problems, however, accepted it. It cost less than the computers then being offered, and it was available. Other southern California aerospace firms besides Northrop carefully evaluated the Card-Programmed Calculator against vendors' claims for electronic computers.[22] Nearly all of them installed at least one CPC.

The Stored-Program Principle

No one who saw a UNIVAC failed to see how much it differed from existing calculators and punched card equipment. It used vacuum tubes—thousands of them. It stored data on tape, not cards. It was a large and expensive system, not a collection of different devices. The biggest difference was its internal design, not visible to the casual observer. The UNIVAC was a "stored program" computer, one of the first. More than anything else, that made it different from the machines it was designed to replace.

The origins of the notion of storing a computer's programs internally are clouded in war-time secrecy. The notion arose as Eckert, Mauchly, and others were rushing to finish the ENIAC to assist the U.S. Army, which was engaged in a ground war in Europe and North Africa. It arose because the ENIAC's creators recognized that while the ENIAC was probably going to work, it was going to be a difficult machine to operate.

Applying the modern term "to program" to a computer probably originated with the ENIAC team at the Moore School. More often, though, they used the phrase "set up" to describe configuring the ENIAC to solve different problems.[23] Setting up the ENIAC meant plugging and unplugging a maze of cables and setting arrays of switches. In effect, the machine had to be rebuilt for each new problem it was to solve. When completed in late 1945, the ENIAC operated much faster than any other machine before it. But while it could solve a complex mathematical problem in seconds, it might take days to set up the machine properly to do that.

It was in the midst of building this machine that its creators conceived of an alternative. It was too late to incorporate that insight into the ENIAC, but it did form the basis for a proposed follow-on machine called the "EDVAC" (Electronic Discrete Variable Computer). In a description written in September of 1945, Eckert and Mauchly stated the concept succinctly: "An important feature of this device was that operating instructions and function tables would be stored exactly in the same sort of memory device as that used for numbers."[24] Six months later, Eckert and Mauchly left the Moore School, and work on the EDVAC was turned over to others (which was mainly why it took five more years to finish building it). The concept of storing both instructions and data in a common storage unit would become basic features of the UNIVAC and nearly every computer that followed.[25]

The stored-program principle was a key to the UNIVAC's success. It allowed Eckert and Mauchly, first of all, to build a computer that had much more general capabilities than the ENIAC, yet required fewer vacuum tubes. It led to the establishment of "programming" (later "software") as something both separate from and as important as hardware design. The basics of this design remained remarkably stable during the evolution of computing from 1945 to 1995. Only toward the end of this period do we encounter significant deviations from it, in the form of "massively parallel" processors or "non–von Neumann" architectures.

John von Neumann's Role

Although Eckert and Mauchly had realized as early as 1944 that computers would need to store information, the "First Draft of a Report on the EDVAC," by John von Neumann, dated June 30, 1945, is often cited as the founding document of modern computing.[26] From it, and a series of reports co-authored by von Neumann a few years later, comes the term "von Neumann Architecture" to describe such a design.[27] According to Herman Goldstine, an army officer assigned to the ENIAC project, John von Neumann (1903–1957) learned of the ENIAC from a chance meeting with him in the summer of 1944 at the

Aberdeen, Maryland, railroad station.[28] Despite his involvement in many other projects, including the design of the atomic bomb, von Neumann was sufficiently intrigued by what was going on at the Moore School to have himself introduced to Eckert and Mauchly and brought onto the project.

Eckert and Mauchly were at that time busy thinking of ways to improve the process of setting up a computer faster.[29] One possibility was to use perforated paper tape to feed instructions, as several relay machines of the 1940s did, but this was too slow for the high speeds of the ENIAC's calculating circuits. So were the decks of cards used by the Card-Programmed Calculator. In Mauchly's words, "calculations can be performed at high speed only if instructions are supplied at high speed."[30]

In the midst of the ENIAC's construction in 1944, Eckert wrote a "Disclosure of a Magnetic Calculating Machine," in which he described the use of "[d]iscs or drums which have at least their outer edge made of a magnetic alloy" on which numbers can be stored.[31] Although it focused on ways of designing a machine that was "speedier, simpler as well as providing features of utility, ruggedness and ease or repair," the disclosure did not articulate the design concepts that later would become known as the stored-program principle.[32] Von Neumann's 1945 Report on the EDVAC went farther—it described a machine in terms of its logical structure rather than its hardware construction. The memorandum that Eckert and Mauchly submitted in September 1945, stated the principle succinctly: they wrote that instructions and numerical data would be stored "in exactly the same sort of memory device."[33]

From the above sequence of reports and memorandums it appears that Eckert and Mauchly had conceived of something like a stored-program principle by 1944, but that it was von Neumann who clarified it and stated it in a form that gave it great force. Von Neumann's international reputation as a mathematician also gave the idea more clout than it might have had coming solely from Eckert and Mauchly, neither of whom were well-known outside the Moore School. Although the term "von Neumann Architecture" is too entrenched to be supplanted, Eckert and Mauchly, who demonstrated such a deep understanding of the nature of electronic computing from an engineering perspective, deserve equal credit.[34]

In the summer of 1946, the Moore School and the U.S. military cosponsored a course on the "Theory and Techniques for Design of Electronic Digital Computers." The course was a recognition of the school's inability to accommodate the numerous requests for information following the public unveiling of the ENIAC.[35] That series of course lectures and the mimeographed reports that appeared a year or two later firmly established the Moore School's approach to computer design. Machines soon appeared that were based on that concept. An experimental computer at the University of Manchester, England, was running test programs by mid-1948. Maurice Wilkes, of Cambridge University, implemented the idea in his EDSAC, operational in the spring of 1949. Eckert and Mauchly completed the BINAC later that year.[36] And of course the UNIVAC would also employ it. Others would continue to propose and build electronic computers of alternate designs, but after the summer of 1946, computing's path, in theory at least, was clear.

The von Neumann Architecture and Its Significance

Before providing a description of the UNIVAC, it is worth a brief look at the essentials of the architecture that von Neumann described in his 1945 report, especially those aspects of it that have remained stable through the past half-century of computer design.

Aside from the internal storage of programs, a major characteristic of a von Neumann computer is that the units that process information are separate from those that store it. Typically there is only a single channel between these two units, through which all transfers of information must go (the so-called von Neumann Bottleneck, about which more later). This feature arose primarily for engineering reasons: it was easier to design storage cells that did not also have to perform arithmetic on their contents.

The main characteristic is that instructions and data are stored in the same memory device, from which any datum can be retrieved as quickly as any other. This concept arose from considering that the processing unit of a computer should not have to sit idle awaiting delivery of the next instruction. Besides that, the ratio of instructions to data usually varies for each problem, so it would not make sense to dedicate separate, expensive storage devices to each. This design implies that one may treat a coded instruction as a piece of data and perform an operation on it, thus changing it into another instruction, but that was not fully understood at first. To give a sense of how this was first implemented, the UNIVAC main store could hold up to 1,000 "words," which could either be numbers (11 digits plus sign), characters (12 characters per word), or instructions (6 characters per instruction; 2 in each word).[37]

Finally, the basic cycle of a von Neumann computer is to transfer an instruction from the store to the processor, decode that instruction, and execute it, using data retrieved from that same store or already present in the processor. Once the processor executed an instruction, it fetched, decoded, and executed another, from the very next position in memory unless directed elsewhere. Having a fast storage device meant that the processor could branch to another stream of instructions quickly whenever it was necessary. Except when explicit branch instructions are encountered, the flow through the instructions stored in the memory was sequential and linear.[38] This concept, of fetching and then executing a linear stream of instructions, is the most lasting of all; even computer designs that purport to be non–von Neumann typically retain the fetch-decode-execute heartbeat of a single-processor machine.[39] As Alan Perils once remarked, "Sometimes I think the only universal in the computing field is the fetch-execute cycle."[40] The UNIVAC could perform this sequence and add two numbers in about half a millisecond.

Since 1990, computer systems with parallel processing structures have become more common, and genuine alternatives to the fetch-execute cycle have been accepted in a few limited markets. Elsewhere the von Neumann architecture, though much modified, prevails. The emergence of practical parallel designs reveals, however, the unifying effect of the von Neumann model as it influenced the computer design of the past five decades.

From ENIAC to UNIVAC: First Transformation[41]

The UNIVAC was going to cut through the Gordian knot of solving complex problems with punched card equipment or plugboard control, and its designers knew that. The ENIAC, though ill-suited for many problems, nevertheless was in such demand that its physical transfer from Philadelphia to Aberdeen had to be put off. With the end of the War there was less urgency to compute firing tables, although the Aberdeen Proving Ground still expected the machine to be moved there for that purpose. After the public unveiling, a flood of interested parties was petitioning to use it. Mauchly reported, for example, that in March of 1948 Pratt & Whitney asked him if they could run an urgent problem "the week of April 17." That gave him a "chuckle"—by 1948 the ENIAC was already fully booked for the next two years![42]

What was less well known was that the Moore School team had carefully evaluated the architecture of the follow-on computer, the EDVAC, in light of the problems it might be expected to solve. Von Neumann found that although it was initially intended for evaluating mathematical expressions, the EDVAC's stored-program design made it "very nearly an 'all-purpose machine'" and that it was better than punched card equipment for sorting data. This was a crucial observation, as sorting was a central task for commercial work, and punched card equipment had been optimized for it.[43]

Still, the climate that surrounded the small group of engineers at the Eckert–Mauchly Computer Corporation was anything but favorable. Many experts were skeptical. Wallace Eckert still felt that modifications to punched card machines, not a radically new and expensive design, would better serve computing's needs. Howard Aiken could not imagine that "the basic logics of a machine designed for the numerical solution of differential equations [could] coincide with the logics of a machine intended to make bills for a department store."[44] Eckert and Mauchly knew otherwise. The UNIVAC's logical structure meant that it could do those things and more. That knowledge drove them and their company through the late 1940s to enter the commercial area, with what eventually became the UNIVAC.

Their drive was matched by an equal, but opposite drive by the University of Pennsylvania to banish commercial interests from the academy. Administrators at Penn did not have the vision of a research university to support technology, which led eventually to the development of areas like Silicon Valley in California and Route 128 in Massachusetts. Irwin Travis, an administrator at the Moore School, asked that members of the staff sign a release form that would prevent them from receiving patent royalties on their inventions. He brooked no discussion. Eckert and Mauchly refused to sign. They resigned on March 31, 1946.[45] The Philadelphia-Princeton region, once a contender for the title of center for computing technology, never recovered.

Eckert and Mauchly could have found work at other universities, or at IBM, but they chose instead the risky course of founding their own company. They formed a partnership, the Electronic Control Company, in 1946; in December 1948 they incorporated as the Eckert–Mauchly Computer Corporation. Added to the engineering problems of designing and building a universal computer and its associated tape drives, memory units, and input-output equipment, was the bigger problem of raising capital. The National Bureau of Standards was encouraging at first; through it Eckert and Mauchly carried out serious discussions with the U.S. Census Bureau. (Census was not allowed to contract for a machine still in development, so the NBS had to be brought in as an intermediary.) The Census Bureau is not usually considered among the technologically astute, but just as it helped inaugurate modern data processing in 1890 by working with Herman Hollerith, Census also helped make electronic computing's transition from the university to the private sector.

Still there were roadblocks. The NBS commissioned a study, which resulted in conservative and skeptical conclusions about electronic computing in general, and Eckert–Mauchly in particular. Another study conducted by the National Research Council in 1947 produced equally negative conclusions, mentioned at the beginning of this chapter. This latter study later became infamous as the source of the statement about how only a few computers would satisfy the world's needs. The search for funds took the fledgling company everywhere: from the American Totalisator Company, who wanted a computer to calculate betting odds at race tracks, to Northrop Aircraft, who wanted an airborne control system for an unmanned, long-range bomber.

Their frantic search for capital makes for a depressing story. But it had a bright side: people wanted this new machine. And as the example of American Totalisator showed, there were many possible customers beyond the obvious ones of the large military or government agencies.

On January 12, 1948, John Mauchly wrote a memorandum to his staff at the Eckert–Mauchly Computer Corporation in which he listed a total of twenty-two industries, government agencies, or other institutions he had contacted. Optimistically he gauged the status of each as a potential customer for a UNIVAC.[46] In the next few years the under-capitalized company would have a great deal of trouble selling UNIVACs. But in the long run, Mauchly was exactly right: each of those industries, and many more, would find compelling reasons to purchase or lease electronic digital computers, if not from Eckert–Mauchly then from someone else. Here are some of the contacts Mauchly listed in his memo:

Prudential. [Edmund C. Berkeley] .... says that considering the number of persons at Prudential who have now expressed themselves in favor of obtaining electronic equipment, he believes there will be no difficulty in getting an order for one UNIVAC.

Oak Ridge .... it was almost 100 percent certain that their purchase order would be approved by Army.

Army Map Service.... Army Map Service has taken an interest in UNIVAC equipment.

Bureau of Aeronautics.... we could possibly obtain a contract.

The Metropolitan Insurance Company has a large problem involving a total file of 18,000,000 policies with 2,000,000 changes per week. There are about twenty digits of information for each policy. It appears that this is a natural application for the UNIVAC.... it would be worthwhile to follow it up.

Presidency College, Calcutta. Professor Mahalanobis...was anxious to contract for a UNIVAC as soon as we were in a position to make definite terms.

Aircraft Companies. A number of aircraft companies are good prospects.... There is no doubt that such companies could use UNIVAC equipment. We have had brief contact with Hughes Aircraft, Glen L. Martin, United Aircraft, North American Aviation, and have been told that Grumman goes in for some rather fancy calculations.

The Information Age had dawned.

UNIVAC

I am pleased that history recognizes the first to invent something, but I am more concerned with the first person to make it work.

—*Grace Hopper*[47]

On March 31, 1951, the Eckert–Mauchly Division of Remington Rand turned over the first UNIVAC to the U.S. Census Bureau. A formal dedication ceremony was held in June at the Division's modest factory in at 3747 Ridge Avenue in Philadelphia. Thus began the era of commercial sales of large-scale stored-program computers in the United States.[48] The event was, however, less of a milestone than it appeared. That first UNIVAC remained at the plant until late December 1952, when it was shipped to Washington. Eckert and Mauchly needed it there: As the only working model of a machine they hoped to sell in quantity, they wanted to show it to other potential customers.[49] And after having gone through heroic efforts to complete and debug the machine, they were apprehensive about dismantling it, moving it, and setting it up again. The first UNIVAC to leave the factory and be installed on a customer's premises was serial #2, installed at the Pentagon for the U.S. Air Force in June 1952.[50] By 1954 about twenty were built and sold, at prices on the order of a million dollars for a complete system.[51] Table 1.1 lists UNIVAC installations from 1951 through 1954.

J. Presper Eckert and John Mauchly, with the help of about a dozen technical employees, designed and built the UNIVAC (figure 1.3). They

Table 1.1
UNIVAC installations, 1951–1954

Date	Customer
Summer 1951	U.S. Census Bureau
late 1952	U.S. Air Force, the Pentagon
late 1952	U.S. Army Map Service
Fall 1953	U.S. AEC, New York, NY (at NYU)
Fall 1953	U.S. AEC, Livermore, CA
Fall 1953	David Taylor Model Basin, Carderock, MD
1954	Remington Rand, New York, NY
1954	General Electric, Louisville, KY
1954	Metropolitan Life, New York, NY
1954	Wright-Patterson AFB, Dayton, OH
1954	U.S. Steel, Pittsburgh, PA
1954	Du Pont, Wilmington, DE
1954	U.S. Steel, Gary, IN
1954	Franklin Life Insurance, Springfield, OH
1954	Westinghouse, Pittsburgh, PA
1954	Pacific Mutual Life Insurance, Los Angeles, CA
1954	Sylvania Electric, New York, NY
1954	Consolidated Edison, New York, NY
1954	Consolidated Edison, New York, NY

Note: This list is compiled from a variety of sources and does not include one or two UNIVACs that were completed but remained with Remington Rand. In some cases the dates are approximate. Depending on how one interprets "installation," the order listed here may be slightly different. UNIVACs were last installed in late 1958 or early 1959.

designed a machine that used four binary digits (bits) to code each decimal digit. In its central processor, four general-purpose accumulators carried out arithmetic. A word was 45 bits long; each word could represent 11 decimal digits plus a sign, or two instructions. The UNIVAC's clock ran at 2.25 MHz, and it could perform about 465 multiplications per second. That was about the same as the ENIAC's multiplication speed; but the UNIVAC's tape system and stored-program architecture made it a much faster machine overall. "Delay Lines" stored 1,000 words as acoustic pulses in tubes of mercury, while magnetic tape units stored up to one million characters on reels of half-inch metal tape.

The UNIVAC was rugged and reliable. Vacuum tube failures, the bane of all early systems, were kept to a reasonably low rate to ensure that the machine would remain useful for practical, day-to-day work. Statistics

Figure 1.3
Grace Murray Hopper and colleagues seated at a UNIVAC console, ca. 1960.
Reels of UNIVAC tape are visible on both sides of the control panel. (*Source*:
Smithsonian Institution photo #83-14878, gift of Grace Murray Hopper.)

gathered by one customer, Metropolitan Life Insurance Company,
showed the central processor was available 81 percent of the time, a
very high figure compared to contemporary vacuum-tube machines.[52]
The Census Bureau said, "We never encountered an incorrect solution
to a problem which we were sure resulted from an internal computer
error."[53] The machine's design reflected Eckert's philosophy of conservative loads on the vacuum tube circuits, plus enough redundancy, to
ensure reliable operation. Its central processor contained over 5,000
tubes, installed in cabinets that were ranged in a 10-foot by 14-foot
rectangle. Inside this rectangle were the mercury delay-line tanks.

Many design features that later became commonplace first appeared
in the UNIVAC: among them were alphanumeric as well as numeric
processing, an extensive use of extra bits for checking, magnetic tapes
for bulk memory, and circuits called "buffers" that allowed high-speed
transfers between the fast delay line and slow tape storage units.[54]

The UNIVAC in Use

A number of UNIVAC customers were private corporations, not military or defense agencies. And of those defense agencies that purchased UNIVACs, many did so for inventory, logistics, and other applications that in many ways were similar to what business customers bought the machine for. In short, and in contrast to the IBM 701 (discussed next), the UNIVAC inaugurated the era of large computers for what is now called "data processing" applications.

For most customers, what was revolutionary about the UNIVAC was not so much its stored-program design or even its electronic processor. It was the use of tape in place of punched cards. To them, the "Automatic" nature of the machine lay in its ability to scan through a reel of tape, find the correct record or set of records, perform some process in it, and return the results again to tape. In a punched card installation, these tasks were performed by people who had to carry large decks of cards from one punched card machine to another. That made punched card processing labor-intensive. Published descriptions of the UNIVAC nearly always referred to it as a "tape" machine. For General Electric, "the speed of computing is perhaps of tertiary importance only."[55] To the extent that its customers perceived the UNIVAC as an "electronic brain," it was because it "knew" where to find the desired data on a tape, could wind or rewind a tape to that place, and could extract (or record) data automatically. Customers regarded the UNIVAC as an information processing system, not a calculator. As such, it replaced not only existing calculating machines, but also the people who tended them.

The Census Bureau, which had been pivotal in getting the fledgling computer company going, hoped to use the UNIVAC for tabulating the 1950 Census. By the time it received its machine in 1951, however, much of the work had already been put on punched card machines for processing. In fact, the Census Bureau had to step aside while the U.S. Air Force and the Atomic Energy Commission commandeered the first machine off the production line, UNIVAC 1, for problems deemed more urgent by the federal government.[56]

Nevertheless, UNIVAC 1 was used for the production of part of the Second Series Population Tables for the states of Alabama, Iowa, Louisiana, and Virginia. This involved classifying individuals into one of several hundred groups, further grouping them by geographic location, and preparing tables showing the number of persons in each group for each local area. The data for this operation, initially punched onto eleven million cards (one for each person), was transferred to tape for processing by the UNIVAC.[57] The machine was also used for tabulating another subset of population involving about five million households. Each problem took several months to complete.

UNIVAC 2, installed at the Pentagon for the Air Comptroller, was intended for use in Project SCOOP (Scientific Computation of Optimum Problems), which grew out of wartime concerns with getting war materials and men across the Atlantic. Following the War, the newly created Air Force was faced with a mathematically similar problem in maintaining and supplying air bases scattered across the globe. Project SCOOP played a key role in the discovery of Linear Programming, a cornerstone of modern applied mathematics.[58]

It was for SCOOP that the Air Force had helped fund construction of a computer called SEAC (Standards Eastern Automatic Computer), but that machine's limited Input/Output facilities made it less than ideal for this problem. Soon after its installation, UNIVAC 2 was put to work on SCOOP around the clock.[59] Although the UNIVAC was superior to the SEAC in many ways, it, too, suffered from a slow output mechanism, which hampered its use for SCOOP. The UNIVAC's UNIPRINTER was based around a standard Remington Rand electric typewriter, and it printed at a rate commensurate with such a machine, about ten characters per second, which was too slow for the data processing applications the UNIVAC was being sold for. In 1954 Remington Rand addressed the problem by introducing the UNIVAC High Speed Printer, which printed a full 130-character line at one time.[60]

The UNIVAC installed in 1954 at Air Force's Air Material Command at Wright-Patterson AFB in Ohio performed similar tasks. One of its first jobs was to calculate "the complete Fiscal 1956 Budget estimate for airborne equipment spare parts, involving approximately 500,000 items."[61] The Air Force noted that the machine did the job in one day, replacing a battery of punched card equipment.

Some UNIVACs performed classified weapons work in the spirit of the one-of-a-kind computers that preceded them. UNIVAC 5, installed at the Lawrence Livermore Labs in April 1953, was one of those. But even that machine did at least one calculation that was not for the purpose of weapons designs. In November 1952, before it was shipped to California, Remington Rand used it to predict Eisenhower's victory over Adlai Stevenson in the 1952 presidential election. Narrated on "live" television, the event inaugurated the intrusion of television into national politics, and of computers into the public's consciousness. For a brief period, the word "UNIVAC" was synonymous with computer, as "Thermos" was for vacuum bottles. That ended when IBM took the lead in the business.[62]

A final example of the UNIVAC in use comes from the experience at General Electric's Appliance Park, outside Louisville, Kentucky. This installation, in 1954, has become famous as the first of a stored-program electronic computer for a nongovernment customer (although the LEO, built for the J. Lyons Catering Company in London, predated it by three years).

Under the direction of Roddy F. Osborn at Louisville, and with the advice of the Chicago consulting firm Arthur Andersen & Co., General Electric purchased a UNIVAC for four specific tasks: payroll, material scheduling and inventory control, order service and billing, and general cost accounting.[63] These were prosaic operations, but GE also hoped that the computer would be more than just a replacement for the punched-card equipment in use at the time. For General Electric, and by implication for American industries, the UNIVAC was the first step into an age of "automation," a change as revolutionary for business as Frederick W. Taylor's Scientific Management had been a half-century earlier.

The term "automation" was coined at the Ford Motor Company in 1947 and popularized by John Diebold in a 1952 book by that title.[64] Diebold defined the word as the application of "feedback" mechanisms to business and industrial practice, with the computer as the principal tool. He spoke of the 1950s as a time when "the push-button age is already obsolete; the buttons now push themselves."[65] Describing the GE installation, Roddy Osborn predicted that the UNIVAC would effect the same kind of changes on business as it had already begun to effect in science, engineering, and mathematics. "While scientists and engineers have been wide-awake in making progress with these remarkable tools, business, like Rip Van Winkle, has been asleep. GE's installation of a UNIVAC may be Rip Van Business's first 'blink.' "[66]

To people at General Electric, these accounts of "electronic brains" and "automation" were a double-edged sword. The Louisville plant was conceived of and built to be as modern and sophisticated as GE could make it; that was the motivation to locate it in Kentucky rather than Massachusetts or New York, where traditional methods (and labor unions) held sway. At the same time, GE needed to assure its stock-holders that it was not embarking on a wild scheme of purchasing exotic, fragile, and expensive equipment just because "longhair" academics—with no concern for profits—wanted it to.

Thus, GE had to emphasize the four mundane jobs, already being done by punched card equipment, to justify the UNIVAC. Once these jobs became routine, other, more advanced jobs would be given to the machine. Although automating those four tasks could have been done with a smaller computer, GE chose a UNIVAC in anticipation of the day when more sophisticated work would be done. These tasks would involve long-range planning, market forecasting based on demographic data, revamping production processes to reduce inventories and shipping delays, and similar jobs requiring a more ambitious use of corporate information.[67] The more advanced applications would not commence until after the existing computerization of "bread and butter" work reached a "break even point... enough to convince management that a computer system can pay for itself in terms of direct dollar savings (people off the payroll) without waiting for the 'jam' of more glamorous applications."[68]

Indeed, the analysis of the UNIVACs benefits was almost entirely cast in terms of its ability to replace salaried clerks and their overhead costs of office space, furnishings, and benefits. Yet at the end of Osborn's essay for the *Harvard Business Review*, the editors appended a quotation from Theodore Callow's *The Sociology of Work*, published that year. That quotation began:

The Utopia of automatic production is inherently plausible. Indeed, the situation of the United States today, in which poverty has come to mean the absence of status symbols rather than hunger and physical misery, is awesomely favorable when measured against the budgetary experience of previous generations or the contemporary experience of most of the people living on the other continents.[69]

It would not be the last time that the computer would be seen as the machine that would bring on a digital Utopia.

On Friday, October 15, 1954, the GE UNIVAC first produced payroll checks for the Appliance Park employees.[70] Punched-card machines had been doing that job for years, but for an electronic digital computer, which recorded data as invisible magnetic spots on reels of tape, it was a milestone. Payroll must be done right, and on time. GE had rehearsed the changeover thoroughly, and they had arranged with Remington Rand that if their machine broke down and threatened to make the checks late, they could bring their tapes to another UNIVAC customer and run the job there.[71] Over the course of the next year they had to exercise this option at least once. There were several instances where the checks were printed at the last possible minute, and in the early months it was common to spend much more time doing the job with UNIVAC than had been spent with punched card equipment. No payrolls were late.

IBM's Response

At the time of the UNIVAC's announcement, IBM was not fully committed to electronic computation and was vigorously marketing its line of punched card calculators and tabulators. But after seeing the competitive threat, it responded with several machines: two were on a par with the UNIVAC; another was more modest.

In May 1952, IBM announced the 701, a stored-program computer in the same class as the UNIVAC. Although not an exact copy, its design closely followed that of the computer that John von Neumann was having built at the Institute for Advanced Study at Princeton. That meant it used a memory device that retrieved all the digits of a word at once, rather than the UNIVAC's delay lines that retrieved bits one at a time. Beginning in January of that year, IBM had hired John von Neumann as a consultant; as with the Institute for Advanced Study computer itself, von Neumann was not involved with the detailed design of the 701. (IBM engineers Jerrier Haddad and Nat Rochester were in charge of the project.) The first unit was installed at IBM's offices in New York in December, with the first shipment outside IBM to the nuclear weapons laboratory at Los Alamos in early 1953.[72]

IBM called the 701 an "electronic data processing machine," a term (coined by James Birkenstock) that fit well with "Electric Accounting Machine," which IBM was using to describe its new line of punched card equipment. IBM deliberately avoided the word "computer," which it felt was closely identified with the UNIVAC and with exotic wartime projects that appeared to have little relevance to business.

For main storage, the 701 used IBM-designed 3-inch diameter vacuum tubes similar to those used in television sets. (They were called "Williams tubes" after their British inventor, F. C. Williams.) Although they were more reliable than those in other contemporary computers, their unreliability was a weak link in the system. One story tells of a 701 behaving erratically at its unveiling to the press despite having been checked out thoroughly before the ceremony. The photographers' flash bulbs were "blinding" the Williams tubes, causing them to lose data.

Another account said that because the memory's Mean Time Between Failure (MTBF) was only twenty minutes, data had to be constantly swapped to a drum to prevent loss.[73]

Each tube was designed to hold 1,024 bits. An array of 72 tubes could thus hold 2,048 36-bit words, and transfer a word at a time by reading one bit from each of 36 tubes.[74] Plastic tape coated with magnetic oxide was used for bulk memory, with a drum for intermediate storage. The processor could perform about 2,000 multiplications/second, which was about four times faster than the UNIVAC.

Within IBM, the 701 had been known as the Defense Calculator, after its perceived market. According to an IBM executive, the name also helped "ease some of the internal opposition to it since it could be viewed as a special project (like the bomb sights, rifles, etc., IBM had built during World War II) that was not intended to threaten IBM's main product line."[75] True to that perception, nearly all of the 19 models installed were to U.S. Defense Department or military aerospace firms.[76] Initial rental fees were $15,000 a month; IBM did not sell the machines outright. If we assume the 701 was a million-dollar machine like the UNIVAC, the rental price seems low; certainly IBM could not have recouped its costs in the few years that the machine was a viable product.

The 701 customers initially used the machine for problems, many still classified, involving weapons design, spacecraft trajectories, and cryptanalysis, which exercised the central processor more heavily than its Input/Output facilities. Punched card equipment had been doing some of that work, but it had also been done with slide rules, mechanical calculators, analog computers, and the Card-Programmed Calculator. Eventually, however, customers applied the 701 to the same kinds of jobs the UNIVAC was doing: logistics for a military agency, financial reports, actuarial reports, payrolls (for North American Aviation), and even predicting the results of a presidential election for network television. (In 1956, the 701 correctly predicted Eisenhower's reelection.)[77]

Unlike the UNIVAC, the 701's central processor handled control of the slow input/output (I/O) facilities directly. All transfers of data had to pass through a single register in the machine's processor, which led to slow operation for tasks requiring heavy use of I/O. However, the 701's lightweight plastic tape could start and stop much faster than the UNIVAC's metal tape and thus speed up those operations. The tape drive also employed an ingenious vacuum-column mechanism, invented by James Wiedenhammer, which allowed the tape to start and stop quickly without tearing.

For scientific and engineering problems, the 701's unbalanced I/O was not a serious hindrance. Computer designers—the few there were in 1953—regarded it as an inelegant design, but customers liked it. The nineteen installations were enough to prevent UNIVAC from completely taking over the market and to begin IBM's transition to a company that designed and built large-scale electronic digital computers.[78]

The 701 became IBM's response to UNIVAC in the marketplace, but that had not been IBM's intention. Before starting on the 701, IBM had developed a research project on a machine similar to the UNIVAC, an experimental machine called the Tape Processing Machine, or TPM. Its design was completed by March 1950.[79] The TPM was a radical departure from IBM's punched card machinery in two ways. It used magnetic tape (like the UNIVAC), and its variable length record replaced the rigid 80-character format imposed by the punched card. Like the UNIVAC, it worked with decimal digits, coding each digit in binary.

IBM chose to market a second large computer specifically to business customers based on the Tape Processing Machine. Model 702 was announced in September 1953 and delivered in 1955. In many ways it was similar to the 701, using most of the same electronic circuits as well as the Williams Tube storage. By the time of the first 702 installations, magnetic core memories were beginning to be used in commercial machines. And 701 customers were finding that their machine, like any powerful general-purpose computer, could be used for business applications as well. IBM received many orders for 702s, but chose to build and deliver only fourteen, with other orders filled by another machine IBM brought out a few years later.[80]

Engineering Research Associates

A third firm entered the field of making and selling large digital computers in the early 1950s: Engineering Research Associates, a Twin Cities firm that had its origins in U.S. Navy-sponsored code-breaking activities during World War II.[81] The Navy gave this work the name "Communications Supplementary Activity—Washington" (CSAW), but it was usually called "Seesaw" after its acronym. It was centered in Washington, on the commandeered campus of a girls school. After the War, two members of this group, Howard Engstrom and William Norris, felt that the talent and skills the Navy had assembled for the war effort were too valuable to be scattered, and they explored ways of keeping the group together. They decided to found a private company, and with financial assistance from John E. Parker, they were incorporated as Engineering Research Associates, Inc., in early 1946. Parker was able to provide space in a St. Paul building that during the war had produced wooden gliders (including those used for the Normandy invasion).

Thus, by one of the coincidences that periodically occur in this history, the empty glider factory gave the Twin Cities an entree into the world of advanced digital computing. The factory was cold and drafty, but ERA had little trouble finding and hiring capable engineers freshly minted from the region's engineering schools. Among them was a 1951 graduate of the University of Minnesota, who went over to "the glider factory" because he heard there might be a job there. His name was Seymour R. Cray.[82] We will encounter Cray and his boss, William Norris, several times in later chapters.

ERA was a private company but was also captive to the Navy, from which it had sprung. (The propriety of this arrangement would on occasion cause problems, but none serious.) The Navy assigned it a number of jobs, or "tasks," that ERA carried out. Most of these were highly classified and related to the business of breaking codes. Task 13, assigned in August 1947, was for a general-purpose electronic computer. ERA completed the machine, code-named "Atlas," and asked the Navy to clear them for an unclassified version they could sell on the open market. In December 1951 they announced it as Model "1101": "13" in binary notation.[83]

As might be expected from a company like ERA, the 1101 was intended for scientific or engineering customers, and its design reflected that. Before it could begin delivering systems, however, ERA found itself needing much more capital than its founders could provide, and like the Eckert–Mauchly Computer Corporation, was purchased by Remington Rand. By mid-1952 Remington Rand could offer not one but two well-designed and capable computer systems, one optimized for science and engineering, the other for commercial use. Installations of the 1103, its successor, began in the fall of 1953. Around twenty were built. As with the IBM 701, most went to military agencies or aerospace companies.

In 1954 the company delivered an 1103 to the National Advisory Committee for Aeronautics (NACA) that employed magnetic core in place of the Williams Tube memory. This was perhaps the first use of core in a commercial machine. The 1103 used binary arithmetic, a 36-bit word length, and operated on all the bits of a word at a time. Primary memory of 1,024 words was supplied by Williams tubes, with an ERA-designed drum, and four magnetic tape units for secondary storage.[84]

Following NACA's advice, ERA modified the machine's instruction set to include an "interrupt" facility—another first in computer design. (Core and interrupts will be discussed in detail in the next chapter.) These enhancements were later marketed as standard features of the 1103-A model.[85] Another aerospace customer, Convair, developed a CRT tube display for the 1103, which they called the Charactron. This 7-inch tube was capable of displaying a 6 × 6 array of characters, which also affected the course of computer history.[86] Overall, the 1103 competed well with the IBM 701, although its I/O facilities were judged somewhat inferior.

The Drum Machines

In the late 1930s, in what may have been the first attempt to build an electronic digital computer, J. V. Atanasoff conceived of a memory device consisting of a rotating drum on which 1,600 capacitors were placed, arrayed in 32 rows.[87] His work influenced the developments of the next decade, although those who followed him did not ultimately adopt his method. In the following years several people continued to work on the idea of rotating magnetic devices for data storage, for example, Perry O. Crawford, who described such a device in his master's thesis at MIT.[88]

After the War, the drum emerged as a reliable, rugged, inexpensive, but slow memory device. Drawing on wartime research on magnetic recording in both the United States and Germany, designers rediscovered and perfected the drum, this time using magnetic rather than capacitive techniques.

The leader in this effort was Engineering Research Associates. Before they were assigned "Task 13," they were asked to research available memory technologies. By 1947 they had made some significant advances in recording speeds and densities, using a drum on which they had glued oxide-coated paper (figure 1.4).[89] Within two years ERA was building drums that ranged from 4.3 to 34 inches in diameter, with capacities of up to two million bits, or 65,000 30-bit words. Access time ranged from 8 to 64 milliseconds.[90] ERA used drums in the 1101; they also advertised the technology for sale to others.

CRC 102A

One of the first to take advantage of magnetic drums was was Computer Research Corporation of Hawthorne, California. This company was

Figure 1.4
Advertisement for magnetic drum memory units, from ERA. (*Source*: *Electronics Magazine* [April 1953]: 397.)

founded by former employees of Northrop Aircraft Company, the company that had built the Card-Programmed Calculator described above. In 1953 they began selling the CRC-102A, a production version of a computer called CADAC that had been built for the Air Force. It was a stored-program, general-purpose computer based on a drum memory. The 102A had a simple design, using binary arithmetic, but a decimal version (CRC 102D) was offered in 1954.[91] In some of the published descriptions, engineers describe its design as based directly on logic states derived from statements of Boolean algebra. This so-called West Coast design was seen as distinct from the designs of Eckert and Mauchly, who thought in terms not of logic states, but of current pulses gated through various parts of a machine. As computer engineering matured, elements of both design approaches merged, and the distinction eventually vanished.[92]

The 102A's drum memory stored 1,024 42-bit words; average access time was 12.5 msec. A magnetic tape system stored an additional 100,000 words. The principal input and output device was the Flexowriter, a typewriter-like device that could store or read keystrokes on strips of paper tape. It operated at about the speeds of an ordinary electric typewriter, from which it was derived. In keeping with its aerospace roots, Computer Research Corporation also offered a converter to enter graphical or other analog data into the machine.[93] It was also possible to connect an IBM card reader or punch to the computer. The computer's operating speed was estimated at about eleven multiplications per second.[94] The 102A was a well-balanced computer and sold in modest numbers. In 1954 the National Cash Register Company purchased CRC, and the 102 formed the basis of NCR's entry into the computer business.[95]

Computer Research's experience was repeated with only minor variations between 1950 and 1954. Typically, a small engineering company would design a computer around a drum memory. I/O would be handled by a standard Flexowriter, or by punched card machines leased from IBM. The company would then announce the new machine at one of the Joint Computer Conferences of the Institute of Radio Engineers/Association for Computing Machinery. They would then get a few orders or development funds from the Air Force or another military agency. Even though that would lead to some civilian orders and modest productions runs, the company would still lack the resources to gear up for greater volume or advanced follow-on designs. Finally, a large, established company would buy the struggling firm, which would then serve as the larger company's entree into computing.

Many of these computers performed well and represented a good value for the money, but there was no getting around the inherent slowness of the drum memory. Their input/output facilities also presented a dilemma. The Flexowriter was cheap, but slow. Attaching punched card equipment meant that a significant portion of the profits would go directly to IBM, and not to the struggling new computer company.

As mentioned, National Cash Register bought CRC. Electronic Computer Corporation, founded by Samuel Lubkin of the original UNIVAC team, merged with Underwood Corporation, known for its typewriters. (Underwood left the computer business in 1957.) Consolidated Engineering of Pasadena, California, was absorbed by Burroughs in 1956. The principal legacy of the drum computers may have been their role as the vehicle by which many of the business machine companies entered the computer business.

Table 1.2 lists several other magnetic drum computers announced or available by mid-1952. For each of these systems, the basic cost was from

Table 1.2
Commercially available small computers, ca. mid-1952

Computer	Word length	Memory capacity (words)	Speed (mult./sec.)	Manufacturer
CE 30-201	10 dec.	4000	118	Consolidated Engineering Pasadena, CA
Circle	40 bits	1024	20	Hogan Labs New York, NY
Elecom 100	30 bits	512	20	Electronic Computer Corp Brooklyn, NY
MINIAC	10 dec.	4096	73	Physical Research Labs Pasadena, CA
MONROBOT	20 dec.	100	2	Monroe Calculating Machine Co Orange, NJ

Source: Data from U.S. Navy, Navy Mathematical Computing Advisory Panel, *Symposium on Commercially Available General-Purpose Electronic Digital Computers of Moderate Price* (Washington, DC, 14 May 1952).

$65,000 to $85,000 for a basic system exclusive of added memory, installation, or auxiliary I/O equipment.

Later Drum Machines, 1953–1956

LGP-30
In the mid-1950s a second wave of better-engineered drum computers appeared, and these sold in much larger quantities. They provided a practical and serious alternative for many customers who had neither the need nor the resources to buy or lease a large electronic computer.

The Librascope/General Precision LGP-30, delivered in 1956, represented a minimum design for a stored-program computer, at least until the minicomputer appeared ten years later. It was a binary machine, with a 30-bit word length and a repertoire of only sixteen instructions. Its drum held 4,096 words, with an average access time of around 2.3 msec. Input and output was through a Flexowriter.

The LGP-30 had only 113 vacuum tubes and 1,350 diodes (unlike the UNIVAC's 5,400 tubes and 18,000 diodes), and looked like an oversized office desk. At $30,000 for a basic but complete system, it was also one of the cheapest early computers ever offered. About 400 were produced and sold.[96] It was not the direct ancestor of the minicomputer, which revolutionized computing in the late 1960s, but many minicomputer pioneers knew of the LGP-30. Librascope offered a transistorized version in 1962, but soon abandoned the general-purpose field and turned to specialized guidance-and-control computers for aerospace and defense customers.

Bendix G-15
The G-15, designed by Harry Huskey and built by Bendix, was perhaps the only computer built in the United States to have been significantly influenced by the design ideas of Alan Turing rather than John von Neumann. Both advocated the stored-program principle, with a provision for conditional branching of instructions based on previously calculated results. For von Neumann, however, the fundamental concept was of a steady linear stream of instructions that occasionally branched based on a conditional test. Turing, on the other hand, felt that there was no fundamental linear order to instructions; for him, *every* order represented a transfer of control of some sort.[97]

Turing's concept (much simplified here) was more subtle than the linear model, and fit well with the nature of drum-based computers.

Turing's model required that every instruction have with it the address where the next instruction was located, rather than assuming that the next instruction would be found in the very next address location. In a drum computer, it was not practical to have instructions arranged one right after the other, since that might require almost a full revolution of the drum before the next one appeared under the read head. Programmers of drum computers often developed complicated "minimum latency coding" schemes to scatter instructions around the drum surface, to ensure that the next instruction would be close to the read head when it was needed. (Note that none of this was required if a memory that took the same amount of time to access each piece of data was used.)

Harry Huskey, who had worked with Turing in 1947 on the ACE project at the National Physical Laboratory in England, designed what became the G-15 while at Wayne State University in Detroit in 1953. First deliveries were in 1956, at a basic price of $45,000. It was regarded as difficult to program, but for those who could program it, it was very fast. Bendix sold more than four-hundred machines, but the G-15's success was not sufficient to establish Bendix as a major player in the computer field.[98] Control Data Corporation later took over Bendix's computer business, and Bendix continued to supply only avionics and defense electronics systems.

IBM 650

Along with the Defense Calculator (a.k.a. IBM 701), IBM was working on a more modest electronic computer. This machine had its origins in proposals for extensions of punched card equipment, which IBM had been developing at its Endicott, New York, plant. IBM's internal management was hesitant about this project, nor was there agreement as to what kind of machine it would be. One proposal, dubbed "Wooden Wheel," was for a plug-programmed machine like the 604 Multiplier.[99] In the course of its development, the design shifted to a general-purpose, stored-program computer that used a magnetic drum for primary memory. (IBM's acquisition, in 1949, of drum-memory technology from Engineering Research Associates was a key element in this shift.[100]) The machine, called the 650, was delivered in 1954 and proved very successful, with eventually around a thousand installations at a rental of around $3,500 a month.[101]

By the time of its announcement, the 650 had to compete with many other inexpensive drum machines. It outsold them all, in part because of

IBM's reputation and large customer base of punched card users, and in part because the 650 was perceived as easier to program and more reliable than its competitors. IBM salesmen were also quick to point out that the 650's drum had a faster access time (2.4 msec) than other drum machines (except the Bendix G-15).[102]

The 650 was positioned as a business machine and continued IBM's policy of offering two distinct lines of products for business and scientific customers. Ironically, it had less impact among business customers, for whom it was intended, than it had at universities. Thomas Watson Jr. directed that IBM allow universities to acquire a 650 at up to a 60 percent discount, if the university agreed to offer courses in business data processing or scientific computing. Many universities took up this offer, making the 650 the first machine available to nascent "computer science" departments in the late 1950s.[103]

Summary

Very few of these machines of anybody's manufacture were *sold* during the period we are talking about. Most of them, and I would guess 80 percent at least, were *bought* by the customer who made the buy, not the salesman who made the sale, although the salesman might get the commission.[104]
—Lancelot Armstrong

The "first generation" began with the introduction of commercial computers manufactured and sold in modest quantities. This phase began around 1950 and lasted through the decade. Computers of this era stored their programs internally and used vacuum tubes as their switching technology, but beyond that there were few other things they had in common. The internal design of the processors varied widely. Whether to code each decimal digit in binary or operate entirely in the binary system internally remained an unsettled question. The greatest variation was found in the devices used for memory: delay line, Williams tube, or drum. Because in one way or another all these techniques were unsatisfactory, a variety of machines that favored one design approach over another were built.

The Institute for Advanced Study's reports, written by Arthur Burks, Herman Goldstine, and John von Neumann, emphasized the advantages of a pure binary design, with a parallel memory that could read and write all the bits of a word at once, using a storage device designed at RCA called the Selectron. By the time RCA was able to produce sufficient quantities of Selectrons, however, core memory was being introduced, and the Selectron no longer looked so attractive. Only the Johnniac, built at the RAND Corporation, used it. Most of the other parallel-word computers used Williams Tubes.[105] In practice, these tubes were plagued by reliability problems.[106]

The result was that memory devices that accessed bits one at a time, serially, were used in most first-generation computers. The fastest computers used mercury delay lines, but the most popular device was the rotating magnetic drum. A drum is fundamentally an electromechanical device and by nature slow, but its reliability and low cost made it the technology of choice for small-scale machines.

Commercial computing got off to a shaky start in the early 1950s. Eckert and Mauchly, who had a clear vision of its potential, had to sell their business to Remington Rand to survive, as did Engineering Research Associates. Remington Rand, however, did not fully understand what it had bought. IBM knew that computers were something to be involved with, but it was not sure how these expensive and complex machines might fit into its successful line of tabulating equipment. Customers took the initiative and sought out suppliers, perhaps after attending the Moore School session in 1946 or visiting a university where a von Neumann type machine was being built. These customers, from a variety of backgrounds, clamored for computers, in spite of a reluctance among UNIVAC or IBM salesmen to sell them.

The UNIVAC and the IBM 701 inaugurated the era of commercial stored-program computing. Each had its drawbacks, but overall they met the expectations of the customers who ordered them. The UNIVAC's memory was reliable but slow; the 701's was less reliable but faster. Each machine worked well enough to establish the viability of large computers. Drum technology was providing storage at a lower cost per bit, but its speed was two orders of magnitude slower, closer to the speeds of the Card-Programmed Calculator (which was capable of reading 125 instruction cards per minute), which had been available since the late 1940s from IBM. Given the speed penalty, drum-based computers would never be able to compete with the others, regardless of price. The many benefits promised in the 1940s by the stored-program electronic computer architecture required high-capacity, high-speed memory to match electronic processing. With the advent of ferrite cores—and techniques for manufacturing them in large quantities—the memory problem that characterized the first generation was effectively solved.

Table 1.3
Selected characteristics of early commercial computers

Computer	Word length	Memory capacity (words)	Access time (microseconds)	Multiplications/ second
CRC-102	9 dec.	1024	12,500	65
ERA 1103	36 bits	1024	10	2500–8000
G-15	29 bits	2160	1,700 avg.	600
LGP-30	30 bits	4096	8,500 avg.	60
IBM 650	10 dec.	1000–2000	2,400 avg.	50–450
IBM 701	36 bits	2048	48	2000
UNIVAC	11 dec.	1000	400 max.	465

Source: Data from Martin Weik, "A Survey of Electronic Digital Computing Systems," Ballistic Research Laboratories Report #971 (Aberdeen Proving Ground, Maryland, December 1955).

Table 1.3 lists memory and processor characteristics of the major computers of this era.

2
Computing Comes of Age, 1956–1964

Computer technology pervades the daily life of everyone in the United States. An airline traveler's tickets, seat assignment, and billing are handled by a sophisticated on-line reservation system. Those who drive a car are insured by a company that keeps a detailed and exacting record of each driver's policy in a large database. Checks are processed by computers that read the numerals written in special ink at the bottom. Each April, citizens file complicated tax returns, which the Internal Revenue Service processes, files, and keeps track of with computers.

It is hard to imagine a world in which computers do not assist with these activities, yet they were not computerized until the late 1950s. This set the stage for further penetration of computing two decades later, in the form of automatic teller machines, bar-coded products scanned at supermarket and retail check-out stations, and massive financial and personal databases maintained by credit-card companies and mail-order houses.

Before 1955, human beings performed all these activities using typewriters, carbon paper, and lots of filing cabinets.[1] Punched-card equipment assisted with some of the work. The preferred aid to arithmetic was the Comptometer, manufactured by Felt and Tarrant of Chicago (figure 2.1).[2] This machine was key-driven: pressing the keys immediately performed the addition, with no other levers to pull or buttons to press. Its use required intensive training, but in the hands of a skilled operator, a Comptometer could perform an addition every few seconds. It could neither multiply nor print the results of a calculation, however.

What these applications had in common was their need to store and retrieve large amounts of data easily and quickly. Required also were a variety of retrieval methods, so that the data could be used later on in different ways. Calculations consisted mainly of additions, subtractions, and less frequently, multiplications. Quantities typically ranged up to a

Figure 2.1
Comptometer. (*Source*: Smithsonian Institution.)

million and required a precision of two decimal places, for dollars and cents. Though similar to the work that punched card installations handled, this activity had the additional requirement of rapid retrieval of individual records from large files, something punched card machines could not easily do. The definition of "data processing" evolved to accommodate this change.

The computers of the early 1950s were ill suited for this work. The inexpensive drum-based machines that proliferated early in the decade lacked the memory capacity, speed, and above all, high-capacity input and output facilities. The larger machines showed more potential, but even the UNIVAC, designed for data processing applications from the start, had a slow printer when first introduced.

By the end of the 1950s, digital electronic computers had begun doing that kind of work. Through the 1950s, computer designers adapted the architecture of a machine developed for scientific problems to applications that required more storage and more voluminous input and output. These were fundamental changes, but computers evolved to accommodate them without abandoning their basic stored-program architecture.

Core Memory

Part of this transformation of computers came from advances in circuit technology. By 1959 the transistor had become reliable and cheap enough to serve as the basic circuit element for processors. The result was increased reliability, lower maintenance, and lower operating costs. Before that, however, an even more radical innovation occurred—the development of reliable, high capacity memory units built out of magnetic cores. These two innovations were able to boost performance to a point where many commercial applications became cost-effective.

Core memory refers to small, doughnut-shaped pieces of material through which several fine wires are threaded to store information (figure 2.2). The wires passing through the core can magnetize it in either direction; this direction, which another wire passing through can sense, is defined as a binary zero or one. The technology exploits the property, known as hysteresis, of certain magnetic materials. A current passing through the center of such a core will magnetize it, but only if it is above a certain threshold.[3] Likewise, a current passing in the other direction will demagnetize such a core if the current is strong enough. A core memory unit arranges cores made of materials having this property in a plane, with wires running vertically and horizontally through the hole in each core. Only when there are currents in both the vertical and the horizontal wires, and both are running in the same direction, will a core be magnetized; otherwise, there is no effect.

A core memory has many advantages over the memories used in the first-generation computers. The cores can be made small. The memory is "nonvolatile": it holds information without having to supply electrical power (as with Williams tubes and mercury delay lines) or mechanical power (as with a rotating drum).

Above all, core provides random access memory, now known as RAM: access to any bit of a core plane is as rapid as to any other. (The term is misleading: it is not really a "random" time, but since the term is in common use it will be retained here.) This overcomes a major drawback to delay lines and drums, where waiting for data to come around can introduce a delay that slows a computer down.

During World War II, the German Navy developed a magnetic material with the property of hysteresis, and they used it in the circuits

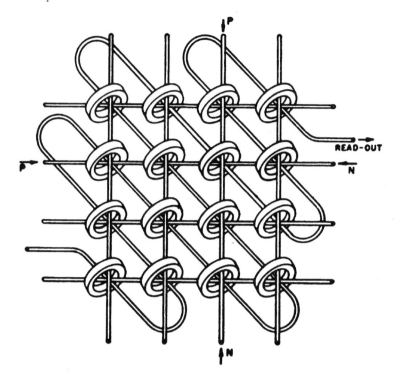

Figure 2.2
Magnetic core memory. (*Source*: From Jan A. Rajchman, "A Myriabit Magnetic-Core Matrix Memory," *IRE Proceedings* (October 1953): 1408.) © 1953 IRE, now known as IEEE

of a fire-control system. After the war, samples were brought to the United States, where it caught the attention of people interested in digital storage. Researchers at IBM, the University of Illinois, Harvard, MIT, and elsewhere investigated its suitability for computers.[4] An Wang, a student of Howard Aiken at Harvard, invented a core memory that was used in the Harvard Mark IV, completed in 1952. Magnetic core memories were installed on both the ENIAC and the Whirlwind in the summer of 1953. The ENIAC's memory, designed by the Burroughs Corporation, used a two-dimensional array of cores; the Whirlwind's memory, designed by Jay Forrester, used a three-dimensional array that offered faster switching speeds, greater storage density, and simpler electronics.[5] One key advantage of Forrester's design was a circuit, developed by Ken Olsen, that reduced the amount of current needed to operate the array.

The core memory made the Whirlwind almost a new machine, so much better was its performance, and commercial systems began appearing with it. As mentioned in the previous chapter, the first commercial delivery was around late 1954, when the ERA division of Remington Rand delivered an 1103A computer to the National Advisory Committee for Aeronautics. ERA had also delivered core memories to the National Security Agency as part of a classified project. At IBM, a team led by Eric Bloch developed a memory unit that served as a buffer between the electrostatic memory of the 702 computer and its card-based input and output units. Deliveries to commercial customers began in February 1955. IBM continued using electrostatic tubes for the 702 but moved to core for machines built after it.[6]

A contract with the U.S. Air Force to build production versions of the Whirlwind was a crucial event because it gave engineers the experience needed for core to become viable in commercial systems. The Air Force's SAGE (Semi-Automatic Ground Environment), a system that combined computers, radar, aircraft, telephone lines, radio links, and ships, was intended to detect, identify, and assist the interception of enemy aircraft attempting to penetrate the skies over the United States. At its center was a computer that would coordinate the information gathered from far-flung sources, process it, and present it in a combination of textual and graphical form. All in all, it was an ambitious design; the Air Force's desire to have multiple copies of this computer in operation round the clock made it even more so.[7] A primary requirement for the system was high reliability, which ruled out mercury delay lines or electrostatic memory.

The design of SAGE's computer had much in common with Whirlwind; some early literature described it as "Whirlwind II." That was especially evident in its core memory, designed to have a capacity of 8,192 words of 32 bits in length. In 1952 the SAGE development team at Lincoln Laboratory asked three companies about the possibility of building production models of the computer then being designed. The team visited the facilities of IBM, Raytheon, and two divisions of Remington Rand. Based on a thorough evaluation of the plants, the team selected IBM.[8] IBM delivered a prototype in 1955, and completed the first production model computer the following year. IBM eventually delivered around thirty computer systems for SAGE. For reliability, each system consisted of two identical computers running in tandem, with a switch to transfer control immediately to the backup if the primary computer failed. Although the computers used vacuum tubes (55,000

per pair), the reliability of the duplexed system exceeded that of most solid-state computers built years later. The last original SAGE computer, operating at a site in North Bay, Ontario, was shut down in 1983.[9]

Initially IBM had contracted with other companies, primarily General Ceramics, to deliver cores. It had also begun a research effort on core production in-house. Among other things, it worked with the Colton Manufacturing Company, which provided machines to the pharmaceutical industry for making pills, to adapt their equipment to press cores of uniform properties. (IBM and other computer companies also used machines modified from those made by General Mills for putting food into consumer packages, and by United Shoe Machinery for making shoes, to insert electronic components onto circuit boards.)[10] As the SAGE project got underway, IBM began to rely more and more on its own expertise. SAGE would require hundreds of thousands of good cores. Given the low yields of cores supplied to IBM at first, it seemed that millions would have to be made and tested.[11] IBM's own research efforts fared much better, producing yields of up to 95 percent by 1954 (figure 2.3).

The SAGE contract generated half a billion dollars in revenue for IBM in the 1950s. Its value in getting IBM into the business of producing cores was probably worth just as much.[12] By 1956, IBM had surpassed UNIVAC in the number of installations of large computers. Already dominant in sales of smaller computers, IBM would continue to dominate the entire computer industry.[13] How it managed to do that has been the subject of many accounts. Most give generous credit to IBM's sales force, and note also that Remington Rand's top management was less forceful in their support of the UNIVAC division. Some accounts believe that IBM took this lead despite the technical superiority of UNIVAC's machines.[14] Also important was the experience in producing reliable core memories that IBM gained from its experience with SAGE.

Figure 2.3
(*top*) Core memory unit developed for the Memory Test Computer, prior to installation on the Whirlwind. (*Source*: Mitre Corporation Archives.) (*bottom*) Core memory unit for the IBM 704 computer. (*Source*: Charles Babbage Institute, University of Minnesota.)

Honeywell, GE, RCA

As IBM and UNIVAC competed for business, other companies took steps to develop and sell large computers. First among them was the Minneapolis Honeywell Corporation, a maker of industrial and consumer controls (including thermostats for the home) and aerospace electronics equipment. In 1955 Honeywell acquired the computer division of Raytheon, which had been the only established company to respond to the U.S. government's request in the late 1940s for large computers for its needs. Raytheon was unable to deliver the machines it promised, although one computer, the RAYDAC, was installed in 1952 at a U.S. Navy base at Point Mugu, California, as part of Project "Hurricane." In 1954 Raytheon established the Datamatic Corporation jointly with Honeywell, but the following year it relinquished all its interest in Datamatic.[15]

Honeywell's first large offering was the Datamatic 1000, delivered in 1957. This machine was comparable to the largest UNIVAC or IBM systems, but it was already obsolete. Among other things, it used vacuum tubes at a time when it was becoming clear that transistors were practical. Honeywell temporarily withdrew from the market and concentrated on designing transistorized machines, which it successfully offered a few years later. That decision laid the grounds for its successful reentry, which began in the mid-1960s.[16]

GE

In 1955, General Electric was the nation's leading electronics firm, with sales of almost $3 billion and over 200,000 employees. (Compare IBM's sales of $461 million and 46,500 employees, or Remington Rand's $225 million and 37,000 employees that year.)[17] In 1953, the company had delivered the OARAC to the U.S. Air Force at Wright-Patterson Air Force Base, and the Air Force had used it for specialized, classified applications. However, the OARAC was a general-purpose electronic computer and GE could have marketed a commercial version of it if its senior management had not decided against entering the computer field. GE engineers later recalled a consistent bias against entering this market throughout the 1950s. GE said that it preferred to concentrate on other products it felt had greater potential, like jet engines and nuclear power. Others dispute that account.[18] One engineer suggested that the fact that IBM was GE's largest customer for vacuum tubes might have been a factor: GE did not want to appear to be in competition with IBM, especially given the perception that GE, with its greater resources, could overwhelm IBM if it chose to do so.

General Electric did, however, produce a computer in the late 1950s for a system called ERMA (Electronic Method of Accounting), an automatic check-clearing system developed with the Stanford Research Institute and the Bank of America. Mindful of the ban by GE chief Ralph Cordiner against general-purpose computers, ERMA's creators sold the project internally as a special, one-time project. A plant was built outside Phoenix, Arizona, and GE engineers, led by Homer R. Oldfield, got to work. While still a major supplier of vacuum tubes, GE had among its sprawling research facilities people who understood the advantages—and problems—of transistors. The ERMA computer would be transistorized. Deliveries began in early 1958. The Bank of America publicly unveiled ERMA in the fall of 1959, at a ceremony hosted by GE spokesman Ronald Reagan.[19]

ERMA sucessfully allowed banks to automate the tedious process of clearing checks, thus avoiding the crisis of paperwork that threatened banks in the booming postwar economy. Among its components was a set of numeric and control characters printed with magnetic ink at the bottom of each check that a machine could read. It was called "MICR." Advertising agencies adopted the typography as a symbol of "computerese," and for a while the type was a symbol of the Age of Automation. Few realize, however, that MICR only specified the shapes of the ten decimal digits and some control characters, not the letters of the alphabet.

ERMA's success emboldened its creators. They continued their risky game by developing other computers, including a system that in 1962 Dartmouth College would adopt for its pioneering time-sharing system (to be discussed at length in chapter 5). As with ERMA, they described their products as special-purpose equipment. But their charade could only go so far. Without full corporate support the company could hardly expect to compete against IBM. In one respect, GE's management had been correct: the computer division never was profitable, despite the high quality and technical innovation of its products. In 1970 GE sold the business to Honeywell for a little over $200 million.[20]

RCA

RCA's entry into commercial computing paralleled GE's. Like GE, RCA had been involved at an early date in computers—it had developed a storage tube for a computer built at Princeton in the early 1950s. RCA was not as large as GE but in 1955 it had over twice the annual sales of IBM and four times the sales of Remington Rand. In November 1955 RCA announced a large-scale computer intended for business and data processing applications, the BIZMAC. (Among the engineers who worked on it was Arnold Spielberg, a talented engineer whose fame among computer circles would be eclipsed by that of his son, Steven, the Hollywood filmmaker. Arnold Spielberg later moved to Phoenix and worked on the GE computers described above.)

The BIZMAC used core memory, which made one of the first commercial machines to do so. Vacuum tubes were used for logic and arithmetic circuits. The BIZMAC did not sell well. Only one full system was installed, at a U.S. Army facility in Detroit. A few smaller systems were installed elsewhere. One reason might have been a too-long development time. By the time of its first installations in 1956, new designs based on simpler architectures and using transistors promised to offer the equivalent performance at a lower cost.[21]

The BIZMAC's architecture was different from the machines it was competing with, which may have been another factor that led to its commercial failure. Unlike the von Neumann design, the BIZMAC had specialized processing units for searching and sorting data on reels of tape. Whereas most contemporary computers had up to a dozen tape drives for mass storage, the BIZMAC was designed to support several hundred drives, all connected to its processor and under machine control. That implied that each reel of tape would be permanently mounted, and there would be little or no need for an operator to mount and demount tapes as there was with other computers.[22] A system of relays connected a particular tape to the BIZMAC's processor. Attached to the main processor was a special-purpose processor whose only function was to sort data.

This design would seem to make the BIZMAC especially suited for business data processing applications, but by 1956 advances in technology had eliminated any advantage gained by this architecture. Other manufacturers were already offering tape drives with much-improved performance, and those drives, combined with advances in core memory and processing speeds, made it cheaper to have only a few high-speed tape drives with human beings employed to mount and demount tapes from a library.

The BIZMAC's failure set back RCA's entry into commercial computing, but it did not end it. After a brief hiatus, the company responded with a line of transistorized computers of more conventional design (the 301 and 501); these were moderately successful.

The BIZMAC was not the only large computer to explore an alternate architecture. In 1956 UNIVAC (now a division of Sperry Rand, formed by a merger of the Sperry Corporation and Remington Rand) introduced the UNIVAC File Computer, which it hoped would be a low-cost successor to the original UNIVAC. As the name implied, the machine was intended for data handling. It used a magnetic drum as its main memory, which lowered costs but also compromised performance. It was programmed by a combination of stored instructions and plugboard panels. Its designers felt that flexible input/output (I/O) facilities were critical for commercial customers.[23] As with the BIZMAC, the UNIVAC File could manipulate data without having to go through its central processor. The UNIVAC File, like the BIZMAC, was a commercial failure, probably for the same reasons. This notion of designing a machine for data storage, retrieval, sorting, and searching reappears from time to time throughout the history of computing. But in nearly every case, a good general-purpose computer has driven specialized machines from the market. Table 2.1 lists the major U.S. computer suppliers and their revenues for 1955.

Table 2.1
Revenues of selected computer and electronic companies, 1955

Company	Annual sales	Net profit	Employees
GE	$2.96 billion	$213 M	210,000
Western Electric*	$1.5 billion	$55 M	98,000
RCA	$940 M	$40 M	70,500
IBM	$461 M	$46.5 M	46,500
NCR	$259 M	$12.7 M	37,000
Honeywell	$229 M	$15.3 M	25,000
Remington Rand**	$225 M	$12.2 M	37,000
Raytheon	$177 M	$3.5 M	18,700
Burroughs	$169 M	$7.8 M	20,000

Source: Data from *Fortune* (July 1955).
* Western Electric was the manufacturing arm of AT&T, which owned and controlled it. AT&T's total revenues for 1955 were greater than GE's, RCA's, and IBM's combined.
** In 1955 Remington Rand merged with the Sperry Corporation, a company with $441 million in sales, mostly defense-related.

A Primer on Computer Architecture

By the end of 1960 there were about 6,000 general-purpose electronic computers installed in the United States.[24] Nearly all of them were descendents of the EDVAC and IAS computer projects of the 1940s, where the concept of the stored program first appeared. One often hears that nothing has really changed in computer design since von Neumann.[25] That is true only in a restricted sense—computers still store their programs internally and separate storage from arithmetic funtions in their circuits. In most other ways there have been significant advances. By 1960 some of these innovations became selling points as different vendors sought to establish their products in the marketplace. The most important architectural features are summarized here.[26]

Word Length   The introduction of reliable core memory made it practical to fetch data in sets of bits, rather than one bit at a time as required by a delay line. For a computer doing scientific work, it seemed natural to have this set correspond to the number of digits required for a typical computation—say, from 7 to 12 decimal digits. That meant a block size, or *word length,* of from 30 to 50 bits. Longer word lengths were preferred for scientific calculations but increased the complexity and cost of the design. By 1960, additional factors made the word-length decision even more difficult.

Computers intended for commercial use did not need to handle numbers with many digits. Money handled in the 1950s seldom exceeded a million dollars, and two digits to the right of the decimal place were sufficient. Business-oriented computers could therefore use a shorter word length or a variable word length, if there was a way to tell the processor when the end of a word was reached. The IBM 702, IBM 1401, RCA 301, and RCA 501 had variable word lengths, with the end of a word set by a variety of means. The 1401 used an extra bit appended to each coded character to indicate whether or not it was the last one of a word; the 702 used a a special character that signified that the end was reached.[27] Although popular in the 1950s, computers with variable word lengths fell from favor in the following decades and are no longer common.

Computers intended for non-numeric applications, especially for controlling other devices, could also use a shorter word length. The most influential of these was the Whirlwind, which had a word length of 16 bits. Commercial machines that exploited the advantages of a short word length began to appear around 1959, and included the Control Data Corporation CDC 160 (12 bits) and the Digital Equipment Corporation PDP-1 (18 bits).

Register Structure   Processing units of early computers contained a set of circuits that could hold a numeric value and perform rudimentary arithmetic on it—usually nothing more than simple addition. This device became known as an *accumulator*, since sums could be built up or "accumulated" in it; a device with more general, though also temporary, storage ability is called a *register* (figure 2.4). Another set of circuits made up the *program counter*, which stored the location of the program instruction that the processor was to fetch from memory and execute.

The typical cycle of a processor was to fetch an instruction from memory, carry out that instruction on data in the accumulator, and update the program counter with the address of the next instruction. In the simplest case, the program counter was automatically incremented by one (hence the term "counter"), but branch instructions could specify that the counter point to a different memory location.

A computer program orders the processor to perform arithmetic or logic (e.g., add or compare), tells the processor where the relevant data are to be found, and tells it where to store results. As with the sequential fetching of instructions, often the processor requires data that are stored close to one another in memory. A program that performs the same operation on such a list of data might therefore consist of a long list of the same operation, followed by only a slight change in the address. Or the program could modify itself and change the address each time an operation is executed. (This idea may have originated with von Neumann in his early reports on computer design.) Neither process is elegant.

Designers began with an experimental computer at the University of Manchester in 1948, and added to the processor an extra *index register* to simplify working with arrays of data. (In early published descriptions of the Manchester computer, its designers called this register a "B-line," and used the symbol "A" for "accumulator," and "C" for "control." These terms persisted into the 1950s in descriptions of processors.) By specifying a value to increment the address field of an instruction, programs no longer had to modify themselves as envisioned by von Neumann and other pioneers. That greatly simplified the already difficult process of programming.[28]

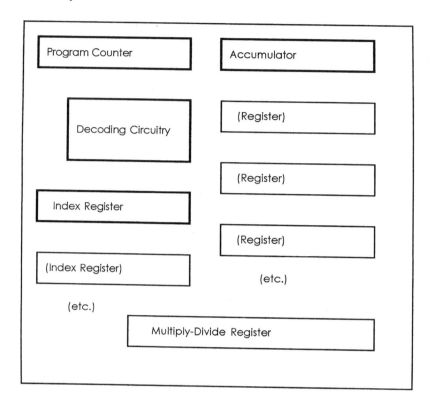

Figure 2.4

Computer central processing unit, or CPU, showing the different registers. The register labeled "Multiply–Divide" typically had twice the word length of the others, because, in general, a multiplication of two n-digit numbers gives a product having $2n$ digits. Below: a single-address instruction consisted of an operation field on one side, and an address field on the other. In between was a set of bits that performed special functions, such as specifying which of the index registers held a desired number.

These three types of registers—accumulator, program counter, and B-line or index register—made up the processing units of most large computers of the 1950s. For example, the IBM 704, announced in 1954, had a 36-bit word length, a core memory holding 4,096 words, and a processor with an accumulator, program counter, and three index registers. Another register was coupled to the accumulator and dedicated to multiplication and division (e.g., to store the extra bits that are generated when two 36-bit numbers are multiplied together).[29]

In 1956 the British firm Ferranti Ltd. announced a machine, called Pegasus, whose processor contained a set of eight registers, seven of which could be used as accumulators or as index registers. That inaugurated the notion of providing general-purpose registers that a program could use for any of the above functions, as needed by a specific program. Other companies were slow to adopt this philosophy, but by the end of the next decade it became the most favored design.[30]

Number of addresses   Instructions for an accumulator-based machine had two parts: The first specified the operation (e.g., add, subtract, or compare) and the second the address of the data to be operated on. If an operation required two pieces of data, the other operand needed to be present in the accumulator. It could be there as the result of the previous operation, or as a result of an explicit instruction to load it into the accumulator from memory. Because each instruction contained one memory reference, this was called a *single-address* scheme. Its virtue was its simplicity; it also allowed the address field of an instruction to be long enough to specify large portions of memory. Many computers built in the 1950s used it, including the original UNIVAC and IBM's series of large scientific computers, the 701, 704, 709, and 7090.

There were alternatives to the single-address scheme. One was to have an operation followed by two addresses, for both operands. A third address field could be added, to store the results of an operation in memory rather than assume they would remain in the accumulator. The UNIVAC 1103, RCA 601, and IBM 1401 used a two-address scheme, while the UNIVAC File Computer and Honeywell H-800 used a three-address scheme.

These schemes all had address fields that told where data were located. One could also include the address of the next *instruction*, instead of going to the program counter for it. Drum computers like the IBM 650 and Librascope LGP-30 used this to minimize the time spent searching the drum for the next instruction—each instruction could direct the computer to the place on the drum where the desired data would be, after executing the previous instruction. Programming this way was difficult, but it got around the inherently slow speeds of drum machinery. With the advent of the magnetic core, this scheme fell from necessity.[31]

Finally, one could design an instruction that specified no addresses at all: both operands were always kept in a specified set of registers, in the correct order. Results likewise went into that place, in the proper order for further processing. That required organizing a set of registers (or memory locations) in a structure called a *stack*, which presented data to the processor as Last-In, First-Out (LIFO). (The concept of LIFO came from the accounting profession; note its similarity to the spring-loaded stacks of plates found in cafeterias.)[32] Computers with this scheme first appeared in the 1960s, but they never seriously challenged the single-address design. Stack architecture resurfaced in the mid-1970s in programmable pocket calculators.

I/O Channels and the "Wheel of Reincarnation"   One of the UNIVAC's innovations was its use of a storage area that served as a "buffer" between the slow input and output equipment, such as card readers and electric typewriters, and the much faster central processor. Likewise the UNIVAC 1103A introduced the concept of the "interrupt," which allowed the machine's processor to work on a problem, stopping to handle the transfer of data to or from the outside world only when necessary. These innovations became well-established and were extended to the large commercial machines of the 1950s.

As the requirements for matching the speeds of I/O with the central processor grew more complex, so too did the devices designed to handle this transfer. With the introduction of the IBM 709, IBM engineers designed a separate processor, called a "channel," to handle input and output. This channel was, in effect, a small computer dedicated to the specific problem of managing a variety of I/O equipment that operated at different data rates.[33] Sometimes, as designers brought out improved models of a computer, they would add to the capabilities of this channel until it was as complex and powerful as the main processor—and they now had a two-processor system. At that point the elegant simplicity of the von Neumann architecture was in danger of being lost. To recapture it, the computer's basic requirements for processing and I/O had to be reconsidered. This so-called wheel of reincarnation, if not broken by a fresh design approach, threatened to lead to a baroque and cumbersome system.[34]

The complexity of I/O channels drove up the cost of systems, but they were necessary for customers who used computers for problems that handled large quantities of data. In time, channels became a defining characteristic of the *mainframe* computer—one that was expensive, physically large, and contained enough memory, flexibility, I/O facilities, and processing speed to handle the needs of large customers. The mainframe computer became the standard of the 1960s, although other classes would arise both at the higher end, where faster processing but simpler I/O was required, and at the lower end, where low cost was a major design goal.

Floating-point Hardware One final design feature needs to be mentioned, which was of concern primarily to scientific applications, but had an impact on commercial customers as well. In the words of one computer designer, whether or not a machine handles floating-point arithmetic in its hardware is the "biggest and perhaps only factor that separates a small computer from a large computer."[35]

Floating-point arithmetic allows users to keep track of the overall scale of a computation. It does so by dividing a quantity into two parts, one of which serves to mark the place of the decimal point (binary point inside the machine). For example, chemists often work with a quantity known as *Avogadro's number,* which they write as 6.02×10^{23}. When written in scientific notation, the number has two parts, the second of which indicates the magnitude. In this example, the 23 indicates that the decimal point belongs 23 places to the right of where it appears in the first part of the number. If written in ordinary notation, Avogadro's number would have to be written as 602 followed by 21 zeroes. It is familiar to scientists and engineers (sometimes under the term "scientific notation"), but almost unknown in the commercial world because commercial calculations do not reach beyond trillions of dollars at the upper end, nor to trillionths of a dollar at the low end. (Inflation has made million- and even billion-dollar figures a lot more common since 1980. At the same time, it has diminished the value of cents; people now do not even bother with pennies at the corner store. Although tax rates are still given in mills (one tenth of a cent), it is unlikely that anyone will calculate the value of a house, as Henry David Thoreau did, to the halfpenny.)

Any computer can be programmed to operate as a floating-point machine, but the programming is complex and slows the machine down. On the other hand, if the electronic circuits of a machine are designed to handle floating point in hardware, the processor will be more complicated and expensive. In the late 1940s there were debates over which approach was better, although it should be noted that the very first electromechanical computers, those built by Konrad Zuse in Germany and by Bell Labs in the United States in the 1940s, had floating point wired in.

Manufacturers felt that commercial customers did not need floating point and would be unwilling to pay for it. They typically offered two separate lines of machines, one for commercial customers and the other for scientific or engineering applications. The former, which included the UNIVAC, the UNIVAC File, and IBM 702 and 705, had only fixed-point arithmetic, and often a variable word length. The latter, like the 704 and the 709, had floating point and a relatively long, fixed word length. I/O facilities were often more elaborate for the business-oriented machines.

This parallel development of two almost-similar lines of equipment persisted through the late 1950s into the next decade. For many customers the distinction was not that clear. For example, the IBM 650, intended for commercial customers, was often installed in university centers, where professors and students labored to develop floating-point software for it. Likewise, the IBM 704 had better performance than the 705, because of what many felt was a superior architecture, and customers who ordered a 704 for scientific work soon found themselves using it for commercial applications as well. The preference for the 704 increased even more as the programming language FORTRAN became available on it. IBM combined both lines with the introduction of the System/360 line of computers in 1964.

The Transistor

Although the transistor as a replacement for the vacuum tube and core memory were both working in the laboratory by the early 1950s, the transistor was not reliable enough to be used in commercial computers until late in that decade, well after cores were in common use.

Bell Laboratories, where the transistor was invented, was not among those considering entering the commercial computer market in the 1950s. The company was a regulated monopoly and weighed every action it took with an eye on the federal courts. In early 1956, after seven years of litigation, it settled a lawsuit brought by the U.S. Justice Department with a consent decree in which it agreed not to enter into any business "other than the furnishing of common carrier communications."[36] AT&T had enough business installing telephones in the booming postwar suburbs anyway. If it did not use transistors for computers, it could use them to replace vacuum tubes in switching equipment and in telephone amplifiers.

AT&T, including Bell Laboratories and its Western Electric manufacturing arm, also had a substantial military business at the time. Bell Labs built several special-purpose computers for the military around 1952, including a digital data transmission set, and a special-purpose computer called TRADIC.[37] Throughout the 1950s and 1960s they provided computing equipment for Air Force and Army missile systems, including the NIKE and Titan. This laid the foundation for other companies, who after a decade of development finally began to supply commercial computers using transistors.

Philco

In part to satisfy federal regulators, Bell Labs made information about transistors available at a nominal cost. Among the many companies that began producing transistors was Philco, an established electronics firm headquartered in Philadelphia. Within a few years Philco pioneered a type of transistor, which they called "surface barrier," that could be made in quantity and that had good performance. Philco's transistors were used in the TX-0 experimental computer at MIT in 1954, and later in Philco's own machines.[38]

Philco's lead in producing surface-barrier transistors catapulted the company to the forefront of computing. In June 1955 it contracted with the National Security Agency to produce a fast computer based on the architecture of the UNIVAC 1103 (itself a commercial version of a computer built for that agency).[39] The result, called "SOLO," was completed sometime between 1956 and 1958, and was probably the first general-purpose transistorized computer to operate in the United States.[40] Philco marketed a commercial version called the TRANSAC S-1000, followed quickly by an upgraded S-2000, in 1958. First deliveries of the S-2000 were in January 1960. These, along with deliveries by UNIVAC of a smaller computer called the Solid State 80, mark the beginning of the transistor era, or "Second Generation."[41]

The S-2000, a large, expensive machine, sold well to those customers, especially in the U.S. Defense Department, who always demanded the top-ranking performance. Several commercial customers, including GE and United Aircraft, bought one as a replacement for the large IBM systems they were already using.[42] Having established a beachhead, Philco found itself under strong pressure to constantly innovate and upgrade its products. That demanded money and manpower the company was unwilling, or unable, to provide. In 1962, shortly after being acquired by Ford Motor Company, Philco dropped out of the computer business.

NCR, Burroughs

The development of transistors also affected the computer business of National Cash Register, which entered the computer industry in 1954 by buying the Computer Research Corporation (CRC), a company founded by a group of engineers from Northrop Aircraft. Although during World War II NCR had built specialized equipment to assist the U.S. military in breaking enemy codes, that had not led it away from its prewar focus on mechanical cash registers. Between 1954 and 1957 the company did not market CRC-inspired electronic computers aggressively and when in 1957 NCR announced the model 304, a large transistorized computer, it turned over production and manufacturing to GE. Modest sales of the 304 (eventually thirty-three were installed) contributed to NCR's later success (and kept GE from leaving the computer field entirely).[43] With its focus on retail sales, banking, and a few other specialized, but large and profitable, commercial niches, the company did not always prosper, but it did survive a painful transition from the era of brass cash registers to one of electronics. AT&T bought the company in 1991, in an attempt to become a competitor in commercial computing. AT&T failed, and it spun off NCR as an independent company in 1995.

The Burroughs Corporation, a manufacturer of adding machines and banking equipment, entered the transistorized era like NCR. In 1956 Burroughs purchased a small firm that made a drum-based scientific computer, Electrodata, a division of Consolidated Engineering of Pasadena. Staffed by engineers who had close ties to the Jet Propulsion Laboratory, Electrodata had built a computer that was slow, like other drum-based computers, but had a superior architectural design. It was among the first American-made computers to have index registers in its processor,[44] which made it a favorite among the more knowledgeable customers, particularly in universities.

Burroughs was already offering a machine it designed in-house, the E-101, which it felt was better suited for the typical commercial customers that bought its accounting machines. The E-101 was only a little bigger than a desk, and inexpensive, selling for about $35,000. It lacked a stored-program architecture.[45] It did not sell very well. Once again, the marketplace chose a well-designed stored-program computer over a less expensive machine intended for specific needs.

Burroughs's failure with the E-101, combined with its failure to adapt the superior Electrodata design to transistors quickly enough, kept Burroughs's influence in the commercial market small. It was able to keep *its* computer expertise intact through military contracts, including a large contract to build specialized computers for the SAGE air defense system. Another contract, for the guidance system for the Atlas inter-continental ballistic missile, led to a transistorized computer that successfully guided Atlas launches from 1957 through 1959. Burroughs claimed that this was "the first operational computer to use transistors rather than vacuum tubes."[46] The SOLO computer, described above, may have preceded it, but SOLO's existence was kept secret for many years. That experience laid a foundation for Burroughs's successful re-entry into commercial computing in the 1960s.

The Rise of IBM

By 1960 IBM dominated the computer industry. With that came an intense interest in its stock, which in turn gave rise to a legion of financial and technical analysts who watched the company and tried to discern its next move. IBM was always careful of its public image, in part a legacy of the humiliation felt by Thomas Watson Sr. over what he felt was an unfair and unwarranted conviction for violating antitrust laws while he was working for National Cash Register in 1912.[47] IBM's public relations department always courteously and promptly supplied outsiders with information, but the information was always carefully structured to reflect an image.

That, in turn, led to groups of people who set out to debunk, disparage, or otherwise dismantle the company. Among them were, of course, IBM's main competitors, but they also included both computer professionals from universities and officials in the U.S. Justice Department. In 1952, before IBM entered the electronic computer business, the Justice Department had alleged that it violated anti-trust laws in conducting its punched card business. That led, in 1956, to a Consent

Decree, which did not break IBM up but deeply affected it and the computer industry. Perhaps its most important provision was that IBM was to allow customers to purchase its machines on terms comparable to those it offered to renters. That signaled a major shift for IBM, and would eventually lead to a large sub-industry of companies that purchased IBM mainframes and leased them out to other customers.[48]

As IBM's stock soared and made millionaires of many employees who had bought shares in earlier decades, "IBM-ologists," at once fascinated and repelled by the company, searched for clues to its inner workings. Like "Kremlin-ologists," who measured how tall various officials appeared in official Soviet photographs, IBM-watchers combed whatever they could find—press clippings, product announcements, figures buried within financial reports—for patterns.[49] Nowhere was this more evident than in the pages of *Datamation*, a trade journal founded as *Research & Engineering* and oriented toward the computer industry in 1957. The magazine's owners at first intended to write it off for tax purposes, but they soon recognized that *Datamation*'s subject was one of growing interest, and also a very good source of advertising revenue.[50] Under the editorial hand of Santo (Sandi) Lanzarotta, and especially Robert Forest after 1963, *Datamation* developed a perfect balance: relentless criticism of IBM for its alleged heavy hand, tempered by a passion for computing and its benefits to society, a passion that most readers recognized was one that IBM shared and furthered in its own way.

Critics charged that IBM was never an innovator but always waited until another, smaller company took the technical risks, and then swept in and took over by questionable marketing practices. IBM was late in recognizing the future of electronic computing, at least compared to Eckert and Mauchly. And some regarded the IBM 701 as inferior to the UNIVAC because of its I/O design. On the other hand, the IBM 704, with its floating-point arithmetic, FORTRAN programming language, and core memory, was technically superior to what UNIVAC offered in 1956, by which time the original UNIVAC was then obsolete. Sales of the 704 were a major factor in IBM's ascendancy, and those sales were not entirely the result of the company's marketing and sales force.

A closer look at this charge reveals that IBM also made up for its lag in technical innovation by superior in-house manufacturing techniques and field service. These efforts merged into marketing and sales, which was always aggressive and which competitors often felt was unfair.

However, with IBM it is not always possible to separate the purely technical dimension from the salesmanship. The following two examples may serve to illustrate this merging of marketing, manufacturing, and technical innovation.

Disk Storage   In 1957 IBM marketed a device that would prove to be one of its most lasting: the spinning disk for random-access storage of large quantities of data. The disk drive was a cousin to the drum store that had been a mainstay of electronic computers since the beginning of the decade, but it had a geometry that exposed more surface area to magnetization and storage. IBM perfected and brought to market an innovation that had originated in the mid-1940s with Presper Eckert, who had suggested using a disk for program and data storage for the ENIAC's successor. And in 1952 the National Bureau of Standards was working on a disk store in response to an order by the Ballistic Research Laboratory at Aberdeen, Maryland.[51]

Using an array of spinning disks offered much greater capacity at lower cost per bit than drums, but it also presented knotty technical problems. Typical drum stores had a set of fixed heads rigidly fastened along a line, one for each track on the drum. The whole mechanism could be made rugged and stable, but it was expensive. A disk array could never be made as rigid as a drum, and it seemed the only way to access the surfaces of the disks was to have heads that could be positioned over the disk like the stylus of a record player. To record and read data, the heads had to be very close to the surfaces. If they touched (as the head of a tape drive touches the tape), the high speeds of the spinning disks would probably cause unacceptable wear.

IBM engineers, working at the newly established laboratory in San Jose, California, came up with the notion of using a thin film of air as a cushion to prevent the heads from touching the disk surface. Their first product used air supplied to the heads from an external compressor. Later IBM disk drives took advantage of an ingenious application of the "boundary layer"—an aerodynamic phenomenon familiar to airplane designers—to float the disk by air set in motion by the disk itself. The labs spent 1953 and 1954 experimenting with a bewildering variety of head geometries and positioning mechanisms.[52]

By 1956 IBM had solved the problems and publicly announced the Model 305 Disk Storage unit. It used a stack of fifty aluminum disks, each 24 inches in diameter, rotating at 1200 rpm. Total storage capacity was five million characters.[53] The press release emphasized its revolutionary quality of "random access": that any piece of data was as accessible as any other, in contrast to the sequential retrieval of a datum from a deck of punched cards or a reel of tape. The machine was later rechristened RAMAC, for Random Access Memory Accounting Machine. IBM also announced that the first commercial customer, United Airlines in Denver, would use it for a reservations system. In the spring of 1958 IBM installed a 305 at the U.S. Pavilion at the Brussels World's Fair, where visitors could query "Professor RAMAC" through a keyboard and receive answers in any of ten languages (including Interlingua, an artificial language like Esperanto).[54] The Brussels exhibit was a masterful piece of IBM public relations. Like the famous demonstration of remote computing staged by George Stibitz at the Dartmouth meeting of the American Mathematical Society in 1940; it foretold a day when direct access to large amounts of data would become essential to the operations of banks, supermarkets, insurance companies, and government agencies. IBM president, Thomas Watson Jr., heralded the RAMAC's introduction as "the greatest product day in the history of IBM," but even he did not fully understand the forces set in motion by Professor RAMAC. In time, the interactive style of computing made possible by random access disk memory would force IBM, as well as the rest of the computer industry, to redefine itself.

The RAMAC was offered as a separate piece of equipment to be attached to the IBM 650, and like the 650 it used vacuum tubes. Within a few years of its introduction it was obsolete and was withdrawn from the market. But the disk technology survived and furthered IBM's dominance of the industry. Eventually direct access to data that allowed users to interact directly with a computer spelled the end of the batch method of processing, on which IBM had built its financial strength ever since the tabulator days.

From Vacuum Tubes to Transistors   A second example of IBM's strength is its introduction of model 7090 only a year after it began deliveries of model 709.

The 7090, a large transistorized computer that many regard as the classic mainframe computer, showed that IBM accepted the obsolescence of vacuum tube technology and that it was willing to take a financial loss on the 709. According to folklore, IBM submitted a bid to the U.S. Air Force to supply solid state computers for the Ballistic Missile Early Warning System (BMEWS) around the Arctic Circle. At the time IBM had announced the 709, but the Air Force insisted on transistorized

machines, perhaps because other, smaller companies had already introduced transistorized products that were getting praise from the trade
press.

IBM planned to meet the Air Force's strict timetable by designing a
computer that was architecturally identical to the 709 and used transistors. They were thus able to use a 709 to develop and test the software
that the new computer would need. The 709 was programmed to
"emulate" the as-yet-unfinished new machine: a program was written
to make the 709 behave as if it were the new computer. (This has since
become a standard procedure whenever a new computer or computer
chip is being designed.) Even that technique did not guarantee that IBM
would meet the Air Force's deadline of installations before 1960. IBM
delivered computers to a site in Greenland in late 1959, but IBM-
watchers claimed that the machines were not finished and that the
company dispatched a cadre of up to 200 engineers to Greenland to
finish the machine as it was being installed.[55]

Whether or not that story is true, the company did deliver a transistorized computer to Greenland, versions of which it marketed commercially as the Model 7090. The 7090 and its later upgrade, called the 7094,
which had four additional index registers, is regarded as the classic
mainframe because of its combination of architecture, performance,
and financial success: hundreds of machines were installed at a price of
around $2 million each (figure 2.5).

A Description of a 7094 Installation   The term "mainframe" probably
comes from the fact the circuits of a mainframe computer were
mounted on large metal frames, housed in the cabinets. The frames
were on hinges and could swing out for maintenance. A typical installation consisted of a number of these cabinets standing on a tiled floor
that was raised a few inches above the real floor, leaving room for the
numerous thick connecting cables that snaked from one cabinet to
another and for the circulation of conditioned air. The entire room
probably had its own climate-control system that not only was separate
from that of the rest of the building, but also kept the room much more
comfortable than anywhere else in the building on a hot summer day.

A cabinet near the operator's console housed the main processor
circuits. These were made up of discrete transistors, mounted and
soldered along with resistors, diodes, jumper wires, inductors, and
capacitors, onto printed circuit boards. The boards, in turn, were
plugged into a "backplane," where a complex web of wires carried

Figure 2.5
Control console of an IBM 7094. Note the small box on top of the console with four rows of lights, which indicate the status of the computer's four additional index registers. The addition of these registers was the principal difference between the 7094 and the IBM 7090. Note also the other rows of lights indicating the bits in the other registers of the CPU. The rows at the bottom are labeled "Storage," "Accumulator," and "M-Q" (for multiply-quotient). (*Source*: IBM.)

signals from one circuit board to another. Some mainframes were laboriously wired by hand, but most used a technique called "wire wrap": it required no soldering, and for production machines the wiring could be done by a machine, thus eliminating errors. In practice, there would always be occasional pieces of jumper wire soldered by hand to correct a design error or otherwise modify the circuits. The density of these circuits was about ten components per cubic inch.

The 7094 was delivered with a maximum of 32,768 words of core memory. In modern terms that corresponds to about 150 Kilobytes (Kbytes, one byte = 8 bits), about what came with the IBM Personal Computer when it first appeared in the early 1980s. Although marketed as a machine for science and engineering, many customers found it well suited for a variety of tasks. It could carry out about fifty to one-hundred-thousand floating-point operations per second, making it among the fastest of its day. Comparisons with modern computers are difficult, as the yardsticks have changed, but it was about as fast as a personal computer of the late 1980s. Its 36-bit word length suited it for scientific calculations that require many digits of precision, and it had the further advantage of allowing the processor to address a lot of memory directly. By comparison, the first personal computers used only an 8-bit word length; the 32-bit length that had become a standard by 1990 was still shorter than what the 7094 machine had.

The console itself was festooned with an impressive array of blinking lights, dials, gauges, and switches. It looked like what people thought a computer should look like. Even into the 1990s some Hollywood movies portrayed computers like this. (A few others, like *Jurassic Park*, showed modest UNIX workstations.) Rows of small lights indicated the status of each bit of the various registers that made up the computer's central processor. In the event of a hardware malfunction or programming error, operators could read the contents of each register directly in binary numbers. They could also execute a program one step at a time, noting the contents of the registers at each step. If desired, they could directly alter the bits of a register by flipping switches. Such "bit twiddling" was exceedingly tedious, but it gave operators an intimate command over the machine that few since that time have enjoyed.

Most of the time, an operator had no need to do those kinds of things. The real controlling of the computer was done by its programmers, few of whom were ever allowed in the computer room. Programmers developed their work on decks of punched cards, which were read by a small IBM 1401 computer and transferred to a reel of tape. The operator took this tape and mounted it on a tape drive connected to the mainframe (although there was a card reader directly attached for occasional use). Many programmers seldom saw the machine that actually ran the programs. In fact, many programmers did not even use a keypunch, but rather wrote out their programs on special coding sheets, which they gave to keypunch operators.[56] The operator's job consisted of mounting and demounting tapes, pressing a button to start a job every now and then, occasionally inserting decks of cards into a reader, and reading status information from a printer. It was not a particularly interesting or high-status job, though to the uninitiated it looked impressive. Over the course of the 7094's heyday, many of the operator's jobs that required judgment were taken over by a control program, appropriately called an "operating system."

A 7094 installation rented for about $30,000 a month, or an equivalent purchase price of about $1.6 million. With that cost it was imperative that the machine never be left idle. Although our personal computers run a screen-saver while we go to a meeting or to lunch, the number of computer cycles wasted by this practice would have been scandalous in 1963. On the 7094, programs were gathered onto reels of tape and run in batches. Programmers had to wait until a batch was run to get their results, and if they then found that they had made a mistake or needed to further refine the problem, they had to submit a new deck and wait once more. However tempting, the idea of gaining direct access to the machine—to submit a program to it and wait a few seconds while it ran—was out of the question, given the high costs of letting the processor sit idle for even a few minutes. That method of operation was a defining characteristic of the mainframe era.

Along with the processor circuit cabinets, magnetic tape drives dominated a mainframe installation. These tapes were the medium that connected a mainframe computer to the outside world. Programs and data were fed into the computer through tapes; the results of a job were likewise sent to a tape. If a program ran successfully, an operator took the tape and moved it to the drive connected to a 1401 computer, which, like a "smart" printer of the Personal Computer era, handled the slower process of printing out results on a chain printer. (Unlike a modern printer, there was typically no direct connection.) Results were printed, in all capital letters, on 15-inch wide, fan-folded paper.

A few mainframes had a video console, but there was none on the 7094's main control panel. Such a console would have been useful only for control purposes, since the sequential storage on tapes prevented direct access to data anyway. In general, they were not used because of their voracious appetite for core memory.

With the advent of personal computers and workstations, mainframes are often viewed as dinosaurs that will not survive the turn of the century. In fact, the mainframe has survived and even prospered, because of its ability to store and move large quantities of data. But the sequential, batch-oriented mode of operation, with its characteristic libraries of tapes, decks of punched cards, and printouts, has ceased to dominate, as it did in the 1960s when there was no alternative.

Small Transistorized Machines

The maturing of transistor technology by the end of the 1950s also had an impact at the low end of the computer industry. It became possible to offer solid state computers at low cost and with much better performance than the drum machines of a only a few years earlier. The use of transistors also made these machines compact, allowing them to be installed in the same rooms that previously housed the array of tabulators and card punches that typically handled a company's data-processing needs.

Once again, the most successful of these machines came from IBM, the model 1401, announced in 1959 (figure 2.6). If the 650 had demonstrated that the market for computers was potentially very large, the 1401, intended for business customers, showed that the market was indeed a real one and included nearly every place where punched card equipment was used. Eventually over ten thousand 1401s were installed—compare this number to the thirty to forty UNIVACs and IBM 701s, or the approximately one thousand installations of the IBM 650. At the same time IBM also introduced the 1620, a small computer with a slightly different architecture, for scientific customers, and it, too, sold well.

The 1401 had a plugboard architecture throughout its early phases of development. As happened with the 650's design, in 1955–1956 IBM engineers redesigned it to incorporate a stored-program architecture which allowed it to be programmed as a general-purpose computer.[57] The 1620 carried this to an extreme: most of its arithmetic was done not by logic circuits wired into the processor, but by references to arithmetic tables stored in the cheaper (but slower) core memory. Some savvy customers even altered the 1620's instruction set and logic by the simple act of storing different numbers in the memory locations the computer looked to for the sums of numbers. They said that the informal name "CADET" was an acronym for "Can't Add; Doesn't Even Try!" But it worked. (The 1620's ability to do arithmetic with such a primitive set of circuits would later inspire one of the inventors of the microprocessor to do the same on a sliver of silicon.)

The 1401 offered modest performance. With its variable word length, its processing speed varied, and on average it was only about seven times faster than a 650. Unlike the 650 it was rarely used for scientific or engineering applications, although most scientific mainframe installations used one or more to transfer data from cards to tape and to print.

Figure 2.6
IBM 1401 production line. The 1401 was the first electronic digital computer to be produced in quantities comparable to IBM's line of punched-card accounting machines. (*Source*: IBM.)

Whether used alone or as an adjunct to a large computer, the 1401's success owed a lot to a piece of peripheral equipment that IBM introduced with it, the type 1403 printer. This printer used a continuous chain of characters that moved laterally across the page. Magnetically driven hammers struck the chain at the precise place where a desired character was to be printed. Capable of printing 600 lines per minute, it was much faster than anything else on the market, and it was rugged enough to stand up under heavy use.[58] The printer used old-fashioned mechanical technology to complement the exotic electronic processing going on in the mainframe. It was a utilitarian device but one that users had an irrational affection for. At nearly every university computer center, someone figured out how to program the printer to play the school's fight song by sending appropriate commands to the printer. The quality of the sound was terrible, but the printer was not asked to play Brahms. Someone else might use it to print a crude image of Snoopy as a series of alphabetic characters. In and out of Hollywood, the chattering chain printer, spinning tapes, and flashing lights became symbols of the computer age.

Conclusion

By 1960 a pattern of commercial computing had established itself, a pattern that would persist through the next two decades. Customers with the largest needs installed large mainframes in special climate-controlled rooms, presided over by a priesthood of technicians. These mainframes utilized core memories, augmented by sets of disks or drums. Backing that up were banks of magnetic tape drives, as well as a library where reels of magnetic tape were archived. Although disks and drums allowed random access to data, most access conformed to the sequential nature of data storage on tapes and decks of cards.

For most users in a university environment, a typical transaction began by submitting a deck of cards to an operator through a window (to preserve the climate control of the computer room). Sometime later the user went to a place where printer output was delivered and retrieved the chunk of fan-fold paper that contained the results of his or her job. The first few pages of the printout were devoted to explaining how long the job took, how much memory it used, which disk or tape drives it accessed, and so on—information useful to the computer center's operators, and written cryptically enough to intimidate any user not initiated into the priesthood.

For commercial and industrial computer centers, this procedure was more routine but essentially the same. The computer center would typically run a set of programs on a regular basis—say, once a week—with new data supplied by keypunch operators. The programs that operated on these data might change slightly from one run to the next, although it was assumed that this was the exception rather than the rule. The printouts were "burst" (torn along their perforations), bound between soft covers, and placed on rolling racks or on shelves. These printouts supplied the organization with the data it needed to make decisions and to run its day to day operations.

Thus the early era of computing was characterized by batch processing. The cost of the hardware made it impractical for users to interact with computers as is done today. Direct interactive access to a computer's data was not unknown but was confined to applications where cost was not a factor, such as the SAGE air defense system. For business customers, batch processing was not a serious hindrance. Reliance on printed reports that were a few days out of date was not out of line with the speeds of transportation and communication found elsewhere in society. The drawbacks of batch processing, especially how it made writing and debugging programs difficult, were more noticed in the universities, where the discipline of computer programming was being taught. University faculty and students thus recognized a need to bring interactive computing to the mainstream. In the following years that need would be met, although it would be a long and difficult process.

Table 2.2 lists the characteristics of some of the machines discussed in this chapter.

Table 2.2
Characteristics of selected computers discussed in this chapter

Name	Year announced or installed	Word length	Words of main memory	Device type
SAGE	1955–1958	32 bits	8 K	Tubes
Philco TRANSAC-2000	1958	—	—	Transistors
RCA 501	1958	12 decimal digits		Transistors
IBM 1401	1959	variable (7 bits/char.)	4–16 K	Transistors
IBM 7090	1960	36 bits	32 K	Transistors

3

The Early History of Software, 1952–1968

He owned the very best shop in town, and did a fine trade in soft ware, especially when the pack horses came safely in at Christmas-time.
—R. D. Blackmore, *Lorna Doone*[1]

There will be no software in this man's army!
—General Dwight D. Eisenhower, ca. 1947[2]

In 1993 the National Academy of Engineering awarded its Charles Stark Draper Prize to John Backus, "for his development of FORTRAN . . . the first general-purpose, high-level computer language."[3] The Academy's choice was applauded by most computer professionals, most of whom knew well the contribution of Backus and FORTRAN to their profession. FORTRAN, although currently still in use, has long been superseded by a host of other languages, like C++ or Visual Basic, as well as by system software such as UNIX and Windows, that reflect the changing hardware environment of personal computers and workstations. In accepting the award, Backus graciously acknowledged that it was a team effort, and he cited several coworkers who had labored long and hard to bring FORTRAN into existence.

The Draper Prize was instituted to give engineers the prestige and money that the Nobel Prize gives scientists. Here it was being awarded for developing a piece of software—something that, by definition, has no physical essence, precisely that which is not "hardware." The prize was being awarded for something that, when the electronic computer was first developed, few thought would be necessary. Not only did it turn out that software like FORTRAN was necessary; by the 1990s its development and marketing overshadowed hardware, which was becoming in some cases a cheap mass-produced commodity. How did the entity now called "software" emerge, and what has been its relationship to the evolution of computer hardware?

A simple definition of software is that it is the set of instructions that direct a computer to do a specific task. Every machine has it. Towing a boat through a lock of a nineteenth-century canal required a performing sequence of precise steps, each of which had to be done in the right order and in the right way. For canal boats there were two sets of procedures: one for getting a boat from a lower to a higher level, and one for going the other way. These steps could be formalized and written down, but no canal workers ever called them "software." That was not because the procedures were simple, but because they were intimately associated with the single purpose of the lock: to get a canal boat from one level stretch to another. A canal lock may have secondary purposes, like providing water for irrigation, but these are not the reasons the lock is designed or installed.

A computer, by contrast, does not specify any single problem to be solved. There is no division into primary and secondary functions: a stored-program digital computer is by nature a general-purpose machine, which is why the procedures of users assume greater importance. These procedures should be considered separate from the machine on which they run.

The word "software" suggests that there is a single entity, separate from the computer's hardware, that works with the hardware to solve a problem. In fact, there is no such single entity. A computer system is like an onion, with many distinct layers of software over a hardware core. Even at the center—the level of the central processor—there is no clear distinction: computer chips carrying "microcode" direct other chips to perform the processor's most basic operations. Engineers call these codes "firmware," a term that suggests the blurred distinction.

If microcode is at one end, at the other one encounters something like an automatic teller machine (ATM), on which a customer presses a sequence of buttons that causes a sophisticated computer network to perform a complex set of operations correctly. The designers of ATMs assume that users know little about computers, but just the same, the customer is programming the bank's computer. Using an ATM shares many of the attributes of programming in the more general sense. Pressing only one wrong key out of a long sequence, for example, may invalidate the entire transaction, and a poorly designed ATM will confuse even a computer-literate customer (like the home video-cassette recorder, which most owners find impossible to program).

Somewhere between these extremes lies the essence of software. One programmer, Scott Kim, said that "there is no fundamental difference between programming a computer and using a computer."[4] For him the layers are smooth and continuous, from the microcode embedded in firmware to the menu commands of an ATM, with his own work lying somewhere in the middle. (Kim is a designer of personal computer software.) Others are not so sure. People who develop complex system software often say that their work has little to do with the kind of computer programing taught in schools. What is worse, they feel that the way computer programming is taught, using simple examples, gives students a false sense that the production of software is a lot simpler than it is.[5] They also point out that developing good software is not so much a matter of writing good individual programs as it is of writing a variety of programs that interact well with each other in a complex system.

The history of software should not be treated separately from the history of computing, even if such a distinction is of value to computer engineers or scientists (figure 3.1). Several of the examples that follow will show innovations in software that had little or no impact until they could mesh well with corresponding innovations in hardware.[6] Likewise, the often-repeated observation that progress in hardware, measured by metrics like the number of circuits on a silicon chip, far outpaces progress in software is probably false.[7] While it is true that hardware technology must face and overcome limits of a physical, tangible nature, both face and overcome the much more limiting barrier of complexity of design.[8]

Beginnings (1944–1951)

In order to program the electromechanical Harvard Mark I, users punched a row of holes (up to 24 in a line) on a piece of paper tape for each instruction.[9] In the summer of 1944, when the machine was publicly unveiled, the Navy ordered Grace Murray Hopper to the Computation Lab to assist Howard Aiken with programming it. Hopper had been a professor of mathematics at Vassar College and had taken leave to attend the Navy's Midshipmen School. According to Hopper, she had just earned her one and one-half stripes when she reported to the lab at Harvard. There, Howard Aiken showed her

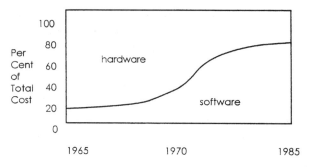

Figure 3.1
Relative costs of software vs. hardware for typical systems, 1965–1985. This famous graph was popularized in the early 1970s by Barry Boehm, then of TRW. The graph has been reprinted in numerous textbooks and articles about software development and has become one of the great myths of software. As with any myth there is much truth in this graph, but more recent studies of software expenditures seem to conclude that over the years the ratio has remained more or less constant. (*Source*: Adapted from Barry Boehm, "Software and its Impact," *Datamation* [May 1973]: 49.)

a large object, with three stripes... waved his hand and said: "That's a computing machine." I said, "Yes, Sir." What else could I say? He said he would like to have me compute the coefficients of the arc tangent series, for Thursday. Again, what could I say? "Yes, Sir." I did not know what on earth was happening, but that was my meeting with Howard Hathaway Aiken.[10]

Thus began the practice of computer programming in the United States. Hopper wrote out the sequence of codes for that series, and later the codes for more complex mathematical expressions—one of the first was a lens design problem for James Baker (a Harvard Fellow known among insider circles for his design of lenses for the top-secret cameras used by U.S. intelligence agencies).[11]

Some sequences that were used again and again were permanently wired into the Mark I's circuits. But these were few and their use did not appreciably extend its flexibility. Since the Mark I was not a stored-program computer, Hopper had no choice for other sequences than to code the same pattern in successive pieces of tape.[12] It did not take long for her to realize that if a way could be found to reuse the pieces of tape already coded for another problem, a lot of effort would be saved. The Mark I did not allow that to be easily done, but the idea had taken root and later modifications did permit multiple tape loops to be mounted.

In the design of a later Harvard calculator (the Mark III), Howard Aiken developed a device that took a programmer's commands, typed on a keyboard in the notation of ordinary mathematics, and translated them into the numerical codes that the Mark III could execute (figure 3.2). These codes, recorded on a magnetic tape, were then fed into the Mark III and carried out. Frequently used sequences were stored on a magnetic drum. In Germany, Konrad Zuse had independently proposed a similar idea: he had envisioned a "Plan Preparation Machine" (*Planfertigungsgeräte*) that would punch tapes for his Z4 computer, built during World War II.[13] Zuse's device would not only translate commands but also check the user's input to ensure that its syntax was correct, that is, that it had the same number of left and right parentheses, that more than one arithmetic operation did not appear between two numbers, and so on.

Figure 3.2
A programming machine attached to the Harvard Mark III, ca. 1952. The operator is Professor Ambros P. Speiser, of the Federal Technical Institute of Zurich. Programs would be keyed into this device in a language similar to ordinary algebra, and the machine would translate it into the codes that the Mark III proper could execute. With a stored program computer this additional piece of hardware is unnecessary. (*Source*: Gesellschaft für Mathematik und Datenverarbeitung [GMD], Bonn, Germany.)

Zuse never completed the Plan Preparation Machine, although he had refined its design (and changed its name to "Programmator") by 1952. Meanwhile, his Z4 computer had been refurbished and installed at the Federal Technical Institute in Zurich. While using it there, Heinz Rutishauser recognized an important fact: that a general-purpose computer could *itself* be programmed to act like such a "Programmator," getting rid of the need for a separate machine. Solving a problem would thus take two steps: one in which the computer is programmed to check and translate the user's commands, and another to carry out these commands, which are now encoded in numerical code on a piece of tape.[14] Rutishauser stated it simply: "Use the computer as its own Plan Preparation Machine."[15]

None of the machines described above stored their programs in internal memory, which meant that programming them to translate a user's commands as Rutishauser envisioned would have been very difficult. The Zuse machine, however, had a flexible and elegant design, which inspired Rutishauser to see clearly how to make computers easier to program. Like Hopper's realization that the tapes she was preparing could be used more than once, Rutishauser's realization that the same computer that solved a problem could prepare its own instructions was a critical moment in the birth of software.

With a stored-program computer, a sequence of instructions that would be needed more than once could be stored on a tape. When a particular problem required that sequence, the computer could read that tape, store the sequence in memory, and insert the sequence into the proper place(s) in the program. By building up a library of sequences covering the most frequently used operations of a computer, a programmer could write a sophisticated and complex program without constant recourse to the binary codes that directed the machine. Of the early stored-program computers, the EDSAC in Cambridge, England, carried this scheme to the farthest extent, with a library of sequences already written, developed, and tested, and punched onto paper tapes that a user could gather and incorporate into his own program.[16] D. J. Wheeler of the EDSAC team devised a way of storing the (different) addresses of the main program that these sequences would have to jump to and from each time they were executed. This so-called Wheeler Jump was the predecessor of the modern subroutine call.[17]

UNIVAC Compilers (1952)

If these sequences were punched onto decks of cards, a program could be prepared by selecting the appropriate decks, writing and punching transitional codes onto cards, and grouping the result on a new deck of cards. That led to the term "to compile" for such activity. By the early 1950s, computer users developed programs that allowed the computer to take over these chores, and these programs were called "compilers." Grace Hopper (1906–1992) played a crucial role in transferring that concept from Howard Aiken's laboratory at Harvard to the commercial world. Even though she had a desire to remain in uniform and the Navy had offered her continued employment in computing, John Mauchly was able to persuade her to join him and work on programming the UNIVAC as it was being built.[18] (She eventually returned to active duty in the Navy and reached the rank of rear admiral at her retirement.)[19]

Hopper defined "compiler" as "a program-making routine, which produces a specific program for a particular problem."[20] She called the whole activity of using compilers "Automatic Programming." Beginning in 1952, a compiler named "A-0" was in operation on a UNIVAC; it was followed in 1953 by "A-1" and "A-2." A version of A-2 was made available to UNIVAC's customers by the end of that year; according to Hopper they were using it within a few months.[21]

The term "compiler" has come into common use today to mean a program that translates instructions written in a language that human beings are comfortable with, into binary codes that a computer can execute. That meaning is not what Hopper had in mind.[22] For her, a compiler handled subroutines stored in libraries.[23] A compiler method, according to Hopper's definition, was a program that copied the subroutine code into the proper place in the main program where a programmer wanted to use it. These subroutines were of limited scope, and typically restricted to computing sines, cosines, logs, and, above all, floating-point arithmetic. Compilers nonetheless were complex pieces of software. To copy a routine that computed, say, the log of a number, required specifying the location of the number it was to calculate the log of, and where to put the results, which would typically be different each time a program used this specific subroutine.[24] The metaphor of "assembling" a program out of building blocks of subroutines, though compelling, was inappropriate, given the difficulty of integrating subroutines into a seamless flow of instructions. The goal for proponents of Automatic Programming was to develop for software what Henry Ford had developed for automobile production, a system based on inter-changeable parts. But just as Ford's system worked best when it was set up to produce only one model of car, these early systems were likewise inflexible, they attempted to standardize prematurely and at the wrong level of abstraction. But it was only in making the attempt that they realized that fact.[25]

Laning and Zierler (1954)

The first programming system to operate in the sense of a modern compiler was developed by J. H. Laning and N. Zierler for the Whirlwind computer at the Massachusetts Institute of Technology in the early 1950s. They described their system, which never had a name, in an elegant and terse manual entitled "A Program for Translation of Mathematical Equations for Whirlwind I," distributed by MIT to about one-hundred locations in January 1954.[26] It was, in John Backus's words, "an elegant concept elegantly realized." Unlike the UNIVAC compilers, this system worked much as modern compilers work; that is, it took as its input commands entered by a user, and generated as output fresh and novel machine code, which not only executed those commands but also kept track of storage locations, handled repetitive loops, and did other housekeeping chores. Laning and Zierler's "Algebraic System" took commands typed in familiar algebraic form and translated them into machine codes that Whirlwind could execute.[27] (There was still some ambiguity as to the terminology: while Laning and Zierler used the word "translate" in the title of their manual, in the Abstract they call it an "interpretive program."[28]

One should not read too much into this system. It was not a general-purpose programming language but a way of solving algebraic equations. Users of the Whirlwind were not particularly concerned with the business applications that interested UNIVAC customers. Although Backus noted its elegance, he also remarked that it was all but ignored, despite the publicity given Whirlwind at that time.[29] In his opinion, it was ignored because it threatened what he called the "priesthood" of programmers, who took a perverse pride in their ability to work in machine code using techniques and tricks that few others could fathom, an attitude that would persist well into the era of personal computers. Donald Knuth, who surveyed early programming systems in 1980, saw another reason in the allegation that the Laning and Zierler system was slower by a factor of ten than other coding systems for Whirlwind.[30] For

Knuth, that statement, by someone who had described various systems used at MIT, contained damning words.[31] Closing that gap between automatic compilers and hand coding would be necessary to win acceptance for compiler systems and to break the priesthood of the programmers.

Assemblers

These systems eventually were improved and came to be known as Programming *Languages*. The emergence of that term had to do with their sharing of a few restricted attributes with natural language, such as rules of syntax. The history of software development has often been synonymous with the history of high-level programming languages— languages that generated machine codes from codes that were much closer to algebra or to the way a typical user might describe a process. However, although these so-called high-level languages were important, programming at many installations continued to be done at much lower levels well into the 1960s. Though also called "languages," these codes typically generated only a single, or at most a few, machine instructions for each instruction coded by a programmer in them. Each code was translated, one-to-one, into a corrresponding binary number that the computer could directly execute. A program was not compiled but "assembled," and the program that did that was called an "assembler." There were some extensions to this one-to-one correspondence, in the form of "macro" instructions that corresponded to more than one machine instruction. Some commercial installations maintained large libraries of macros that handled sorting and merging operations; these, combined with standard assembly-language instructions, comprised soft- ware development at many commercial installations, even as high-level languages improved.

A typical assembler command might be "LR" followed by a code for a memory address. The assembler would translate that into the binary digits for the operation "Load the contents of a certain memory location into a register in the central processor." An important feature was the use of symbolic labels for memory locations, whose numerical machine address could change depending on other circumstances not known at the time the program was written. It was the job of the assembler program to allocate the proper amount of machine storage when it encountered a symbol for a variable, and to keep track of this storage through the execution of the program. The IBM computer user's group

SHARE (described next) had a role in developing an assembler for the IBM 704, and assembly language continued to find strong adherents right up into the System/360 line of IBM computers.

SHARE (1955)

While computer suppliers and designers were working on high-level languages, the small but growing community of customers decided to tackle software development from the other direction. In 1955, a group of IBM 701 users located in the Los Angeles area, faced with the daunting prospect of upgrading their installations to the new IBM 704, banded together in the expectation that sharing experiences was better than going alone. That August they met on the neutral territory of the RAND Corporation in Santa Monica. Meeting at RAND avoided problems that stemmed from the fact that users represented competing companies like North American Aviation and Lockheed. Calling itself SHARE,[32] the group grew rapidly and soon developed an impressive library of routines, for example, for handling matrices, that each member could use.

IBM had for years sponsored its own version of customer support for tabulator equipment, but the rapid growth of SHARE shows how different was the world of stored-program digital computers. Within a year the membership—all customers of large IBM systems—had grown to sixty-two members. The founding of SHARE was probably a blessing for IBM, since SHARE helped speed the acceptance of IBM's equipment and probably helped sales of the 704. As SHARE grew in numbers and strength, it developed strong opinions about what future directions IBM computers and software ought to take, and IBM had little choice but to acknowledge SHARE's place at the table. As smaller and cheaper computers appeared on the market, the value and clout of the groups would increase. For instance, DECUS, the users group for Digital Equipment minicomputers, had a very close relationship with DEC, and for personal computers the users groups would become even more critical, as will be discussed in chapter 7.

Sorting Data

Regardless of what level of programming language they used, all commercial and many scientific installations had to contend with an activity that was intimately related to the nature of the hardware—

namely, the handling of data in aggregates called *files*, which consisted of records stored sequentially on reels of tape. Although tape offered many advantages over punched cards, it resembled cards in the way it stored records one after the other in a sequence. In order to use this data, one frequently had to sort it into an order (e.g., alphabetic) that would allow one to find a specific record. Sorting data (numeric as well as non-numeric) dominated early commercial computing, and as late as 1973 was estimated to occupy 25 percent of all computer time.[33] In an extreme case, one might have to sort a very large file after having made only a few changes to one or two records—obviously an inefficient and costly use of computer time. Analysts who set up a company's data processing system often tried to minimize such situations, but they could not avoid them entirely.

Computer programming was synchronized to this type of operation. On a regular basis a company's data would be processed, files updated, and a set of reports printed. Among the processing operations was a program to sort a file and print out various reports sorted by one or more keys. These reports were printed and bound into folders, and it was from these printed volumes that people in an organization had access to corporate data. For example, if a customer called an insurance company with a question about his or her account, an employee would refer to the most recent printout, probably sorted by customer number. Therefore sorting and merging records into a sorted file dominated data processing, until storage methods (e.g., disks) were developed that allowed direct access to a specific record. As these methods matured, the need to sort diminished but did not go away entirely—indeed, some of the most efficient methods for sorting were invented around the time (late 1960s) that these changes were taking place.[34]

As mentioned in chapter 1, John von Neumann carefully evaluated the proposed design of the EDVAC for its ability to sort data. He reasoned that if the EDVAC could sort as well as punched-card sorting machines, it would qualify as an all-purpose machine.[35] A 1945 listing in von Neumann's handwriting for sorting on the EDVAC is considered "probably the earliest extant program for a stored-program computer."[36] One of the first tasks that Eckert and Mauchly took on when they began building the UNIVAC was to develop sorting routines for it.

Actually, they hired someone else to do that, Frances E. (Betty) Holberton, one of the people who followed Eckert and Mauchly from the ENIAC to the Eckert–Mauchly Computer Corporation. Mauchly gave her the responsibility for developing UNIVAC software (although that word was not in use at the time).[37] One of Holberton's first products, in use in 1952, was a routine that read and sorted data stored on UNIVAC tape drives. Donald Knuth called it "the first major 'software' routine ever developed for automatic programming."[38]

The techniques Holberton developed in those first days of electronic data processing set a pattern that was followed for years. In the insurance company mentioned above, we can assume that its customer records have already been placed on the tape sequentially in customer number order. If a new account was removed or changed, the computer had to find the proper place on the tape where the account was, make the changes or deletions, and shuffle the remaining accounts onto other positions on the tape. A simple change might therefore involve moving a lot of data. A more practical action was to make changes to a small percentage of the whole file, adding and deleting a few accounts at the same time. These records would be sorted and written to a small file on a single reel of tape; then this file would be merged into the master file by inserting records into the appropriate places on the main file, like a bridge player inserting cards into his or her hand as they are dealt. Thus for each run a new "master" file was created, with the previous master kept as a backup in case of a mechanical failure.[39]

Because the tape held far more records than could fit in the computer's internal memory, the routines had to read small blocks of records from the tape into the main memory, sort them internally, write the sorted block onto a tape, and then fetch the next block. Sorted blocks were merged onto the master file on the tape until the whole file was processed. At least two tape drives were used simultaneously, and the tapes were read both forward and backward. The routines developed by Holberton and her team at UNIVAC were masterpieces of managed complexity; but even as she wrote them she recognized that it would be better if one could organize a problem so that it could be solved without recourse to massive sorts.[40] With the advent of disk storage and a concept of "linked lists," in which each record in a list contained information about where the next (or previous) record was, sorting lost its dominance.

FORTRAN (1957)

The programming language FORTRAN ("Formula Translation"—the preferred spelling was all capitals) was introduced by IBM for the 704 computer in early 1957. It was a success among IBM customers from the beginning, and the language—much modified and extended—continues to be widely used.[41] Many factors contributed to the success of FORTRAN. One was that its syntax—the choice of symbols and the rules for using them—was very close to what ordinary algebra looked like, with the main difference arising from the difficulty of indicating superscripts or subscripts on punched cards. Engineers liked its familiarity; they also liked the clear, concise, and easy-to-read users manual. Perhaps the most important factor was that it escaped the speed penalty incurred by Laning and Zierler's system. The FORTRAN compiler generated machine code that was as efficient and fast as code written by human beings. John Backus emphasized this point, although critics have pointed out that FORTRAN was not unique among high-level languages.[42] IBM's dominant market position obviously also played a role in FORTRAN's success, but IBM's advantage would not have endured had the Model 704 not been a powerful and well-designed computer on which to run FORTRAN. Backus also noted that the provision, in the 704's hardware, of floating-point arithmetic drove him to develop an efficient and fast compiler for FORTRAN, as there were no longer any cumbersome and slow floating-point routines to "hide" behind.[43]

FORTRAN's initial success illustrates how readily users embraced a system that hid the details of the machine's inner workings, leaving them free to concentrate on solving their own, not the machine's, problems. At the same time, its continued use into the 1990s, at a time when newer languages that hide many more layers of complexity are available, reveals the limits of this philosophy. The C language, developed at Bell Labs and one of the most popular after 1980, shares with FORTRAN the quality of allowing a programmer access to low-level operations when that is desired. The successful and long-lasting computer languages, of which there are very few, all seem to share this quality of hiding some, but not all, of a computer's inner workings from its programmers.

COBOL

FORTRAN's success was matched in the commercial world by COBOL ("Common Business Oriented Language"), developed a few years later. COBOL owed its success to the U.S. Department of Defense, which in May 1959 convened a committee to address the question of developing a common business language; that meeting was followed by a brief and concentrated effort to produce specifications for the language, with preliminary specifications released by the end of that year. As soon as those were published, several manufacturers set out to write compilers for their respective computers. The next year the U.S. government announced that it would not purchase or lease computer equipment that could not handle COBOL.[44] As a result, COBOL became one of the first languages to be standardized to a point where the same program could run on different computers from different vendors and produce the same results. The first recorded instance of that milestone occurred in December 1960, when the same program (with a few minor changes) ran on a UNIVAC II and an RCA 501. Whether COBOL was well designed and capable is still a matter of debate, however.

Part of COBOL's ancestry can be traced to Grace Hopper's work on the compilers for the UNIVAC. By 1956 she had developed a compiler called "B-0," also called in some incarnations "MATH-MATIC" or "FLOW-MATIC," which unlike her "A" series of compilers was geared toward business applications. An IBM project called Commercial Translator also had some influence. Through a contract with the newly formed Computer Sciences Corporation, Honeywell also developed a language that many felt was better than COBOL, but the result, "FACT," did not carry the imprimatur of the U.S. government. FACT nevertheless had an influence on later COBOL development; part of its legacy was its role in launching Computer Sciences Corporation, one of the first commercial software companies.[45]

It was from Grace Hopper that COBOL acquired its most famous attribute, namely, the ability to use long character names that made the resulting language look like ordinary English. For example, whereas in FORTRAN one might write:

```
IF A > B
```

the corresponding COBOL statement might read:

```
IF EMPLOYEE-HOURS IS GREATER THAN MAXIMUM
```
[46]

Proponents argued that this design made COBOL easier to read and understand, especially by "managers" who used the program but had little to do with writing it. Proponents also argued that this made the program "self-documenting": programmers did not need to insert comments into the listing (i.e., descriptions that were ignored by the compiler but that humans could read and understand). With COBOL, the actual listing of instructions was a good enough description for humans as well as for the machine. It was already becoming known what later on became obvious: a few months after a code was written, even the writer, never mind anyone else, cannot tell what that code was supposed to do.

Like FORTRAN, COBOL survived and even thrived into the personal computer era. Its English-like syntax did not achieve the success its creators hoped for, however. Many programmers felt comfortable with cryptic codes for variables, and they made little use of the ability to describe variables as longer English words. Not all managers found the language easy to read anyway. Still, it provided some documentation, which was better than none—and too many programs were written with none. In the years that followed, researchers explored the relationship between machine and human language, and while COBOL was a significant milestone, it gave the illusion that it understood English better than it really did. Getting computers to "do what I meant, not what I said" is still at the forefront of computer science research.

Although the year 2001 is fast approaching, computer languages are still a long way from the level of natural language understanding shown by HAL, the computer that was the star of the Stanley Kubrick movie, *2001: A Space Odyssey* (figure 3.3).[47] Before 2001 arrives, the world will also have to face a more serious issue: the inability of many computer programs to recognize that when a year is indicated by the decimal digits "00," it means the year 2000, not 1900. The programs causing these troubles were written in the 1960s and 1970s, mostly in COBOL. Today, program managers are desperately searching for programmers who can wade through old COBOL programs and find and correct the offending lines of code. A story circulating around Internet discussion groups is that companies are visiting retirement communities in Florida, where they are coaxing old-time COBOL programmers off the golf course. The "Year-2000 Bug" gives ample evidence that, although it was *possible* to write COBOL programs that were self-documenting, few ever did. The programs that need to be corrected are incomprehensible to many of the best practitioners of modern software development.

The word "language" turned out to be a dangerous term, implying much more than its initial users foresaw. The English word is derived from the French *langue,* meaning tongue, implying that it is spoken. Whatever other parallels there may be with natural language, computer languages are not spoken but written, according to a rigidly defined and precise syntax.[48] Hopper once recounted how she developed a version

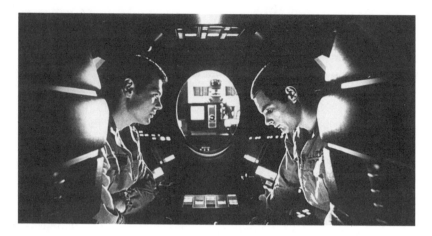

Figure 3.3
A scene from *2001: A Space Odyssey*. A camera eye of HAL, the on-board computer, is visible between the two astronauts. Publicity that accompanied the film's release in 1968 stated that its creators depicted a level of technology that they felt was advanced but not unreasonably so for the year 2001. Computers have become more powerful and more compact, but it seems unlikely that machine understanding of natural language, on a level shown by HAL, will be attained by 2001. (*Source: 2001: A Space Odyssey.* © 1968 Turner Entertainment Co.)

of FLOW-MATIC in which she replaced all the English terms, such as "Input," "Write," and so on, with their French equivalents. When she showed this to a UNIVAC executive, she was summarily thrown out of his office. Later on she realized that the very notion of a computer was threatening to this executive; to have it "speaking" French—a language he did not speak—was too much.[49]

Languages Versus Software

From the twin peaks of FORTRAN and COBOL we can survey the field of software through the 1960s. After recognizing the important place of assembly language and then looking at the histories of a few more high-level languages, we might conclude that we have a complete picture.

Among the high-level languages was ALGOL, developed mainly in Europe between 1958 and 1960 and proposed as a more rigorous alternative to FORTRAN. ALGOL was intended from the start to be independent of any particular hardware configuration, unlike the original FORTRAN with its commands that pointed to specific registers of the IBM 704's processor. Also unlike FORTRAN, ALGOL was carefully and formally defined, so that there was no ambiguity about what an expression written in it would do. That definition was itself specified in a language known as "BNF" (Backus-Normal-Form or Backus-Naur-Form). Hopes were high among its European contributors that ALGOL would become a worldwide standard not tied to IBM, but that did not happen. One member of the ALGOL committee ruefully noted that the the name ALGOL, a contraction of algorithmic language, was also the name of a star whose English translation was "the Ghoul." Whatever the reason for its ill fate, ALGOL nonetheless was influential on later languages.

Of the many other languages developed at this time, only a few became well known, and none enjoyed the success of FORTRAN or COBOL. JOVIAL (Jules [Schwartz's] Own Verison of the International Algebraic Language) was a variant of ALGOL developed by the Defense Department, in connection with the SAGE air-defense system; it is still used for air-defense and air-traffic-control applications. LISP (List Processing) was a language oriented toward processing symbols rather than evaluating algebraic expressions; it has been a favorite language for researchers in artificial intelligence. SNOBOL (StriNg-Oriented symBOlic Language) was oriented toward handling "strings"— sequences of characters, like text. A few of the many other languages developed in the late 1960s will be discussed later.[50]

Somewhere between assemblers and COBOL was a system for IBM computers called RPG (Report Program Generator; other manufacturers had similar systems with different names). These were in common use in many commercial installations throughout the 1960s and after. Textbooks on programming do not classify RPG as a language, yet in some ways it operated at a higher level than COBOL.[51] RPG was akin to filling out a preprinted form. The programmer did not have to specify what operations the computer was to perform on the data. Instead the operations were specified by virtue of where on the form the data were entered (e.g., like on an income tax return). Obviously RPG worked only on routine, structured problems that did not vary from day to day, but in those situations it freed the programmer from a lot of detail. It is still used for routine clerical operations.

System Software

Besides programs that solved a user's problem, there arose a different set of programs whose job was to allocate the resources of the computer system as it handled individual jobs. Even the most routine work of preparing something like a payroll might have a lot of variation, and the program to do that might be interspersed with much shorter programs that used the files in different ways. One program might consist of only a few lines of code and involve a few pieces of data; another might use a few records but do a lot of processing with them; a third might use all the records but process them less, and so on.

Early installations relied on the judgment of the operator to schedule these jobs. As problems grew in number and complexity, people began developing programs that, by the 1990s, would dominate the industry. These programs became known as *operating systems* (figure 3.4). The most innovative early work was done by users. One early system, designed at the General Motors Research Laboratories beginning in 1956 was especially influential.[52] Its success helped establish batch computing—the grouping of jobs into a single deck of cards, separated by control cards that set the machine up properly, with only minimal work done by the human operator. A simple but key element of these systems was their use of special "control" cards with specific codes punched into reserved columns. These codes told the computer that the cards that followed were a FORTRAN program, or data, or that a new job was starting, and so on. That evolved into a system known at IBM as Job Control Language (JCL). Many a novice programmer has a vivid memory of JCL cards, with their distinctive double slash (//) or slash-asterisk (/*) punched into certain fields. Many also remember the confusion that resulted if a missing card caused the computer to read a program deck as data, or vice versa.[53]

MAD

In university environments there arose a similar need to manage the workflow efficiently. Student programs tended not to be uniform from week to week, or from one student to another, and it was important that students received clear messages about what kinds of errors they made. (In fact, every installation needed this clarity but few recognized that at the time.) In 1959 a system called MAD (Michigan Algorithmic Decoder) was developed at the University of Michigan by Bernie Galler, Bob

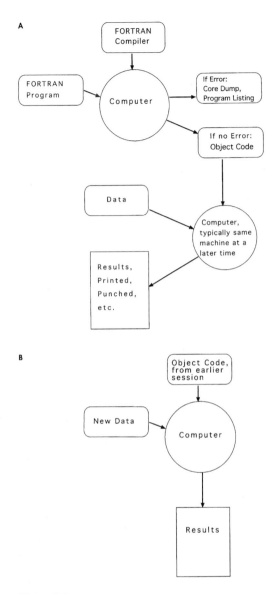

A

FORTRAN
Compiler

FORTRAN
Program

Computer

If Error:
Core Dump,
Program Listing

If no Error:
Object Code

Data

Computer,
typically same
machine at a
later time

Results,
Printed,
Punched,
etc.

B

Object Code,
from earlier
session

New Data

Computer

Results

Figure 3.4
The origins and early evolution of operating systems. (*a*) Simple case of a single
user, with entire computer's resources available. Note that the process requires a
minimum of two passes through the computer: the first to compile the program
into machine language, the second to load and execute the object code (which
may be punched onto a new deck of cards or stored on a reel of tape). If the
original program contained errors, the computer would usually, though not
always, print a diagnostic message, and probably also a "dump," rather than
attempt to generate object code. (*b*) If at a later date the user wants to run the
same program with different data, there is no need to recompile the original
program.

C

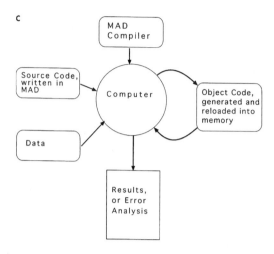

Figure 3.4 (Continued)
The origins and early evolution of operating systems. (*c*) "Load and Go": The Michigan Algorithmic Decoder (MAD) system collapsed the generation of object code and execution, so that students and other users could more quickly get results, or diagnostic messages if there were errors in their programs.

Graham, and Bruce Arden. MAD was based on ALGOL, but unlike ALGOL it took care of the details of running a job in ways that few other languages could do. MAD offered fast compilation, essential for a teaching environment, and it had good diagnostics to help students find and correct errors. These qualities made the system not only successful for teaching, but also for physicists, behavioral scientists, and other researchers on the Michigan campus. One feature of MAD that may have helped win its acceptance among students was that it printed out a crude picture of Alfred E. Newman, the mascot of *Mad* Magazine, under many error conditions. (Bob Rosin, who coded the image on a deck of punched cards, recalled that this "feature" had eventually to be removed because students were deliberately making errors in order to get the printout.)[54]

Both industrial and teaching installations had the same goal of accommodating programs of different lengths and complexity. For economic reasons, another goal was to keep the computer busy at all times. Unfortunately, few industrial and commercial installations realized, as MAD's creators did, the importance of good error diagnosis. And since the commercial systems did not have such diagnostics, many teaching environments did not, either, reasoning that students would

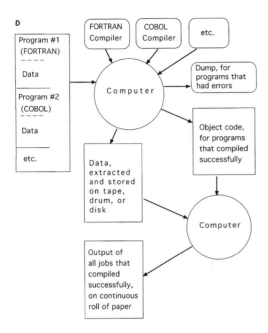

Figure 3.4 *(Continued)*
The origins and early evolution of operating systems. (*d*) Batch processing. The economics of a large computing system made it unlikely that a single user could have exclusive use of the machine, as in (*a*). In practice his or her deck of cards would be "batched" with other users. The program and data would be separated by a specially punched card; likewise, each person's job would be separated from the next person's job by one or more "job control" cards. The operating system would load the appropriate compiler into the computer as each user required it; it might also extract the data from the deck of cards and load it onto a faster medium such as tape or disk, and handle the reading and printing of data and results, including error messages. The computer operator might be required to find and mount tapes onto drives, as indicated by a signal from the console; the operator would also pull the printout from the printer and separate it for distribution to each user.

sooner or later have to get used to that fact. In these batch systems, if a program contained even a simple syntax error, the operating system decided whether the computer should continue trying to solve the problem. If it decided not to, it would simply transfer the contents of the relevant portion of memory to a printer, print those contents as rows of numbers (not even translating the numbers into decimal), suspend work on that program, and go on to the next program in the queue. The word for that process was "dump." Webster's definition, "to throw down

E

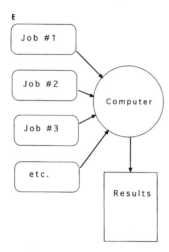

Figure 3.4 (Continued)
The origins and early evolution of operating systems. (*e*) Multiprogramming. A
mix of programs as in (*d*) would use different portions of the computer in a
different mix. One program might make heavy use of the tape drives but little
use of the central processor's advanced mathematical powers. Operating systems
thus evolved to support "multiprocessing": the ability of a computer to run more
than one program at the same time, each using portions of the machine that the
other was not using at a given instant. The operating system took care that the
programs did not interfere with one another, as in two programs attempting to
write data at the same time to the same memory location.

or out roughly," was appropriate. The hapless user who received a "core
dump" was in effect being told, rather rudely, that the computer had
decided that someone else's work was more important. Trying to find
the error based on row upon row of numbers printed by a core dump
was intimidating to the lay person and difficult even for the expert.

As operating systems evolved, they tended to consume more precious
amounts of memory, until there was little left for running the programs
the computer was installed for in the first place. The evolution of the
name given to them reflects their growing complexity: They were called
"monitors," then "supervisor systems," and finally "operating systems."
In the early days, simple and lean systems were developed by customers,
for example, the Fortran Monitor System for the IBM 7090 series.
Scaling up to more complex systems proved difficult. SOS (Share
Operating System), developed by SHARE for IBM mainframes, was
more complex but less efficient. When IBM decided to combine its
line of scientific and business systems into a series called System/360, the

company also set out to develop an operating system, OS/360, to go with it.[55] System/360 was a success for IBM and redefined the industry, as subsequent chapters will show, but the operating system OS/360, available by 1966, was a failure and its troubles almost sank the company.[56] Fred Brooks, who had been in charge of it, wrote a book on OS/360's development, *The Mythical Man-Month*, which has become a classic description of the difficulty of managing large software projects. Among Brooks's many insights is that committees make poor structures for software projects; this was also a factor in the problems with the Share Operating System noted above, as well as with the languages PL/I and ALGOL-68 (discussed later).

IBM eventually developed workable system software for its 360 Series, but when the minicomputer was invented in the mid 1960s, the history of operating systems started over again: the first minis did not have the internal memory capacity to do more than support a simple monitor, and individuals wrote Spartan monitors that worked well. As minicomputers got more powerful, so did their operating systems, culminating in Digital Equipment Corporation's VMS for its VAX line of computers (1978). The phenomenon was repeated yet again with the personal computer. The first personal computers had rudimentary monitors that loaded data from tape cassettes, and these were followed by more complex but still lean disk operating systems. Finally, "windows"-based systems appeared, whose complexity required teams of programmers working on subsections. As expected, some of these projects have been accompanied by the same management problems discussed by Brooks for the System/360. Computers seem to be cursed with having to go through this painful wheel of reincarnation every other decade or so.

Computer Science

These examples show that there was more to software than the development and evolution of programming languages. But programming languages came to dominate the academic discipline that arose during this period, namely, computer science. The discipline first appeared in the late 1950s at pioneering institutions, including Stanford and Purdue, under different names and often as a division of the Mathematics or Electrical Engineering Departments. It established a beachhead based on inexpensive drum-based computers, including the Librascope LGP-30 and especially the IBM 650. Summer school sessions at several top universities in the mid-1950s further legitimized the discipline.[57]

In the early 1960s computer science struggled to define itself and its purpose, in relation not only to established disciplines of electrical engineering and applied mathematics, but also in relation to—and as something distinct from—the use of computers on campus to do accounting, record keeping, and administrative work.[58] Among those responsible for the discipline that emerged, Professor George Forsythe of Stanford's mathematics faculty was probably the most influential. With his prodding, a Division of Computer Science opened in the mathematics department in 1961; in 1965 Stanford established a separate department, one of the first in the country and still one of the most well-regarded.[59]

In the fall of 1967 Herbert Simon, Alan Perlis, and Allen Newell, all of the Carnegie Institute of Technology in Pittsburgh, wrote an eloquent and influential letter to the editor of *Science*, in which they defined computer science as "the study of computers."[60] Implicit in that definition was the notion of a computer not as a static artifact but as a system that carried out dynamic processes according to a set of rules. They defended that definition and the legitimacy of computer science against a number of objections, including the obvious one that computers were a man-made phenomenon and hence their study could not be one of the natural sciences. Simon (who won a Nobel Prize for his work in what might be called management science) argued that many natural sciences studied phenomena that were not totally divorced from human creation, and anyway, there was nothing wrong with making a science of the study of the artificial. The following year he delivered a series of lectures that further developed this argument, published as *The Sciences of the Artificial*.[61]

One objection that the trio raised and then refuted was that computer science was really "the study of algorithms (or programs), not computers."[62] They felt that such a definition was too restrictive. Their refutation was weak—among the reasons they gave was that their professional society was called the Association for Computing Machinery. In any event, computer science evolved in subsequent years to mean precisely what they said it was not—the study of algorithms, with a focus on the even narrower field of programming languages. The ACM's name notwithstanding, hardware issues remained with electrical engineering. Six months after the appearance of the letter in *Science*, the ACM published Curriculum '68, a set of courses that the association felt would provide an intellectually defensible grounding in computer science for undergraduates.[63] Curriculum '68 emphasized a mathematical and theoretical basis for computer science. The study of computer hardware was absent. An earlier version published in 1965 recommended (as electives) courses in electronics and Analog Computers; these were dropped from the 1968 curriculum. A student wanting to study "computers," as Newell, Simon, and Perlis defined them, would have to study Electrical Engineering as well. The ACM chose to emphasize algorithmic procedures, programming languages, and data structures. One critic called Curriculum '68 as influential on the discipline of computer science as the 1945 EDVAC Report.[64] By 1968 computer science had gained respectability. An undergraduate could obtain a degree in it at one of one hundred U.S. universities, up from only twelve in 1964. By the 1980s it had become one of the most popular undergraduate majors on nearly every campus in the United States.[65]

Other Events of 1968 and 1969

In 1968 and 1969 a cluster of similar events further established the place of software, its relationship to computer science, and its relationship to industrial, commercial, and military computer users.

Donald E. Knuth

In 1968 Donald E. Knuth published the first of a projected seven volumes of a series on *The Art of Computer Programming. Fundamental Algorithms*, in his words, set down in print techniques of programming that had been passed along as "folklore,...but [for which] comparatively little theory had been developed."[66] Others had also attempted to place computer programming on solid theoretical ground, but often these attempts did not offer much practical help in solving actual programing problems. Most teaching of programming was intimately bound up with the idiosyncrasies of a particular machine, including the binary codes for certain registers, the timing of signals to and from disk or drum stores, and so on. Knuth provided a theoretical basis for computing that was practical, and his books established it as the *algorithm*, a formal procedure that one can use to solve a problem in a reasonable length of time, given the constraints of actual computing machines.

Structured Programming

In March 1968, in the same issue of the ACM journal that published Curriculum '68, there appeared a letter to the editor with the curious title "Go-To Statement Considered Harmful."[67] The letter opened with the statement,

For a number of years I have been familiar with the observation that the quality of programmers is a decreasing function of the frequency of **go to** statements in the programs they produce.

The author was Edsger Dijkstra of the Technical University of Eindhoven, the Netherlands. His letter set off an argument that continued for the next few years. It was only a minor part of his long effort to move computer science toward a more formal theoretical basis, but the letter's bold assertion became a symbol for the more complex work he was trying to do to move complex software systems from foundations of sand to the bedrock of basic theory.

There followed a long and acrimonious debate over Dijkstra's assertion about "go to." Few participants seemed to realize that Dijkstra was concerned with something more profound than one particular command. Critics in the industrial world saw the fuss as one more example that proved the irrelevance of academic computer science. In the short term, the letter gave a push to the concept of "structured programming," a method that its adherents believed would advance programming from an art (as Knuth called it in the title of his books) to a science. Whether Dijkstra's letter was the impetus or not, programming did move in this direction in the following years.

Intellectual Property Issues

The very next exchange of letters to the editor of the *Communications* of the ACM concerned another issue, that also came to have a profound effect on the quality of programming.[68] The exchange was between Professor Bernard Galler of the University of Michigan (one of the creators of MAD), and Calvin Mooers, a developer of a language he called "the TRAC language." The Rockford Research Institute, of which Mooers was a founder, had sought legal protection for the TRAC language, protection which would have prevented anyone from altering, modifying, or extending it. Galler pointed out that the best and most successful languages were those that had benefitted from users, including graduate students, who had improved or modified them in use. Both sides had a point. Without some protection, one version of a language, albeit "improved," would be incompatible with another, and this alone could vitiate whatever improvements might be claimed. But as the examples of SHARE and MAD proved, software development needed the user to progress. In part because of Mooers's policy, his language found little use.[69]

Eight years later, during the early development of the personal computer, the question of ownership and control of commercial software became crucial. As the Microsoft Corporation clarified that question and established its rights, not only did Microsoft become a dominant software provider, it also set the stage for the production of large amounts of inexpensive software from a myriad of vendors.[70] (This will be discussed further in chapter 7.)

Software Engineering

In October 1968 a conference was convened in Garmisch, Germany, with the provocative title "Software Engineering."[71] The conference marked the end of the age of innocence, a realization that a "crisis" in software production would not end soon but could—and had to—be managed. The name given to that activity—engineering—was deliberately provocative, suggesting that what lay behind the software crisis was the fact that programmers lacked the theoretical foundations and disciplines of daily practice that one found in traditional fields of engineering. That the conference was sponsored by NATO further revealed software engineering's distance from computer science, which was centered in the universities.[72] Conference organizers had recognized that computers were responsible for systems that put human lives at risk, including the military systems employed at NATO. These systems could not tolerate a "bug" that in a batch data processing environment might be only a minor irritant. That suggested a parallel with other forms of engineering, especially civil engineering, where people routinely trusted a bridge because they trusted the people who designed it. In other respects, however, the analogy broke down. In civil engineering, a tradition of certification and a chain of legal responsibility had evolved over the years; no such tradition had been established in computing and none would emerge. Attempts to control who might claim the title "computer programmer" seem always to be futile. Part of the reason is that new technology, like the personal workstation, offers an avenue for new entrants into the field, and established members of the profession cannot control that. A 1996 conference on the history of software engineering, also held in Germany, came to the unintended conclusion that the attempt to establish software engineering on the whole had failed.[73]

Unbundling

In December 1968, under pressure from the U.S. government, IBM announced that the following year it would "unbundle" its software; that is, charge separately for it instead of combining its costs with that of the hardware systems.[74] One of the first products it began to sell was also one of the most successful in the history of computing, its Customer Information Control System (CICS), which it offered on a tape beginning in July 1969 for $600 a month.[75] Software remained ethereal, but now it could be bought and sold. The effect of that decision was to open up the field of software to commercial vendors, who would now be driven by the powerful and unforgiving forces of the free marketplace.

The effects of what was going on in the academic world eventually made their way to the software houses. Programming became more structured and more firmly based on theory, although the software crisis became a permanent fixture. Meanwhile, the computer industry was going through one of its most innovative periods. It was in the late 1960s that the integrated circuit began to show up in commercial systems, which immediately created a new class of inexpensive computers that had limited memory requirements. These computers had no room to implement the highly structured languages, like Pascal, that critics of FORTRAN preferred. So programmers resorted to unstructured machine or assembly language, "go to" and all. Putting software on a more formal basis, which so many had hoped for, would arrive late, if ever.

In 1969 Ken Thompson and Dennis Ritchie at the Bell Telephone Laboratories in New Jersey began work on what would become the UNIX operating system. The computer they used was a Digital Equipment Corporation PDP-7, a machine with extremely limited memory even by the standards of that day.[76] Thompson wrote the earliest version of this system in assembler, but soon he and his colleagues developed a "system programming language" called "B," which by 1973 had evolved into a language called C.[77] C is a language that is "close to the machine," in Ritchie's words, a characteristic that reflected its creators' desire to retain the power of assembly language.[78] To that extent it went against the tenets of structured programming that were then being espoused. That quality also made it one of the most popular languages of the personal computer era, as practiced at the giant software houses of the 1990s such as Microsoft.

The success of UNIX and C balanced two failed programming languages of the late 1960s, ALGOL-68 and PL/I. These failed languages illustrated, on the one hand, the worst of the gulf between academic computer science and the users, and on the other hand, the already familiar realization that committees are ill-suited to good software development.

ALGOL-68 was an attempt to improve upon ALGOL-60, which outside the academy only the Burroughs Corporation supported. The reasons for ALGOL-60's failure was the subject of much discussion at the time. Many believed that IBM's support for FORTRAN doomed it, but we shall see that IBM's support for PL/I did not have the power to save it. More serious was ALGOL-60's lack of I/O specifications. For computer users who were still debating whether to use any high-level language at all, that was a fatal omission.[79] In the mid-1960s, the International Federation for Information Processing (IFIP) established a working group to extend the ALGOL language, and they released a new version of it, which came to be known as ALGOL-68.[80] It, too, failed in the marketplace, but for different reasons. Whereas ALGOL-60 was based on a formal structure and was very lean, ALGOL-68 was burdened by an attempt to do too much, with the effects that some features interfered with the clean implementation of others. It was hard to understand. In an attempt to satisfy a broad range of users worldwide, the committee produced something that satisfied few.[81] It was implemented in a few places, notably on ICL computers in the U.K. Otherwise, its chief legacy may have been the language Pascal, the tightly structured language that Nicholas Wirth of the Swiss Federal Technical Institute developed in reaction to ALGOL-68's complexity.[82]

At IBM there was a similar effort to develop a language to replace the venerable FORTRAN and COBOL. When in 1964 IBM announced its System/360 series of computers as a replacement for both its business and scientific computers (*see* chapter 5), the company assumed that a new language could likewise be developed for both applications. A joint IBM-SHARE committee concluded in early 1963 that the new language would not be an extension of FORTRAN, even though the existing version of that language, FORTRAN-IV, was very popular and heavily used. The new language, PL/I (Programming Language, One), drew from FORTRAN, COBOL, and ALGOL. Preliminary versions were released in 1964, but by the time the full language was ready, COBOL and FORTRAN-IV had established a foothold on the System/360 series and could not be dislodged. PL/I's complexity overwhelmed its many advantages, including the advantage that IBM was supporting it and that it was suitable for both business and science. Many IBM installations made PL/I available, but it never became very popular.

Conclusion

The activity known as computer programming was not foreseen by the pioneers of computing. During the 1950s they and their customers slowly realized: first, that it existed; second, that it was important; and third, that it was worth the effort to build tools to help do it. These tools, combined with the applications programs, became collectively known as "software," a term that first came into use around 1959.[83] People like Grace Hopper and Maurice Wilkes initially focused on building up libraries of subroutines, and then getting the computer to call these up and link them together to solve a specific problem. That gave way to a more general notion of a high-level computer language, with the computer generating fresh machine code based on a careful analysis of what the programmer specified, in something that resembled a combination of algebra and English.

Despite great strides in software, programming always seemed to be in a state of crisis and always seemed to play catch-up to the advances in hardware. This crisis came to a head in 1968, just as the integrated circuit and disk storage were making their impact on hardware systems. That year, the crisis was explicitly acknowledged in the academic and trade literature and was the subject of a NATO-sponsored conference that called further attention to it. Some of the solutions proposed were a new discipline of software engineering, more formal techniques of structured programming, and new programming languages that would replace the venerable but obsolete COBOL and FORTRAN. Although not made in response to this crisis, the decision by IBM to sell its software and services separately from its hardware probably did even more to address the problem. It led to a commercial software industry that needed to produce reliable software in order to survive. The crisis remained, however, and became a permanent aspect of computing. Software came of age in 1968; the following decades would see further changes and further adaptations to hardware advances.

4

From Mainframe to Minicomputer, 1959–1969

The room could have been designed by Hollywood producers of dystopian films like *Blade Runner* or *Brazil*. As far as the eye could see were rows of IBM Model 027 keypunches—machines that punched rectangular holes into 80-column cards, each of them a standard 3-1/4 × 7-3/8 inches. Seated at each station was a woman, her head tilted to the left to scan a piece of paper mounted on a frame, her right hand deftly floating over the keys of the machine. Each press of a key caused it to punch a hole in a card with a solid "thunk." When the room was in full swing, said one operator, "there was a certain rhythm, a beat, a sound" that let each operator know she was getting the job done.[1] A data processing manager had a slightly different opinion: he said the sound was "like you had a helmet on and someone was hitting it with a hammer."[2] A film was made of the operation; from its soundtrack one might conclude that, if anything, the second opinion was conservative.

The room was in one of several regional centers set up in the mid-1960s by the U.S. Internal Revenue Service to process tax returns. By then the IRS had embraced the electronic digital computer and owned one of the most sophisticated and complex systems in the world. At its heart was a set of IBM mainframes at a national center in Martinsburg, West Virginia. In 1964, around the time the film was made, the Center was using a set of transistorized IBM 7070s, business versions of the 7090 discussed in chapter 2.[3]

At the same time that the women were keypunching tax returns, the Worcester (Massachusetts) *Telegram and Gazette* was also entering the computer age. A few big-city newspapers had already installed mainframes to replace the Mergenthaler Linotypes that set type in hot lead. The Worcester paper was able to join this movement by purchasing a much smaller, but very capable, $30,000 "Computer Typesetting

System" from the Digital Equipment Corporation of Maynard, Massachusetts. The system was not much bigger than a couple of office desks. At its heart was a new type of computer offered by Digital, a PDP-8.[4]

Digital computers began the decade of the 1960s with a tentative foothold; they ended with an entrenched position in many business, accounting, and government operations. They were also now found in a host of new applications.

The forces driving this movement were both technical and social. Among the former were the introduction of transistors in place of vacuum tubes and the development of languages like FORTRAN and COBOL that made programming easier. Among the latter was the increased demand for record-keeping by the federal government, brought on by programs of the "Great Society." President John Kennedy's challenge, in May 1961, to put a man on the moon and return him safely by the end of the decade transformed the U.S. space program into a complex of research and production centers with unlimited budgets and insatiable appetites for computing power. The United States was entering a decade of economic growth and prosperity, accompanied by major investments in interstate highways, suburban housing, and jet aircraft. All of these put a strain on information-processing procedures that were based on punched card tabulators, mechanical adding machines, and calculators.

Computing in the 1960s was not just a story of existing firms selling new machines in increasing volume to existing customers, although that did occur. That decade also saw the nature of the industry transformed. In a sense the computer was reinvented yet again. Just as Eckert and Mauchly transformed a fast calculator, the ENIAC, into a general-purpose data processing and scientific device, so now did new companies like the Digital Equipment Corporation rework the computer's internal architecture, its programming, the way it was marketed, and the applications it was used for.

The rate of technological advance in computing, and the rapid obsolescence of existing products, had few counterparts in other industries. It was a fact that was well understood by IBM, whose market share hovered around 70 percent from the late 1950s onward.[5] IBM built up a large research department, with major laboratories on both coasts as well as in Europe, to ensure that it could stay abreast of developments in solid-state electronics, tape and disk storage, programming languages, and logic circuits. Some of that research was conducted at fundamental levels of solid-state physics and mathematics, which offered little chance of a quick payoff, but the nature of computing dictated that such research be done. Applied research might have produced a faster vacuum tube or a better card punch (and such devices were indeed invented), but it would not produce the kinds of advances in computing that are regarded in the popular press as revolutionary.[6]

Supporting such research, though expensive, gave IBM an advantage over its competitors, who had to work from a smaller customer base. And it created a high barrier to any new firm wishing to enter the industry. If a newcomer wanted to exploit a radically new piece of technology IBM had developed, it would have to build and market a balanced system, including software—all the pieces of which IBM was probably also better at producing. One such company was Philco, discussed earlier. The surface-barrier transistors that Philco developed put it a year ahead of IBM, which in the late 1950s had just introduced the vacuum-tube 709.[7] But once IBM countered with its transistorized 7090, Philco could not maintain the pace of competition and left the business by 1964.[8] In order to survive, a new entrant into the field had to have, in addition to superior technology, a niche that was poorly served by IBM and the other mainframe companies. IBM's dominant position meant that it could mete out technical advances at a pace that did not render its installed base obsolete too quickly. Almost no manufacturers save IBM made a profit selling large computer systems in the late 1950s.

Despite what its critics charged, IBM did not always succeed in controlling the pace of innovation. It abandoned the vacuum-tube 709 faster than it wanted to, for example; ten years later it made an uncomfortably quick transition from the System/360 to the System/370. One place where IBM did succeed was in keeping viable the basic input medium of the punched card, and with that the basic flow of data through a customer's installation. The same card, encoded the same way and using a keypunch little changed since the 1930s, served IBM's computers through the 1960s and beyond. The sequential processing and file structure, implicit in punched card operations, also survived in the form of batch processing common to most mainframe computer centers in the 1960s. That eased the shock of adopting the new technology for many customers, as well as ensuring IBM's continued influence on computing at those sites.

IBM thus created a state of equilibrium in the industry. Its 70 percent market share, some economists felt, was "just enough" to maintain innovation, stability, and profits, but not so much as to bring on stagnation and the other evils of monopolization.[9] Were it not for a radical change in an external factor, the computer industry might have gone on this way for decades, just as the U.S. auto industry, dominated by three firms, achieved an equilibrium into the 1980s.

The Influence of the Federal Government

The external factor was the U.S. Defense Department, whose funding was crucial to the advance of computer technology. Military support was nothing new to computing, beginning with the ENIAC. What changed was the nature of research done under defense support, especially after the onset of the war in Korea in 1950. Military support for basic research in physics, electrical engineering, and mathematics increased dramatically after 1950. The nature of that research also changed, from one where the military specified its needs in detail, to one where the researchers themselves—professors and their graduate students at major universities—took an active role in defining the nature and goals of the work.[10] Military funding, channeled into research departments at prestigious universities, provided an alternative source of knowledge to that generated in large industrial laboratories. This knowledge, in turn, allowed individuals outside established corporations to enter the computer industry. Its effect on computing was dramatic.

This chapter begins by looking at case studies that illustrate how other branches of the federal government affected computing as customers; that is, how their heavy demands for computation combined with generous budgets spurred the growth of large mainframe installations, dominated by the IBM 7000 series of computers. Following that, the chapter looks at computing from the other side and describes how research in solid-state physics and electronics produced a fundamentally different type of computer, and how a key group of individuals leveraged that research to redefine the the industry.

Massachusetts Blue Cross

Massachusetts Blue Cross was a typical commercial customer. In December 1960, after three years of planning and analysis, Massachusetts Blue Cross/Blue Shield acquired an IBM 7070 computer to process work it had been doing on tabulating machines.[11] Blue Cross intended to place, on twenty-four reels of tape, the records of 2,500,000 subscribers. By June 1961 the transfer was completed, and the 150 file cabinets that held the punched cards and printed records were retired.[12] The 7070 was upgraded three years later to a 7074; Blue Cross also acquired a smaller

IBM 1401 dedicated solely to input and output. Although by this time COBOL and FORTRAN were both widely available and supported, Blue Cross chose to use instead the more primitive language AUTOCODER, because, in the words of a former employee, "FORTRAN and COBOL...used up too much main memory and took too much processing time for compilation."[13] (See figure 4.1.)

The fortunes of the medical insurance business took a dramatic turn in 1965, with the passage of amendments to the Social Security Act that established Medicare for Americans age sixty-five and over. Blue Cross/Blue Shield of Massachusetts successfully bid for the job of administering the program in that commonwealth and managed to computerize the account by the fall of 1966. It claimed to be the first in the country to have fully computerized Medicare. However, the thousand-fold increase in processing speeds that Blue Cross got from using a 7070, revolutionary in 1961, was now inadequate. The company rented computer time on another 7070 located in the Boston suburbs, with Blue Cross employees driving a car loaded with decks of cards out to Southbridge every evening, running the programs overnight, and driving back to Boston with the output in the morning.

In 1967 Blue Cross acquired one of IBM's new-generation System/360 computers to handle the workload. By the end of that decade there were *three* System/360s on site, as well as the 7074/1401 system. Forty-three tape drives handled the records, and computer operations went on twenty-four hours a day, seven days a week. COBOL was now the preferred language, although the AUTOCODE programs continued to be used by running an emulator on one of the 360s that made it "look like" a 7074/1401 system.[14] The company continued to rely on IBM mainframe systems.

NASA-Ames Research Center

The NASA-Ames Research Center, located in Mountain View, California, had been a center for high-speed aerodynamics research since its founding in 1940, as part of what was then the National Advisory Committee for Aeronautics. The shock of the Soviet's Sputnik in 1957, followed by President Kennedy's challenge to the nation to put a man on the Moon before the end of the 1960s, gave the center a sense of urgency not seen since the Second World War. Its focus on aerodynamic research meant that the laboratory had been involved with numerical calculations from its beginning. In 1955 Ames had acquired its first stored-program electronic computer, an IBM 650. In 1958, shortly after

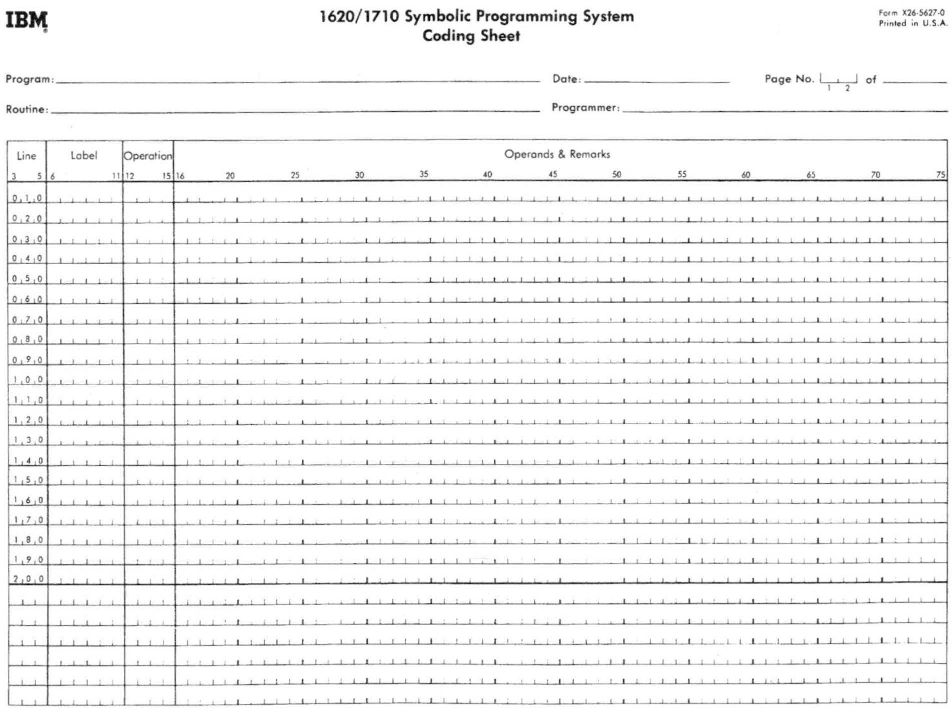
IBM

1620/1710 Symbolic Programming System
Coding Sheet

Form X26-5627-0
Printed in U.S.A.

Program: Date: Page No. of

Routine: Programmer:

Line (3–5)	Label (6–11)	Operation (12–15)	Operands & Remarks (16–75)
010			
020			
030			
040			
050			
060			
070			
080			
090			
100			
110			
120			
130			
140			
150			
160			
170			
180			
190			
200			

Figure 4.1
IBM coding forms. Not only would a programmer hardly ever see the computer, he or she might never even see the keypunch on which the programs were entered into the computer. (*Source*: Thomas E. Bergin.)

the center became part of the newly founded National Aeronautics and Space Administration (NASA), it acquired an IBM 704, replaced in 1961 by an IBM 7090.[15] These were all used for scientific calculations, for example, satellite trajectories, heat transfer, and particle physics.

In the fall of 1961 the center acquired a medium-size Honeywell H-800 for processing of wind tunnel data. That was followed by the acquisition of other similar machines for dedicated purposes: controlling experiments, operating a flight simulator, and reducing wind tunnel data. These computers came from Honeywell, Digital Equipment Corporation, Scientific Data Systems (SDS), and EAI. IBM was also represented by its 1800, which was used for controlling an experiment in real time, but IBM's presence at the Center was mainly in the large, centralized computing system used for "general scientific" work.[16]

Throughout the 1960s the demands on the central IBM installation grew at a compounded rate of over 100 percent a year. Meeting that demand was a never-ending headache for the Ames Computation Division. Beginning in 1963 the 7090 was upgraded to a complex of machines called a Direct Couple System. At its heart was an IBM 7094, acquired in July. To keep this machine from being diverted from its main scientific work, an IBM 7040—itself a large mainframe—was coupled to it to handle input/output (I/O). Although the 7094 had channels, the Direct Couple System allowed the 7040 to handle I/O instead, putting less strain on the 7094. Each machine had a core memory of 32,768 36-bit words. An IBM 7740 communications computer handled a connection to several remote terminals.

The Direct Couple System communicated to the outside world mainly through reels of magnetic tape, which were prepared by an IBM 1401 connected to a keypunch; the 1401 also handled printing results, from tapes brought to it from the 7074. The 1401 could also operate as a stand-alone computer, a very capable one at that. NASA did use it this way to handle the center's administrative work, such as budgeting.[17] In 1974, when it was declared surplus government property, the complete DCS was valued at $1.6 million. Monthly costs were in the range of around $35,000. A variety of other equipment rounded out the configuration.

This Direct Couple System served the Ames laboratory through the dramatic years of the space race. Similar configurations appeared at other large research centers, especially among West Coast aerospace firms. By 1968 the system was working around the clock (except for weekends), with about twenty-seven hours a month reserved for maintenance, and it was running well over capacity. That year an IBM System/360 Model 50 replaced it, at a rental of $21,000 a month for the processor alone, and $45,000 a month for the whole system. NASA was still reluctant to move administrative work off the 1401 to the 360, however, and eventually acquired another small computer to handle those tasks. Nevertheless, one of the main programs installed on the 360 was a piece of software that allowed it to behave as if it were a 1401, to run programs developed for that machine.[18]

The large systems at Ames operated in batch, using decks of punched cards, tape, and line printers. Some of the smaller computers, especially those connected to a wind tunnel, simulator, or other piece of apparatus, operated in real time—processing data as fast as data were presented to it. These systems might also have a provision for direct, interactive control by a user. By the end of the 1960s people at Ames wanted to extend that kind of interactive access to its mainframe system, through a technique called time-sharing. Time-sharing will be discussed at length in the next chapter, but for now it is worth noting how it first appeared at places like NASA.[19]

In 1969 the Ames laboratory installed an IBM 360 Model 67, IBM's attempt to provide for time-sharing on that product line. But the Model 67 proved a disappointment,[20] and by 1971 its "heavy compute load" was shifted to other machines. The Model 67 was retained but reconfigured to serve as a "communications center," mainly for connection to the newly established ARPA-Network.[21] The Model 67's failure was due to deficiencies in its design and to difficulties in incorporating time-sharing into work patterns at NASA-Ames. NASA engineers were using computers to analyze wind-tunnel data. In these problems the basic program remained unchanged, with new data arriving with each test. Wind tunnels were scheduled well in advance, the programs were debugged, and the engineers had established a rhythm between the two large and expensive systems: tunnel and computer. With the full resources of the main computer brought to bear on the data they generated each evening, they were confident that they would have useful results the next morning. With time-sharing that was not the case: how fast or slow one person's job ran depended on who else was using the machine and what kinds of jobs they were doing. That was unpredictable and not under the wind tunnel team's control.[22] In the case studies that follow we shall encounter variants of both these technical and social issues as interactive computing became more of

Table 4.1
Selected computer installations at NASA-Ames, 1955–1969

Date acquired	Date released	Equipment	Use
2/55	6/63	Electrodata, Datatron	On-line processing of wind tunnel data.
5/55	9/58	IBM 650	General scientific computing.
9/58	7/61	IBM 704	General scientific computing and for satellite trajectory and heat transfer calculations. Replaced the IBM 650.
7/61	3/63	IBM 7090	General scientific computing. Added additional capacity for inlet design calculations and plasma particle studies. Replaced the IBM 704.
11/61		Honeywell 800	On-line processing of wind tunnel data.
3/63	7/64	IBM 7094	Replaced the IBM 7090.
7/64		IBM 7094/7040 Direct-Coupled System (DCS)	General scientific computing, large-scale data reduction and administrative data processing. Included an IBM 1401. Replaced the IBM 7094.
8/64		Honeywell 200	Served as an I/O control unit to the H-800 system; replaced associated HON-800 system.
7/65	12/67	IBM 7740, 4 IBM 1440s	Communications unit and terminals to provide remote access to Central Facility.
12/67		IBM 360/50 and IBM 1800	General scientific computing. Acquired to supplement capacity of the DCS and assume remote job entry function. Replaced IBM 7740, four IBM 1440s, and two IBM 1401 systems.
1968		ILLIAC IV	Advanced parallel-processing. Early supercomputer.
7/69		IBM 360/67	Time-sharing operation.

Source: Data from NASA Ames Research Center, "ADPE Acquisition Plan: Proposed Central Computer Facility" (October 1969): 4–5; NASM Archives.

an option. Table 4.1 lists the computing facilities at NASA-Ames during this period.

The IRS

The U.S. Internal Revenue Service performs such a monumental job of processing data that one can scarcely imagine that they ever did their work without computers. As with Blue Cross and NASA-Ames, its computer needs increased by orders of magnitude from 1959 to 1969. We shall also see that it followed a similar trajectory.

The basic operation of this agency is familiar to most Americans. In contrast to the work of NASA, it involves mainly simple arithmetic and the quantities involved are seldom more than a million. But unlike scientific calculations, these calculations have to be accurate to the penny. What really distinguishes this work is the huge number of tax returns the IRS must process, year in and year out, with no slack time.

The need to raise revenue to wage the Second World War set in motion events that would transform the Internal Revenue Service.[23] At that time the number of Americans who were required to file returns and pay taxes increased from around eight million to sixty million; the practice of withholding tax from a paycheck also became common. The IRS handled this work with a combination of Friden calculators, Burroughs or National accounting machines, and pencil and paper. Punched card equipment was not installed until 1948. In 1955 the agency installed an IBM 650 in its regional center in Kansas City, where it helped process 1.1 million returns on a test basis. But keypunching was still the main activity at each regional center, where around 350 employees keyed in basic information from each return.[24]

In 1959 the U.S. Treasury Department authorized the IRS to computerize its operations fully. The IRS selected IBM after soliciting bids from forty manufacturers. An IBM 1401 with a 4K core memory, a card reader, punch, line printer, and two tape drives was installed in each regional center. An IBM 7070 mainframe, the first of several, was purchased for a National Center established in Martinsburg, West Virginia. (Note that thus far the IRS was following the same path as Massachusetts Blue Cross.) The changeover to electronic processing was complete by 1967: Honeywell H-200s had replaced the 1401 and IBM 360s had replaced the 7070s.[25]

Although stored-program computers were now processing the returns, the first step in the process was still keypunchers entering data from returns onto punched cards—over 400 million cards a year, for over 100 million taxpayers, by the mid-1960s. Rooms full of mostly women worked at a steady, unflagging pace, each woman's eye focused on a return propped up to her left, her right hand floating over a keypunch. The 1401s at each regional center read the cards, verified that the required information was there, did some simple data reduction, and transferred the results to tape. Couriers flew these tapes to Martinsburg, where the 7070 processed the returns. The National Center then sent a tape to the

Treasury Department to issue refund checks; for others less fortunate, a tape was sent back to a regional center to send out a bill or otherwise ask for more information. Since the topic of computer networking will arise later, it is worth noting that a courier carrying these reels of tape was moving data at a rate of about 30,000 bits per second on a cross-country flight. That was equivalent to what a personal computer in the 1990s could handle over ordinary telephone lines.[26]

By 1965 the IRS was identifying each taxpayer by a unique number—his or her Social Security number—eliminating the confusion of handling persons with the same names. That had required an act of Congress; one easily forgets the modest origins of the Social Security number when it was established in the 1930s. Requiring one to put this number on all forms, plus the attendant publicity about "electronic brains," led to a few nasty letters to congresspeople about "Big Brother." Few realized the social watershed they had just crossed. (They would later on, as will be discussed.) The punching of cards ended in 1967, when machines were installed that allowed direct entry of data onto a drum (later a disk), but otherwise this division of labor among field centers and the National Center remained into the 1990s. (When the keypunch machines were retired, managers found that productivity did not go up as they expected. By reintroducing some of the sound that was lost, the operators were able to reestablish the rhythm necessary to maintain high rates of data entry.)[27]

As at Blue Cross and NASA-Ames, the IRS processed data sequentially. To find or change a particular record, one mounted the appropriate tape and ran it through a tape drive until that record appeared. A taxpayer who had a problem would have to wait for his or her record to be delivered to a regional center. By the mid-1970s the operation had settled into a pattern whereby the master file was updated once a week, producing an updated file on microfilm that could be flown to a regional center to address questions about a specific return. This kind of data retrieval was not due to any bureaucratic inertia on the part of the IRS; it was built into the structure of the system.

By 1967, with the computerized processing in place and operating smoothly, the agency began looking for an improved system. It hoped not only to eliminate punching cards but also to eliminate manually keying in data altogether—whether by using machines that could read handwritten figures, or by having the taxpayer fill out his or her return in some sort of machine-readable form. The agency also intended to move away from sequential toward random, on-line access to data.

An ambitious plan, projected to cost 650 to 850 million dollars, called the Tax Administration System (TAS), was conceived to implement these goals.[28] Processing would be dispersed to ten service centers across the United States, instead of only to the one in Martinsburg. A combination "batch and realtime [sic] transaction-oriented computer network employing a decentralized database" would be installed, with direct access to taxpayer information available at one of over 8,000 terminals.[29] Taxpayer data would be stored at the centers on "random access storage devices" (probably magnetic disks), instead of on tapes. Other terminals would allow data entry directly into the network, without the need for punching cards.

The planners of the TAS gave much thought to making the system secure—from physical damage, from malicious intrusion, and from simple human errors. But the seed of mistrust in computers that had lain dormant now sprouted. The late 1960s was a time when many citizens questioned the federal government's truthfulness. During the Watergate hearings, which led to President Nixon's resignation in 1974, it was revealed that the White House had breached the wall of integrity that the IRS had carefully built up to shield its operations from political interference. Although the IRS had not yet adopted an interactive data-retrieval system, White House operatives had been able to obtain the tax records of those not in their favor. Trust, without which no system of taxation can function, had eroded.

This time there were more than a few irate letters to Congresspeople. Congress directed the General Accounting Office (GAO) to look at the privacy implications of the proposed TAS; the GAO's preliminary report, issued in 1976 and early 1977, criticized the system for not addressing security and privacy issues thoroughly enough.[30] A copy of the report was leaked to the trade journal *Computerworld*, which ran a lead story under the headline "Proposed IRS System May Pose Threat to Privacy."[31] In the spring of 1977 there were hearings in both the Senate and the House, at which IRS officials were asked questions such as whether someone could "attach a terminal in parallel" with the existing terminal network and thereby be capable of "pulling all this information out."[32] Some IRS employees recall members of Congress dictating what types of computer architecture the agency was allowed to bid on.[33]

Under pressure from Congress, the IRS dropped its plans for the TAS in January 1978. In its place the IRS proposed an "equipment replacement and enhancement program." (Congress made them drop the word "enhancement.") The old architecture in which a centralized master file was kept on magnetic tape was retained. Patrick Ruttle of the IRS called this "a way of moving into the future in a very safe fashion."[34] Instantaneous on-line access to records was verboten. Hamstrung by a hostile Congress, the agency limped along. In 1985 the system collapsed; newspapers published lurid stories of returns being left in dumpsters, refund checks lost, and so on.[35] Congress had a change of heart and authorized money to develop a new data-handling architecture.

NASA's Manned Space Program

Both NASA-Ames and the IRS made attempts to move away from batch processing and sequential access to data, and both failed, at least at first. But the failures revealed advantages of batch operation that may have been overlooked otherwise. Batch operation preserved continuity with the social setting of the earlier tabulator age; it also had been fine-tuned over the years to give the customer the best utilization of the machine for his or her dollar. The real problem with batch processing was more philosophical than technical or economic. It made the computer the equivalent of a horseless carriage or wireless telegraph—it worked faster and handled greater quantities than tabulators or hand calculations, but it did not alter the nature of the work.

During this period, up to the late 1960s, direct, interactive access to a computer could exist only where cost was not a factor. NASA's Manned Space Program was such an installation where this kind of access was developed, using the same kind of hardware as the IRS, NASA-Ames, and Blue Cross.[36] In the late 1950s a project was begun for which cost was not an objection: America's race to put men on the Moon by the end of the decade.

Most of a space mission consists of coasting in unpowered flight. A lot of computing must be done during the initial minutes of a launch, when the engines are burning. If the craft is off-course, it must be destroyed to prevent its hitting a populated area. If a launch goes well, the resulting orbit must be calculated quickly to determine if it is stable, and that information must be transmitted to tracking stations located around the globe. The calculations are formidable and must be carried out, literally, in a matter of seconds.

In 1957 the Naval Research Laboratory established a control center in Washington, D.C., for Project Vanguard, America's first attempt to orbit a satellite. The Center hoped to get information about the satellite to its IBM 704 computer in real time: to compute a trajectory as fast as the telemetry data about the booster and satellite could be fed to it.[37] They did not achieve that goal—data still had to be punched onto cards. In November 1960 NASA installed a system of two 7090 computers at the newly formed Goddard Space Flight Center in Greenbelt, Maryland. For this installation, real-time processing was achieved. Each 7090 could compute trajectories in real time, with one serving as a backup to the other. Launch data were gathered at Cape Canaveral and transmitted to Greenbelt; a backup system, using a single IBM 709, was located in Bermuda, the first piece of land the rocket would pass over after launch. Other radar stations were established around the world to provide continuous coverage.[38]

The system calculated a predicted trajectory and transmitted that back to NASA's Mission Control in Florida. Depending on whether that trajectory agreed with what was planned, the flight controller made a "Go" or "No Go" decision, beginning ten seconds after engine cut-off and continuing at intervals throughout the mission.[39] At launch, a special-purpose Atlas Guidance computer handled data at rates of 1,000 bits per second. After engine cut-off the data flowed into the Goddard computers at a rate of six characters a second.[40] For the generation of Americans who remember John Glenn's orbital flight in February 1962, the clipped voice of the Mercury Control Officer issuing periodic, terse "Go for orbit!" statements was one of the most dramatic aspects of the flight.

In a typical 7090 installation, its channels handled input and output between the central processor and the peripheral equipment located in the computer room. In this case the data was coming from radar stations in Florida, a thousand miles away from Greenbelt. IBM and NASA developed an enhancement to the channels that further conditioned and processed the data. They also developed system software, called Mercury Monitor, that allowed certain input data to interrupt whatever the processor was doing, to ensure that a life-threatening situation was not ignored. Like a busy executive whose memos are labeled urgent, very urgent, and extremely urgent, multiple levels of priority were permitted, as directed by a special "trap processor." When executing a "trap," the system first of all saved the contents of the computer's registers, so that these data could be returned after the interruption was handled.[41]

The Mercury Monitor represented a significant step away from batch operation, showing what could be done with commercial mainframes not designed to operate that way.[42] It evolved into one of IBM's most ambitious and successful software products and laid the foundation for the company's entry into on-line systems later adopted for banking, airline reservations systems, and large on-line data networks.[43]

In the mid-1960s Mission Control moved to Houston, where a system of three (later five) 7094 computers, each connected to an IBM 1401, was installed. In August 1966 the 7094s were replaced by a system based on the IBM 360, Model 75. The simple Mercury Monitor had evolved into a real-time extension of the standard IBM 360 operating system. IBM engineers Tom Simpson, Bob Crabtree and three others called the program HASP (Houston Automatic Spooling Priority—SPOOL was itself an acronym from an earlier day). It allowed the Model 75 to operate both as a batch and real-time processor. This system proved effective and for some customers was preferred over IBM's standard System/360 operating system. HASP was soon adopted at other commercial installations and in the 1970s became a fully supported IBM product.[44]

These modifications of IBM mainframes could not have happened without the unique nature of the Apollo mission: its goal (to put a man on the Moon and return him safely) and its deadline ("before the decade is out"). Such modifications were neither practical nor even permitted by IBM for most other customers, who typically leased and did not own equipment.[45] NASA's modifications did show that a large, commercial mainframe could operate in other than a batch mode. NASA's solution involved a lot of custom work in hardware and software, but in time other, more traditional customers were able to build similar systems based on that work.

The Minicomputer

Having described changes in computing from the top down, changes caused by increased demands by well-funded customers, we'll now look at how these changes were influenced by advances in research into solid-state physics, electronics, and computer architecture. The result was a new type of machine called the "minicomputer." It was not a direct competitor to mainframes or to the culture of using mainframes. Instead the minicomputer opened up entirely new areas of application. Its growth was a cultural, economic, and technological phenomenon. It introduced large groups of people—at first engineers and scientists, later others—to direct interaction with computing machines. Minicomputers, in particular those operated by a Teletype, introduced the notion of the computer as a personal interactive device. Ultimately that notion would change our culture and dominate our expectations, as the minicomputer yielded to its offspring, the personal computer.

Architecture

A number of factors define the minicomputer: architecture, packaging, the role of third-parties in developing applications, price, and financing. It is worth discussing the first of those, architecture, in some detail to see how the minicomputer differed from what was prevalent at the time.

A typical IBM mainframe in the early 1960s operated on 36 bits at a time, using one or more registers in its central processor. Other registers handled the addressing, indexing, and the extra digits generated during a multiplication of two 36-bit numbers. The fastest, most complex, and most expensive circuits of the computer were found here. A shorter word length could lower the complexity and therefore the cost, but that incurred several penalties. The biggest penalty was that a short word length did not provide enough bits in an instruction to specify enough memory addresses. It would be like trying to provide telephone service across the country with seven-digit phone numbers but no area codes. Another penalty of using a short word was that an arithmetic operation could not provide enough digits for anything but the simplest arithmetic, unless one programmed the machine to operate in "double precision." The 36-bit word used in the IBM 7090 series gave the equivalent of ten decimal digits. That was adequate for most applications, but many assumed that customers would not want a machine that could not handle at least that many.

Minicomputers found ways to get around those drawbacks. They did that by making the computer's instruction codes more complex. Besides the operation code and memory address specified in an instruction, minicomputers used several bits of the code to specify different "modes" that extend the memory space. One mode of operation might not refer directly to a memory location but to another register in which the desired memory location is stored. That of course adds complexity; operating in double precision also is complicated, and both might slow the computer down. But with the newly available transistors coming on the market in the late 1950s, one could design a processor that, even with these added complexities, remained simple, inexpensive, and fast.

The Whirlwind had a word length of only 16 bits, but the story of commercial minicomputers really begins with an inventor associated with very large computers: Seymour Cray. In 1957, the Control Data

Corporation was founded in the Twin Cities by William Norris, the cofounder of Engineering Research Associates, later part of Remington Rand UNIVAC, as mentioned in chapter 1. Among the many engineers Norris persuaded to go with him was Cray. While at UNIVAC Cray had worked on the Navy Tactical Data System (NTDS), a computer designed for Navy ships and one of the first transistorized machines produced in quantity.[46] Around 1960 CDC introduced its model 1604, a large computer intended for scientific customers. Shortly thereafter the company introduced the 160, designed by Cray ("almost as an after-thought," according to a CDC employee) to handle input and output for the 1604. For the 160 Seymour Cray carried over some key features he pioneered for the Navy system, especially its compact packaging. In fact, the computer was small enough to fit around an ordinary-looking metal desk—someone who chanced upon it would not even know it was a computer.

The 160 broke new ground by using a short word length (12 bits) combined with ways of accessing memory beyond the limits of a short address field.[47] It was able to directly address a primary memory of eight thousand words, and it had a reasonably fast clock cycle (6.4 micro-seconds for a memory access). And the 160 was inexpensive to produce. When CDC offered a stand-alone version, the 160A, for sale at a price of $60,000, it found a ready market. Control Data Corporation was concentrating its efforts on very high performance machines (later called "supercomputers," for which Cray became famous), but it did not mind selling the 160A along the way. What Seymour Cray had invented was, in fact, a minicomputer.[48]

Almost immediately new markets began to open for a computer that was not tied to the culture of the mainframe. One of the first customers, which provides a good illustration of the potential of such designs, was Jack Scantlin, the head of Scantlin Electronics, Inc. (SEI). When he saw a CDC 160A in 1962, he conceived of a system built around it that would provide on-line quotations from the New York Stock Exchange to brokers across the country. By 1963 SEI's Quotron II system was operational, providing stock prices within about fifteen seconds, at a time when trading on the NYSE averaged about 3.8 million shares a day.[49] SEI engineers resorted to some ingenious tricks to carry all the necessary information about stock prices in a small number of 12-bit words, but ultimately the machine (actually, two 160As connected to a common memory) proved fully capable of supporting this sophisticated application.

The Digital Equipment Corporation

In the same year that CDC was founded, 1957, Kenneth H. Olsen and Harlan Anderson founded the Digital Equipment Corporation (DEC, pronounced "deck"). Financing came from the American Research and Development Corporation, a firm set up by Harvard Business School Professor Georges Doriot, whose goal was to find a way to commercialize the scientific and technical innovations he had observed during the Second World War as an officer in the U.S. Army. They set up operations in a corner of a woolen mill astride the Assabet River in Maynard, Massachusetts. As a student at MIT, Olsen had worked on fitting the Whirlwind with core memory in place of its fragile and unreliable storage tubes, and in the mid-1950s he had worked for MIT's Lincoln Laboratory in suburban Lexington. He had represented the Lincoln Lab to IBM when it was building computers for the SAGE air-defense system. In 1955 Olsen had taken charge of a computer for Lincoln Lab called TX-0, a very early transistorized machine.[50] Under his supervision, the TX-0 first operated at Lincoln Lab in 1956.[51]

The TX-0 had a short word length of 18 bits. It was designed to utilize the new surface-barrier transistors just then being produced by Philco (it used around 3,600 of them). These transistors were significantly faster and of higher quality than any transistors available previously. Although each one cost $40 to $80 (compared to about $3 to $10 for a tube), and their long-term reliability was unknown, the TX-0 designers soon learned that the transistors were reliable and did not need any treatment different from other components.[52] Reflecting its connections to the interactive SAGE system, the TX-0 had a cathode-ray tube display and a light-pen, which allowed an operator to interact directly with a program as it was running. The designer of that display was Ben Gurley, who left Lincoln Labs in 1959 to become one of Digital Equipment Corporation's first employees.

When completed in 1957, the TX-0 was one of the most advanced computers in the world, and in 1959 when Digital Equipment Corporation offered its PDP-1 designed by Gurley, it incorporated many of the TX-0's architectural and circuit innovations. Recall that the IBM 7090 was a transistorized machine that employed the same architecture as the vacuum tube 709, with transistors replacing the individual tubes. The PDP-1 owed nothing to tube design; it was intended to take full advantage of what transistors had had to offer from the start. It was capable of 100,000 additions per second, not as fast as the IBM 7090, but respectable and much faster than the drum-based computers in its price class. Its basic core memory held four thousand, later expanded to sixty-four thousand, 18-bit words.

The PDP-1 was not an exact copy of the TX-0, but it did imitate one of its most innovative architectural features: foregoing the use of channels, which mainframes used, and allowing I/O to proceed directly from an I/O device to the core memory itself. By careful design and skillful programming, this allowed fast I/O with only a minimal impact on the operation of the central processor, at a fraction of the cost and complexity of a machine using channels.[53] In one form or another this "direct memory access" (DMA) was incorporated into nearly all subsequent DEC products and defined the architecture of the minicomputer. It is built into the microprocessors used in modern personal computers as well. To allow such access to take place, the processor allowed interrupts to occur at multiple levels (up to sixteen), with circuits dedicated to handling them in the right order. The cost savings were dramatic: as DEC engineers later described it, "A single IBM channel was more expensive than a PDP-1."[54] The initial selling price was $120,000.

Digital Equipment Corporation sold about fifty PDP-1s. It was hardly a commercial success, but it deserves a place in the history of computing for its architectural innovations—innovations that were as profound and long-lasting as those embodied in John von Neumann's 1945 report on the EDVAC.

The modest sales of the PDP-1 set the stage for Digital's next step. That was to establish a close relationship between supplier and customer that differed radically from those of IBM and its competitors. From the time of its founding, IBM's policy had been to lease, not sell, its equipment. That policy gave it a number of advantages over its competitors; it also required capital resources that DEC did not have. Although IBM agreed to sell its machines as part of a Consent Decree effective January 1956, leasing continued to be its preferred way of doing business.[55] That policy implied that the machine on the customer's premises was not his or hers to do with as he wished; it belonged to IBM, and only IBM was allowed to modify it. The kinds of modifications that NASA made at its Houston center, described above, were the rare exceptions to this policy.

The relationship DEC developed with its customers grew to be precisely the opposite. The PDP-1 was sold, not leased. DEC not only permitted, it encouraged modification by its customers. The PDP-1's customers were few, but they were sophisticated. The first was the Cambridge consulting firm Bolt Beranek and Newman (BBN), which later became famous for its role in creating the Internet. Others included the Lawrence Livermore Laboratory, Atomic Energy of Canada, and the telecommunications giant, ITT.[56] Indeed, a number of improvements to the PDP-1 were suggested by Edward Fredkin of BBN after the first one was installed there. Olsen donated another PDP-1 to MIT, where it became legendary as the basis for the hacker culture later celebrated in popular folklore. These students flocked to the PDP-1 rather than wait their turn to submit decks of cards to the campus IBM mainframe. Among its most famous applications was as a controller for the Tech Model Railroad Club's layout.[57] Clearly the economics of mainframe computer usage, as practiced not only at commercial installations but also at MIT's own mainframe facility, did not apply to the PDP-1.

DEC soon began publishing detailed specifications about the inner workings of its products, and it distributed them widely. Stan Olsen, Kenneth Olsen's brother and an employee of the company, said he wanted the equivalent of "a Sears Roebuck catalog" for Digital's products, with plenty of tutorial information on how to hook them up to each other and to external industrial or laboratory equipment.[58] At Stan's suggestion, and in contrast to the policy of other players in the industry, DEC printed these manuals on newsprint, cheaply bound and costing pennies a copy to produce (figure 4.2). DEC salesmen carried bundles of these around and distributed them liberally to their customers or to almost anyone they thought might be a customer.

This policy of encouraging its customers to learn about and modify its products was one borne of necessity. The tiny company, operating in a corner of the Assabet Mills, could not afford to develop the specialized interfaces, installation hardware, and software that were needed to turn a general-purpose computer into a useful product. IBM could afford to do that, but DEC had no choice but to let its customers in on what, for other companies, were jealously guarded secrets of the inner workings of its products. DEC found, to the surprise of many, that not only did the customers not mind the work but they welcomed the opportunity.[59]

The PDP-8   The product that revealed the size of this market was one that was first shipped in 1965: the PDP-8 (figure 4.3). DEC installed over 50,000 PDP-8 systems, plus uncounted single-chip implementations developed years later.[60]

The PDP-8 had a word length of 12 bits, and DEC engineers have traced its origins to discussions with the Foxboro Corporation for a process-control application. They also acknowledge the influence of the 12-bit CDC-160 on their decision.[61] Another influence was a computer

Figure 4.2
DEC manuals. DEC had these technical manuals printed on cheap newsprint, and the company gave them away free to anyone who had an interest in using a minicomputer. (*Source*: Mark Avino, NASM.)

designed by Wes Clark of Lincoln Labs called the LINC, a 12-bit machine intended to be used as a personal computer by someone working in a laboratory setting.[62] Under the leadership of C. Gordon Bell, and with Edson DeCastro responsible for the logic design, DEC came out with a 12-bit computer, the PDP-5, in late 1963. Two years later they introduced a much-improved successor, the PDP-8.

The PDP-8's success, and the minicomputer phenomenon it spawned, was due to a convergence of a number of factors, including performance, storage, packaging, and price. Performance was one factor. The PDP-8's circuits used germanium transistors made by the "micro-alloy diffused" process, pioneered by Philco for its ill-fated S-2000 series. These transistors operated at significantly higher speeds than those made by other techniques. (A PDP-8 could perform about 35,000 additions per second.)[63] The 12-bit word length severely limited the amount of memory a PDP-8 could directly access. Seven bits of a word comprised the address field; that gave access to 2^7 or 128 words. The

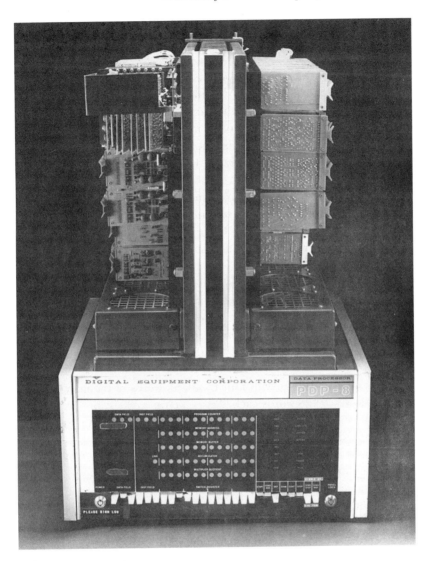

Figure 4.3
Digital Equipment Corporation PDP-8. The computer's logic modules were mounted on two towers rising from the control panel. Normally these were enclosed in smoked plastic. Note the discrete circuits on the boards on the left: The original PDP-8 used discrete, not integrated circuits. (*Source*: Laurie Minor, Smithsonian.)

PDP-8 got around that limitation in two ways. One was to use "indirect addressing," to specify in the address field a memory location that contained not the desired piece of data but the *address* of that data. (This allowed for the full 12-bits of a word instead of only seven to be used for an address.) The other was to divide the memory into separately-addressed "pages," exploiting the fact that most of the time one is accessing data from a small portion of memory; only occassionally would the computer have to jump to another page. That process was not as simple as addressing memory directly, but it did not slow things down if it did not happen too often.

Improvements in logic and core memory technology reduced the memory cycle time to 1.6 microseconds—slightly faster than the IBM 7090, four times faster than the CDC 160, and over a thousand times faster than the Bendix G-15, the fastest drum computer of the late 1950s.[64] The PDP-8's short word length meant that it could not compete with its mainframe competitors in doing arithmetic on 10-digit decimal or floating-point numbers, but for many other applications it was as fast as any computer one could buy at any price.[65] That kind of performance made the PDP-8 and the minicomputers that followed it fundamentally different from the G-15, the LGP-30, the IBM 1401, and other "small" computers.

The basic PDP-8 came with four thousand words of memory, divided into 32 blocks of 128 words each. Access across a block, or "page," was possible by setting one of two bits in the operation code of an instruction word. For external memory DEC provided a simple, inexpensive, but capable tape system derived from the LINC. They called it "DECtape." Again in contrast to mainframe tape systems, a reel of DECtape was light and portable; the drive was compact and could fit into the same equipment rack as the computer itself. Data could be read or written in either direction, in blocks of 128 words, not just appended at the end of a record. DECtape acted more like the floppy disk drives on modern personal computers, than like the archival storage style of mainframe tape drives.[66]

The physical packaging of the PDP-8, a factor that mattered less for large systems, played a key role in its success. The PDP-8 used a series of compact modules, on which transistors, resistors, and other components were mounted. Each module performed a well-defined logic function (similar to the functions that the first integrated circuits performed). These in turn were plugged into a hinged chassis that opened like a book. The result was a system consisting of processor, control panel, and core memory in a package small enough to be embedded into other equipment. The modules themselves were interconnected by wire-wrap (see chapter 2). DEC used automatic wire-wrapping machinery from the Gardner-Denver Corporation to wire the PDP-8. This eliminated wiring errors and allowed DEC to handle the large orders it soon received. The computer occupied eight cubic feet of volume and weighed 250 pounds.[67]

There was the matter of pricing the PDP-8. A low price would generate sales, but it might also prevent DEC from generating enough revenue to support research and development, which it would need to keep its lead in technology and (avoid the fate of many of the start-up computer companies of the mid-1950s, which ended up being bought by established companies like Burroughs or NCR). Executives at DEC decided to take the risk, and they priced the PDP-8 at $18,000, including a teletype terminal for I/O. Within a few years one could be bought for less than $10,000. The low price shocked the computer industry and generated a flood of orders. Once again all estimates of the size of the market for computers turned out to be too timid.[68] Established companies, including IBM, eventually entered this market, but DEC continued to grow and prosper. It found a way, first of all, to stay at the forefront of computer technology by continuing to draw from the knowledge and skills of the MIT research community. It also continued to keep the cost of its operations low. Being based in an old woolen mill certainly helped, but even more important was the relationship DEC developed with its customers, who took responsibility for development work and associated costs. (This will be discussed shortly.)

For loading and editing programs the PDP-8 used a new device from the Teletype Corporation, the Model 33 ASR ("automatic send-receive").[69] It was cheaper, simpler, and more rugged than the Flexowriter used by earlier small computers (figure 4.4). Like the Flexowriter, it functioned as a typewriter that could print onto a roll of continuous paper, send a code indicating what key was pressed directly to a computer, or punch that code onto a paper tape. Data were transmitted at a rate from six to ten characters per second. Introduced in the mid-1960s, the Model 33 was one of the first to adopt the standard for coding bits then being promulgated by the American Standards Association, a code known as ASCII (American Standard Code for Information Interchange). The Flexowriter's code was popular with some business equipment companies, but its code was rejected as a basis for the computer industry when ASCII was developed.[70] Just as the Chain Printer symbolized the mainframe computing environment, the Model 33 came to symbolize the minicomputer era and the beginnings of the personal computer era that followed it. It had a far-reaching effect on personal computing, especially on the keyboard: the *control* and *escape* keys, for example, first made their general appearance on the Model 33. Many other key codes peculiar to this machine found their way into personal computer software fifteen years later, with few people realizing how they got there.

Finally, there was the computer's name. "Minicomputer" was catchy, it fit the times, and it gave the PDP-8 an identity. One could obtain a minicomputer and not feel obliged also to get a restrictive lease

Figure 4.4
An ASR-33 Teletype, the standard input/output device for early minicomputers, although it was not originally designed for that purpose. Note the "Control" (CTRL) and "Escape" (ESC) keys, which later became standard for desktop computer keyboards. The "X-ON" (CTRL-Q) and "X-OFF" (CTRL-S) commands also became embedded into personal computer operating systems. The "@" symbol (Shift-P) was later adopted for indicating addresses on the Internet. (*Source*: Charles Babbage Institute, University of Minnesota.)

agreement, a climate-controlled room, or a team of technicians whose job seemed to be keeping users away. The miniskirt happened to come along (from Britain) at the time the PDP-8 was beginning to sell, and no doubt some of its glamour was transferred to the computer. It may have been a DEC salesman stationed in Europe who gave the PDP-8 that name.[71] (Given Kenneth Olsen's conservative religious upbringing, it was unlikely that he would have come up with it. Of Scandinavian descent, he neither smoked nor drank nor used profanity.) Another source of the name, one that fits the PDP-8 perfectly, was also a British export—the Morris Mini-Minor, designed by the legendary automobile engineer Alec Issigonis, in response to the Suez Canal Crisis that cut off Persian Gulf oil to Britain in 1956. Issigonis's design was lightweight, responsive, and economical to operate. Most important, it outperformed most of the stodgy, bloated British cars with which it competed. The British exported Mini-Minors and miniskirts around the world. Digital Equipment Corporation did the same with minicomputers.

Programming a PDP-8 to do something useful required no small amount of skill. Its limited memory steered programmers away from high-level programming languages and toward assembly or even machine code. But the simplicity of the PDP-8's architecture, coupled with DEC's policy of making information about it freely available, made it an easy computer to understand. This combination of factors gave rise to the so-called original equipment manufacturer (OEM); a separate company that bought minicomputers, added specialized hardware for input and output, wrote specialized software for the resulting systems, and sold them (at a high markup) under its own label. The origin of the term "OEM" is obscure. In some early references it implies that the computer manufacturer, not the third party, is the OEM, which seems a logical definition of "original equipment." Eventually, however, the meaning attached entirely to the party that built systems around the mini.[72]

Dealing with an OEM relieved the minicomputer manufacturer of the need to develop specialized software. DEC developed some applications of its own, such as the computerized typesetting system, but that was the exception.[73] A typical OEM product was the LS-8 from Electronics Diversified of Hillsboro, Oregon, which it was used to operate theatrical stage lighting, controlling a complex of lights through programmed sequences. The LS-8's abilities were cited as a key element in the success of the long-running Broadway hit *A Chorus Line*.[74] Inside the LS-8 was a PDP-8A, a model that DEC had introduced in 1975. Users of the LS-8 did not necessarily know that, because the LS-8 had its own control panel, tailored not to computer users but to theatrical lighting crews. OEM applications ranged across all segments of society, from medical instrumentation to small business record keeping, to industrial controllers. One PDP-8–based system was even installed in a potato-picking machine and carried on the back of a tractor (figure 4.5).[75]

The DEC Culture   Alec Issigonis believed that the key to the success of the Morris Mini-Minor was that it was designed by a capable engineering team of no more than six persons, which was permitted by management to operate with little or no outside interference.[76] That is about as good a description of the culture at Digital Equipment as one could hope to find.[77] Though growing fast, DEC retained the atmosphere of a small company where responsibility for product development fell to small groups of engineers. In 1965 it had revenues of $15 million and 876 employees. By 1970 DEC had revenues of $135 million and 5,800

Figure 4.5
A PDP-8 mounted on a tractor and controlling a potato-picker. Although an awkward installation, it foreshadowed the day when microprocessors were embedded into nearly all complex machinery, on the farm and elsewhere. (*Source*: Digital Equipment Corporation.)

employees.[78] That was a small fraction of IBM's size, although DEC was shipping as many PDP-8 computers as IBM was shipping of its 360 line.

As Digital grew into one of IBM's major competitors, it remained Spartan—excessively so. Digital gradually took over more and more of the Assabet Mills, until it eventually bought it all (figure 4.6). Finding one's way through the complex was daunting, but the "Mill rats" who worked there memorized the location of the corridors, bridges, and passageways. Digital opened branch facilities in neighboring towns, but "the Mill" remained the spiritual center of the company. Customers were continually amazed at its simplicity and lack of pretension. One Wall Street analyst said, with unconcealed scorn, that the company had only "barely refurbished" the nineteenth-century mill before moving in.[79] An administrator from the Veterans Administration, who was adapting DEC equipment for monitoring brain functions during surgery, expressed similar surprise:

I don't know if you've ever been to the original factory, but it is (or was) a nice old nineteenth-century mill that was used to make wool blankets during the civil war, so the wooden floors were soaked with lanolin and had to be swabbed occasionally. It was a huge building, and a little spooky to work in at night when no one else was around.[80]

Figure 4.6
The Mill, Maynard, Massachusetts. Headquarters for Digital Equipment Corporation. (*Source*: Digital Equipment Corporation.)

A professor of English from a small midwestern college, who wanted to use a PDP-8 to sort and classify data on the London Stage in the seventeenth and eighteenth centuries, described his first visit to the Mill this way:

Maynard is still rural enough to remind one that Thoreau once roamed its woods. Like many New England towns it has a dam in its river just above the center and a jumble of old red brick mills mellowing toward purple beneath the dam. DEC apparently occupied all the mill buildings in Maynard Center, and they were all connected by abutment at some angle or another by covered bridges, and the river got through them somehow.

The main entrance from the visitors' disintegrating asphalt parking lot was a wooden footbridge across a gully into an upper floor of one of the factory buildings. One entered a fairly large, brightly lighted, unadorned, carpetless section of a loft with a counter and a door at the far end. At the counter a motherly person helped one write down one's business on a card and asked one to take a seat in a row of about seven chairs down the middle of the room. There were a few dog-eared magazines to look at. It was impossible to deduce the principle of their selection or the series of accidents by which they had arrived here. *Colorado Municipalities, Cat-Lover's Digest, Psychology Today*.[81]

A cult fascination with Digital arose, and many customers, especially scientists or fellow engineers, were encouraged to buy by the Spartan image. DEC represented everything that was liberating about computers, while IBM, with its dress code and above all its punched card, represented everything that had gone wrong.[82] Wall Street analysts, accustomed to the trappings of corporate wealth and power, took the Mill culture as a sign that the company was not a serious computer company, like IBM or UNIVAC.[83] More to the point, DEC's marketing strategy (including paying their salesmen a salary instead of commissions) was minimal. Some argued it was worse than that: that DEC had "contempt" for marketing, and thus was missing chances to grow even bigger than it did.[84] DEC did not grow as fast as Control Data or Scientific Data Systems, another company that started up at the same time, but it was selling PDP-8s as fast as it could make them, and it was opening up new markets for computers that neither CDC nor SDS had penetrated. It was this last quality that set the company apart. One could say from the perspective of the 1990s that DEC was just another computer company that grew, prospered, and then was eclipsed by events. But that would miss the fact that DEC reoriented computing toward what we now assume is the "natural" or obvious way to define computing. It is impossible to understand the state of computing at the end of the twentieth century without understanding computing's debt to the engineers at the Assabet Mills.

But whatever its image, DEC did not see *itself* as a company that built only small computers. Simultaneously with the PDP-8 it introduced a large system, the 36-bit PDP-6. Only twenty-three were sold, but an improved version, the PDP-10, became a favorite of many university computer science departments and other sophisticated customers. First delivered in 1966, the PDP-10 was designed from the start to support time-sharing as well as traditional batch processing. Outside the small though influential group of people who used it, however, the PDP-10 made only a small dent on the mainframe business that IBM dominated with its 7090 and 360-series machines.

DEC did eventually became a serious contender in the large systems market with its VAX line, beginning in the late 1970s. By that time it had also smoothed the rougher edges off of the Mill culture. Its sales force continued to draw a salary, but in other respects DEC salesmen resembled IBM's. Digital remained in the Mill but refurbished the visitors' reception area so it resembled that of any other large corporation. (Because of its location in the middle of Maynard, however, there still was limited parking; visitors simply parked on a downtown street, being careful to put a few dimes into the meter to keep from getting a ticket. Maynard still was a thrifty New England town.) The brick walls were still there, adorned with a few well-chosen pieces of a loom or carding machine leftover from the woolen mill days. A visitor could announce his or her name to a receptionist seated at a well-appointed security desk, settle into a comfortable and modern chair, and peruse the *Wall Street Journal* while waiting for an appointment. By the late 1980s the manufacturing had moved overseas or to more modern and utilitarian buildings scattered throughout Massachusetts and New Hampshire. The Mill was now a place for office workers seated at desks, not for engineers at workbenches. Olsen's successor, Robert B. Palmer, decided in 1993 to move the company's headquarters out of the Mill and into a smaller, modern building in Maynard. Around the same time word went out that the company was to be called Digital, not DEC—a small change but somehow symbolic of the passing of an age. The era of the minicomputer came to an end, but only after it had transformed computing.

The MIT Connection   The Mill was one clue to DEC's approach to entering the computing business. A more revealing clue is found in a corporate history that the company published in 1992 (when the personal computer was challenging DEC's business). The first chapter of *Digital at Work* is a discussion not of the Mill, the PDP-1, or of Olsen, but of "MIT and the Whirlwind Tradition."[85] The chapter opens with a photograph of MIT's main building. The first photographs in the book of people are of MIT students; next are photos of professors and of the staff (Jay Forrester, Robert Everett, and J. A. O'Brien) of Project Whirlwind.

The Whirlwind computer was operational in 1950, and by the time DEC was founded it was obsolete. But the foundations laid by Project Whirlwind were stong enough to support DEC years later. The most visible descendant of Whirlwind was the SAGE air-defense system. DEC, the minicomputer, and the other computer companies that sprouted in suburban Boston were other, more important offspring. Ken Olsen, allied with Georges Doriot, found a way to carry the MIT atmosphere of engineering research, whose greatest exponent was Jay Forrester, off the campus, away from military funding, and into a commercial company. It was so skillfully done, and it has been repeated so often, that in hindsight it appears natural and obvious. Although there have been parallel transfers to the private sector, few other products of World War II and early Cold War weapons labs (radar, nuclear fission, supersonic aerodynamics, ballistic missiles) have enjoyed this trajectory. Computing, not nuclear power, has become "too cheap to meter."

That new culture of technical entrepreneurship, considered by many to be the main force behind the United States's economic prosperity of the 1990s, lasted longer than the ambience of the Mill. It was successfully transplanted to Silicon Valley on the West Coast (although for reasons yet to be understood, Route 128 around Boston, later dubbed the Technology Highway, faded). In Silicon Valley, Stanford and Berkeley took the place of MIT, and the Defense Advanced Research Projects Agency (DARPA) took over from the U.S. Navy and the Air Force.[86] A host of venture capital firms emerged in San Francisco that were patterned after Doriot's American Research and Development Corporation. Many of the popular books that analyze this phenomenon miss its university roots; others fail to understand the role of military funding. Some concentrate on the wealth and extravagant lifestyles adopted by the millionaires of Silicon Valley—hardly applicable to Ken Olsen, whose plain living was legendary.

IBM represented the perfection of what John Kenneth Galbraith called the "technostructure": a large, highly organized, vertically integrated firm that controlled, managed, and channeled the chaos of technical innovation into market dominance. Central to smooth operations at IBM was a character from a best-seller from that era, *The Organization Man*, by William Whyte.[87] People made fun of the IBM employee, with his white shirt and conservative suit, who followed the "IBM way" so closely. Yet who among them was not jealous of the company's profits and the generous commissions earned by IBM salesmen? A closer reading of Whyte's book reveals a genuine admiration for such people, without whom a company could hardly survive, let alone prosper. Olsen tapped into an alternate source of knowledge; he had no choice. Olsen and his young engineers just out of MIT were "organization men," too, only of a different stripe. They, too, shared a set of common values, only theirs came from the old temporary buildings on the MIT campus, the ones where the Radiation Lab was housed during the War. Those values seemed very different from IBM's, but they were strong enough to mold DEC employees into a competitive organization. These engineers refuted the wisdom of the day, which stated that the era of the lone pioneer was over, that start-up companies could never compete against the giants.

The modest appearance of the PDP-8 concealed the magnitude of the forces it set into motion. Mainframe computing would persist, although its days of domination were numbered. As long as the economics were in its favor, many would continue to use a computer by punching decks of cards. IBM would continue to dominate the industry. The computer business was not a zero-sum game; DEC's gain was not automatically IBM's loss—at least not for a while. The mini showed that with the right packaging, price, and above all, a more direct way for users to gain access to computers, whole new markets would open up. That amounted to nothing less than a redefinition of the word "computer," just as important as the one in the 1940s, when that word came to mean a machine instead of a person that did calculations. Fulfilling that potential required two more decades of technical development. Ultimately Digital Equipment Corporation, as well as IBM and the other mainframe companies, would be buffeted by the forces unleashed in the Assabet Mills, forces that would prove impossible to restrain.

The "Go-Go" Years and the System/360, 1961–1975

IBM, the Seven Dwarfs, and the BUNCH

As the minicomputer established its markets in the mid-1960s, most computer dollars continued to be spent on large mainframes sold by IBM and a few competitors. IBM held about a 70% share of the commercial market, with 1963 revenues of $1.2 billion, growing to over $3 billion in 1965, and $7.5 billion by 1970.[1] Second to IBM was Sperry Rand, inheritor of the original UNIVAC and ERA developments of the 1940s, with $145 million in revenue. Other players in the U.S. market were Control Data, Honeywell, Philco, Burroughs, RCA, General Electric, and NCR. (AT&T also manufactured computers, but as a regulated monopoly its figures are not comparable here.)[2]

With the partial exception of Control Data, all the above companies focused on the same model of computing espoused by IBM: large, centralized mainframe installations, running batches of programs submitted as decks of punched cards.[3] Those who wished to compete in this business provided everything from bottom to top—hardware, peripherals, system and applications software, and service. They sought further to compete with IBM by offering to lease as well as sell their computers outright. That required enormous amounts of capital, and profits for everyone except IBM were low or nonexistent.

The status of the players at the time led IBM-ologists to call them "Snow White and the Seven Dwarfs." The term was ironic: "Snow White" was periodically the target of lawsuits either from one of the "Dwarfs" (e.g., Control Data) or the Federal government itself, for monopoly practices. By the 1970s General Electric and RCA had left the business, leading to a new term for IBM's competitors, the "BUNCH" (Burroughs, UNIVAC, NCR, Control Data, and Honeywell). This constellation remained stable into the 1980s—remarkably so in an industry as volatile as computers. The advent of personal computers in the 1980s changed the nature of the entire business, and the simple grouping of mainframe suppliers unraveled.

IBM System/360

As DEC began shipping its PDP-8 in early 1965, IBM delivered the first of a series of mainframes that would propel that company into an even more commanding position in the industry. That was the System/360, announced in April 1964 (figure 5.1). It was so named because it was aimed at the full circle of customers, from business to science, at customers who did a lot of mathematical calculation and at those who did simpler arithmetic on large sets of data. System/360's primary selling point was that IBM was offering not one but a whole line of computers,

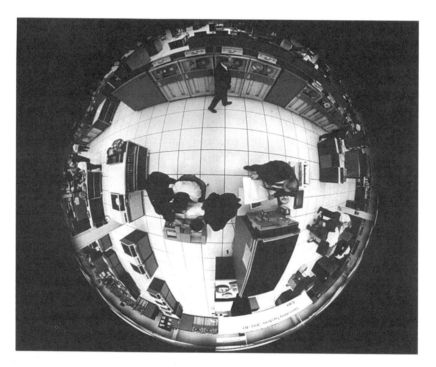

Figure 5.1
IBM System/360. A publicity photo from IBM, showing the vast size and scope of products announced in 1964 (*Source*: IBM.)

with a promise that programs written for one model would work on larger models, thus saving a customer's investment in software as business grew. IBM announced six models on April 7, 1964. Later on it announced others, while dropping some of the original six by the time deliveries began. The idea was not entirely new: computer companies had tried to preserve software compatibility as they introduced newer models, as IBM had done with its 704, 709, and 7090 machines. But the 360 was a *series* of computers, all announced at the same time, offering about a 25 : 1 performance range. Except for a small run of machines delivered to the Army in the late 1950s, that had never been attempted before.[4]

In an often-repeated phrase, first used in a *Fortune* magazine article, an IBM employee said "you bet your company" on this line of computers.[5] Besides the six computer models, IBM introduced "over 150 different things—new tapes, new disks, the 029 card punch" on the same day.[6] Had the 360 failed, it would have been a devastating blow, although IBM would still have survived as a major player in the business. The company could have introduced newer versions of its venerable 1401 and 7090-series machines, and it still had a steady stream of revenue from precomputer punched card installations. But such a failure would have restructured the computer industry.[7]

System/360 did not fail. Within weeks of the product announcement in April 1964 orders began coming in. "Orders for System/360 computers promptly exceeded forecasts: over 1100 were received in the first month. After five months the quantity had doubled, making it equal to a fifth of the number of IBM computers installed in the U.S."[8] The basic architecture served as the anchor for IBM's product line into the 1990s.

Manufacturing and delivering the line of computers required enormous resources. The company expanded its production facilities, but delivery schedules slipped, and shortages of key components arose. The success of the 360 threatened the company's existence almost as much as a failure might have. For those employees driven to the breaking point— and there were many—the jump in revenues for IBM may not have been worth the physical and mental stress. From 1965 to 1970, thanks mostly to System/360, IBM's gross income more than doubled. Net earnings also doubled, surpassing $1 billion by 1971. IBM had led the U.S. computer industry since the mid 1950s. By 1970 it had an installed base of 35,000 computers, and by the mid-1970s it made sense to describe the U.S. computer industry as having two equal parts: IBM on one side and everyone else *combined* on the other.[9]

The problems IBM faced in trying to meet the demand—employee burnout, missed shipping dates, quality control on the production lines—were problems its competitors might have wished for. Obviously many customers found this line of machines to their liking. Most NASA centers, for example, quickly switched over to 360 (Model 65 or higher) from their 7090 installations to meet the demands of putting a man on the Moon. Commercial firms that used computers for business data processing likewise replaced their 7030s and other systems with models of the System/360. There was some resistance to replacing the venerable 1401 with the low-end 360, but in general the marketplace gave overwhelming approval to the notion of a compatible family of machines suitable for scientific as well as business applications (figure 5.2).

The decisions that led to System/360 came from an IBM committee known as SPREAD, which met daily in the Sheraton New Englander motel in Cos Cob, Connecticut, for two months in late 1961. Their

Figure 5.2
A small-scale System/360 installation. Note the vacuum-column tape drives in the background and a typewriter with the Selectric mechanism in the front. In the extreme foreground is a disk drive. (*Source*: IBM Archives.)

report, issued internally on December 28, 1961, and published twenty-two years later, reveals much about the state of computing, as it then existed, and as key engineers and executives at IBM thought it would become.[10]

Their deliberations began with a survey of the company's existing products. In 1961 IBM was fielding a confusing tangle of machines, few of which were compatible with one another. Two of them stood out and have already been described. The 1401 was a small, character-oriented computer that rented at a low price and was well liked. Intended for business customers, it was also popular for scientific use, mainly due to the excellent Model 1403 chain printer that came with it. Sales of the 1401 were measured in the thousands. The other outstanding machine was the 7090/7094 scientific computer. It was expensive, but its performance made it popular with customers like NASA and the large aerospace firms. Its sales measured in the low hundreds. IBM's large business computer, the 7070, had had disappointing sales, while the small scientific machine, the 1620, was doing well, although not as well as the 1401.[11]

All were transistorized machines, although the 7090 was a transistorized version of the vacuum-tube 709, hastily introduced to win an Air Force contract. Because both the 7090 and the 1401 were built on an old foundation, it would have been difficult to achieve an order-of-magnitude increase in performance for either. And they were incompatible with each other.[12] Meanwhile, the notion of what divided business from scientific use was not holding up. According to that notion, business customers handled large sets of data, on which they performed simple arithmetic, while scientific customers did the opposite, advanced calculation on small sets of data. In fact, however, along with a need for floating-point arithmetic, scientists and engineers also needed to handle large data sets in applications like finite-element analysis, for instance, a technique developed for building complex aerospace structures.[13] And routine business transactions like payroll had increasing complexity, as federal programs like Medicare spread through the workplace. The SPREAD Committee, composed of members from both of IBM's product lines, did not agree at first on a unified product line, but eventually they recognized its advantages and incorporated that as a recommendation in their final report. As with many great ideas, the notion of having a unified product line seems obvious in retrospect, but that was not the way it seemed at first to those assembled in the rooms of the motel.[14]

Less obvious was scalability. Even though the SPREAD Committee agreed that this was needed, at the early stages both Fred Brooks and Gene Amdahl—later two of the 360's principal architects—argued that "it couldn't be done."[15] Few other technologies, if any, scale simply. Civil engineers, for example, use different criteria when designing large dams than they use for small ones. The engine, transmission, power train, and frame of a large sedan are not simply bigger versions of those designed for a subcompact. What the SPREAD Committee was proposing was a range of 25 : 1 in computing—more like comparing a subcompact to an 18-wheeler. By 1970, however, after IBM had announced an upgrade to the 360 line, it was offering compatible computers with a 200 : 1 range.[16]

What changed Brooks's and Amdahl's mind was the rediscovery of a concept almost as old as the stored-program computer itself. In 1951, at a lecture given at a ceremony inaugurating the Manchester University digital computer, Maurice Wilkes argued that "the best way to design an automatic calculating machine" was to build its control section as a little stored-program computer of its own, wherein each control operation (say, the command to add two numbers) is broken down into a series of "micro-operations" directed by a matrix of components that stored a "micro-programme [sic]."[17] By adding a layer of complexity to the design, Wilkes in fact simplified it. The design of the control unit, typically the most difficult, could now be made up of an array of simpler circuits, like those for the computer's memory unit.[18] Wilkes made the bold assertion that this was the "best way" because he felt it would give the design more logical regularity and simplicity; almost as an afterthought he mentioned that "the order code need not be decided on finally until a late stage in the construction of the machine."[19] He did not say anything about a series of machines or computers having a range of power.

The idea was kept alive in later activity at Manchester, where John Fairclough, a member of the SPREAD Committee, studied electrical engineering. Through him came the notion of using microprogramming (adopting the American spelling) as a way of implementing a common set of instructions across the line of 360s, while allowing the engineers charged with the detailed design of each specific model to optimize the design for low cost and adequate performance. The microprogram, in the form of a small read-only memory built into the control unit of each model's processor, would be written to ensure compatibility. Microprogramming gave the 360's designers "the ability to separate the design process...from the control logic that effectively embodied the instruction-set characteristics of the machine we were trying to develop."[20]

IBM's adoption of this concept extended Wilkes's original insight. In essence it is a restatement of the fundamental property of a general-purpose, stored-program computer—that by accepting complexity at one level (computers require very large numbers of components), one gains power and simplicity at another level (the components are in the form of regular arrays that can be analyzed by tools of mathematics and logic). Some understanding of this concept appears inchoate in the earliest of the digital machines. Wilkes himself may have been inspired by the Bell Labs relay computer Model VI, which he probably inspected during a visit to America in 1950. On the Model VI a set of coils of wire stored information that allowed the machine to execute complex sub-sequences upon receiving one simple instruction from a paper tape.[21]

By adopting microprogramming, IBM gained one further advantage, which some regard as the key to the 360's initial success.[22] That was the ability to install a microprogram that would allow the processor to understand instructions written for an earlier IBM computer. In this way IBM salesmen could convince a customer to go with the new technology without fear of suddenly rendering an investment in applications software obsolete. Larry Moss of IBM called this ability *emulation*, implying that it was "as good as" (or even better than) the original, rather than mere "simulation" or worse, "imitation." The 360 Model 65 sold especially well because of its ability to emulate the large business computer 7070, and IBM devoted extra resources to the low-end models 30 and 40 to emulate the 1401.[23]

In theory, any stored-program computer can be programmed to act as if it were another—a consequence of its being a "Universal Turing Machine," named after the mathematician Alan M. Turing, who developed this concept in the 1930s. In practice, that usually implies an unacceptable loss of performance, as the extra layers of code slow things down. Trying to emulate one computer with another usually lands the hapless designer in the "Turing Tar-Pit," where anything is possible but nothing is practical.[24] The 360 avoided that pit because its emulation used a combination of software and the microprogram of each machine's control unit (figure 5.3). When combined with the faster circuits it also used, the combination permitted the new machines to run the old programs as much as *ten times* faster than the same program would have run on, say, a 1401. By 1967, according to some estimates, over half of all 360 applications were emulations of older hardware.

Figure 5.3
IBM 9020 Configuration Control Console. The 9020 consisted of a set of three System/360, Model 50 computers, configured to operate in real time. The system was designed to run correctly during the failure of any one or even two individual computers. The 9020 systems were used for en route civilian air traffic control operations throughout the United States until the summer of 1997. (*Source*: Terry McCrae, Smithsonian.)

1401 emulation was especially crucial to IBM's bet-the-company gamble: In December 1963 Honeywell introduced the H-200 computer, with a program they called "Liberator" that allowed it to run 1401 programs. H-200 sales were immediately brisk, just as IBM was announcing the 360 line with its implied incompatibility with the 1401. The IBM division that sold the 1401 went through a Slough of Despond in early 1964, but it climbed out after orders for the lower-end models of the 360 came rolling in.[25] The success of emulation demonstrated a paradox of computer terminology: software, despite its name, is more permanent and hard to modify than hardware. To this day there are 1401 programs running routine payroll and other data-processing jobs, on modern computers from a variety of suppliers. When programmers coded these jobs in the early 1960s using keypunch machines, they had no idea how long-lived their work would be. (The longevity of 1401 software is a major cause of the "Year-2000" bug.)

The System/360 had other architectural features worth mentioning. Many of these were first introduced in a system called STRETCH, designed for the Los Alamos National Laboratory and completed in the early 1960s. The name came from its goal, to "stretch" the state of the art in processing speed. After delivering the STRETCH to Los Alamos in 1961, IBM marketed a commercial version as the IBM 7030, but after eight deliveries it withdrew it and wrote off a large financial loss. Later on IBM realized that perhaps it was not a failure after all, since so many concepts first explored in STRETCH found their way into the System/360.[26]

Every System/360 except for the smallest Model 20 contained sixteen general-purpose registers in its central processor. Nearly all previous computer designs specified one register, the accumulator, where simple arithmetic and logical operations took place; another register, the index register or "B-line," held indexing information for memory access. Still other registers might be dedicated to other special functions. In the 360, any of the sixteen registers could be used for any operation (with a few exceptions, like extra registers for floating-point numbers).

The 360's word length was 32 bits—4 bits shorter than word length of the 7090/7094 scientific computers, but because 32 was a power of 2, it simplified the design. Most early computers used sets of 6 bits to encode characters; System/360 IBM used 8 bits, which Werner Buchholz of IBM called a "byte," in 1956.[27] Because eight is also a power of 2, this further simplified the machine's logic. It also allowed 2^8 or 256 different combinations for each character; which was more than adequate for upper- and lowercase letters, the decimal digits 1 to 10, punctuation, accent marks, and so on. And since 4 bits were adequate to encode a single decimal digit, one could "pack" two decimal digits into each byte, compared to only one decimal digit in a 6-bit byte. (The 360's memory was addressed at the byte level; one could not fetch a sequence of bits that began in the middle of a byte.)

To encode the 256 different combinations, IBM chose an extension of a code they had developed for punched card equipment. This Extended Binary Coded Decimal Interchange Code (EBCDIC) was well designed, complete, and offered room for future expansion. It had one unfortunate characteristic—incompatibility with the ASCII standard being developed at the same time. ASCII, supported by the American National Standards Institute in 1963, standardized only seven bits, not eight. One reason was that punched paper tape was still in common use, and the committee felt that punching eight holes across a standard piece of tape would weaken it too much. (There were a few other reasons as well.) The lack of an 8-bit standard made it inferior to EBCDIC, but because of its official status, ASCII was adopted everywhere but at IBM. The rapid spread of minicomputers using ASCII and Teletypes further helped spread the code. With the dominance by IBM of mainframe installations, neither standard was able to prevail over the other.[28] IBM had had representatives on the committee that developed ASCII, and the System/360 had a provision to use either code, but the ASCII mode was later dropped as it was little used.[29] The adoption of two incompatible standards within a few years of each other was unfortunate but probably not surprising. Similar events would occur later on.[30]

There have been only a few consequences of the spread of these two standards. In ASCII, the ten decimal digits were encoded with lower numerical values than the letters of the alphabet; with EBCDIC it was the opposite. Therefore a sorting program would sort "3240" before "Charles" if the data were encoded in ASCII, but "Charles" before "3240" if EBCDIC had been used. In EBCDIC, possibly reflecting its punched card ancestry, the eight bits of a byte were numbered sequentially from left to right, with the leftmost representing the most significant bit; for example, the bit representing the 2^8th value was bit #0, 2^7 bit 1, and so on. With ASCII it was the reverse. From an engineering standpoint this is a trivial difference, and most users never have to worry about it.[31] Because of its beachhead in minicomputers, ASCII would prevail in the personal computer and workstation environment beginning in the 1980s.

The 360's designers allowed for 4 bits of a word to address the 16 general-purpose registers, and 24 bits to address the machine's core memory. That allowed direct access to 2^{24}, or 16 million addresses, which seemed adequate at the time. Like nearly every other computer design, the address space was eventually found to be inadequate, and in 1981 IBM extended the number of address bits to 31, allowing for access to 2 billion addresses.[32]

For the cheaper models, even allowing 24 bits was extravagant, as these were intended to do their work with a much smaller memory space. Carrying the extra address bits would impose an overhead penalty that might allow competitors like Honeywell to offer machines that were more cost-effective. IBM's solution was to carry only 12 of the possible 24 address bits in an instruction. This number would then be added to another number stored in a "base" address register to give the full 24-bit address.[33] If a program required fewer than 2^{12} or four thousand bytes of memory, going to the base register was not necessary. That was the case for many smaller problems, especially those that the cheaper models of the 360 were installed for. For longer problems there was of course the additional penalty incurred when going to the base register to obtain an address, but in practice this was not a severe problem.

Finally, the System/360 retained the concept of having channels to handle input and output. With a standard interface, IBM could offer a single line of tape, card, and printing equipment that worked across the whole line of machines—a powerful selling point whose advantages easily offset whatever compromises had to be made to provide compatibility. The trade press called I/O devices "peripherals," but they were central to the System/360 project—a new model keypunch, new disk and tape drives, and even the Selectric typewriter with its famous golf-ball print head and classic keyboard layout. All of these devices defined the 360-era of mainframe computing as much as the beige, slanted control panel.[34]

The architectural design of the 360 used creative and sometimes brilliant compromises to achieve compatibility across a range of performance. Initially it had a fairly simple design, but over the years it grew ever more complex, baroque, and cumbersome. The fact that it could grow as it did, enough to remain viable into the 1990s, is testimony to the strength of the initial effort.

System/360 and the Full Circle of Computing   The orders that began streaming in for models of the 360 shortly after it was announced validated the decision to offer such an ambitious, unified line of products. But did those orders mask any sectors in the "360 degrees" that the machines did not cover well? They did, though not in obvious ways.

Chapter 4 discussed the rise of the minicomputer, led by the PDP-8, which was introduced just as deliveries of the 360 began. As minicomputers grew in capability, they began to compete with IBM's mainframe line, but initially there was little overlap or competition. The PDP-8 was not intended for the jobs the mainframes were being used for, such as processing large payrolls, and the System/360 was ill-suited as a controller for laboratory experiments, real-time data collection, and other uses that the PDP-8 was especially good for. For those applications, IBM offered an incompatible line of hardware.

Time-Sharing and System/360

There was, however, one very important sector that System/360 did not cover—using a large computer interactively or "conversationally." For economic reasons one could not dedicate a mainframe to a single user, so in practical terms the only way to use a large machine interactively was for several users to share its computational cycles, or "time," simultaneously.

The term "time-sharing" for computers has had several meanings. The earliest meaning was fairly restricted, and referred only to the ability of programmers to debug a program without having to prepare and submit a new deck of cards into the queue of jobs for a system. The time the computer spent going through the batch of jobs would be shared by a programmer making a few changes to a program, which would then be inserted into the original program that was already on the machine. By the late 1960s the term had a more general definition, one that was especially advocated by Professor John McCarthy, then of MIT, and which will be used in the following discussion of the System/360. By that definition, each user had the illusion that a complete machine and its software was at his or her disposal. That included whatever programming languages the computer supported, and any data sets the user wanted to use, whether supplied by others or by the user. The only constraint was the physical limits of the machine. That went far beyond the notion of time-sharing as a tool for programmers, as well as beyond the interactive nature of SAGE, which allowed multiple users of one and only one data set, and beyond NASA's real-time systems, which restricted users to both specialized data sets and programming languages.

What made such a concept thinkable was the disparity between the few milliseconds (at least) between a typist's keystrokes and the ability of a computer to fetch and execute dozens, perhaps hundreds, of simple instructions. The few seconds a person might pause to ponder the next command to type in was time enough for a computer to let another user's creation emerge, grow, and die—at least in theory. In practice, instructions that directed a computer to switch from one user to another required many cycles of the computer's processor just to keep track of things. The time required by those instructions could easily take up all the time—and more—between a user's keystrokes. But the rewards for a successful time-sharing system were great enough to lead many to try to build one.

By the mid-1960s, time-sharing seemed an idea whose time had come. An experimental system was operating at MIT on an IBM 7090 by 1962. It evolved into Compatible Time-Sharing System (CTSS) using the upgraded 7094.[35] CTSS supported only a few users simultaneously, but it did successfully address many concerns about time-sharing's viability. It led in part to a proposal for a more ambitious system, which would become the centerpiece for Project MAC ("Man and Computer," a.k.a. "Machine-Aided Cognition").[36] With support from the Defense Advanced Research Projects Agency, Project MAC sought a computer on which to base its system. System/360s were among those considered, but in May 1964 Project MAC informed IBM that the just-announced System/360 was not suitable. The project chose a General Electric machine instead.[37] Shortly after that, Bell Laboratories, another of IBM's most favored customers, spurned IBM and chose a GE system for its time-sharing work. By 1967, GE seemed to be on its way to a position of leadership in the computer business, based on its successful GE-635 line of computers that seemed to support time-sharing better than IBM's products.

IBM's own history of those events describes an air of crisis, a feeling that after such hard work the company was rapidly losing its place in the field and sinking in the face of competition.[38] The 360's basic architecture was not hostile to time-sharing applications, but neither was it optimum.[39] Some of the participants in Project MAC recall that the System/360's most serious deficiency was its lack of dynamic address translation: an ability to stop the execution of a program, move it out of core memory to a disk, then at a later time move it back (probably into a different section of core) and resume execution of it. Other members of the Project MAC team have a different recollection—that it was not so much a matter of dynamic address translation as it was of the inherent "processor-oriented" architecture of the 360.[40] In any event, time-sharing requires the ability to swap programs to and from core quickly, if each of several users is to have the illusion that the entire resources of the computer are brought to bear on his or her problem.

The problem seemed to be that IBM simply did not see time-sharing as as important as some customers wanted the company to see it. Amdahl and Brooks had focused their attention on the need to introduce a unified line of processors, which may have caused them to miss the depth of feeling among IBM's best customers for intrinsic hardware features that supported time-sharing. Amdahl's and Brooks's response, that the 360 would meet these customers' needs, was technically true. But it did not convince MIT or Bell Labs that IBM shared their vision of the future of computing.

With the heroic rededication of resources that IBM was famous for, the company announced the Model 67 with address translation hardware in August 1965. Historians have accepted the view that the Model 67 was a failure for IBM, perhaps because IBM's announcement of it was later introduced as evidence in an antitrust trial. This hasty announcement was cited as an example of the company's alleged policy of announcing products more with an aim of cutting off competition than of introducing genuine value. The Model 67's time-sharing system software (TSS) did not work well and IBM's announcement of it did not stop MIT and Bell Labs from going through with their purchase of computers from General Electric. One systems programmer who lived through that era stated:

Losing Project MAC and Bell Labs had important consequences for IBM. Seldom after that would IBM processors be the machines of choice for leading-edge academic computer science research. Project MAC would go on to implement Multics [a time-sharing operating system] on a GE 645 and would have it in general use at MIT by October, 1969. Also in 1969, the system that was to become UNIX would be begun at Bell Labs as an offshoot and elegant simplification of both CTSS and Multics, and that project, too, would not make use of IBM processors.[41]

Comments like these are clouded as much as they are clarified by hindsight. Early time-sharing systems worked well enough to demonstrate the feasibility of the concept, but it took years before any of them fulfilled their initial promise of supporting many users and running different types of jobs with a quick response time. The Model 67, although a failure, laid the basis for a revamping of the entire

System/360 line, in which time-sharing was available and did work well: the System/370, announced at the end of the decade as an upgrade to the 360 line. Throughout the 1970s and 1980s, System/370s using the Conversational Monitoring System (CMS) software handled demanding time-sharing applications.

Nor did General Electric vault to a position of leadership. GE sold its computer business to Honeywell in 1970, a sale that allowed Honeywell to make solid profits and gain customer loyalty for a few years.[42] Bell Laboratories found the GE time-sharing system wanting, and dropped out of the MULTICS project in 1969. Two researchers there, Ken Thompson and Dennis Ritchie, ended up developing UNIX, in part because they needed an environment in which to do their work after their employer removed the GE system. They began work on what eventually became UNIX on a Digital Equipment Corporation PDP-7, a computer with far less capability than the GE mainframe and already obsolete in 1969. They later moved to a PDP-11. For the next decade and a half, UNIX's development would be associated with DEC computers. The name implies that "UNIX" is a simplified form of "MULTICS," and it did borrow some of MULTICS's features. UNIX was also inspired by the earlier, and simpler, CTSS as well. UNIX's impact on mainstream computing would occur in the 1980s, and it will be discussed again in a later chapter.

Whatever advantage Honeywell had in obtaining GE's business, it did not last the decade. We shall see that Honeywell also squandered a lead in minicomputers. Still, Honeywell's and GE's loss was not IBM's gain. Successful commercial time-sharing systems became common in the late 1960s—too common, as venture capital firms funded too many companies for the market. Most of these companies used neither IBM nor GE/Honeywell products, but rather PDP-10s from Digital Equipment Corporation or SDS-940s from Scientific Data Systems. The revolutionary breakthroughs in interactivity, networking, and system software that characterized computing in the late 1970s and early 1980s would not be centered on IBM equipment.

The System/360 Model 67 repeated IBM's experience with STRETCH. Both were commercial failures, but both laid the groundwork for the line of successful products that followed, whose success overcame the initial lost revenues. The issues raised in litigation have obscured the lesson of System/360 architecture—because it was so well designed, it could absorb a major enhancement five years later and still maintain software compatibility and its customer base. As for the crisis precipitated by MIT's defection to General Electric, one can say that IBM should have been more accommodating to MIT. To be fair, one could also say, in hindsight, that MIT would have been better off basing Project MAC around a different computer from the one it chose.

As the 1960s drew to a close, the minicomputer—especially new models like the PDP-11 and Data General Nova—was breaking out of its laboratory and OEM setting and moving into territory that IBM and the "Seven Dwarfs" had considered their own. Unfortunately for IBM, the 360's architecture did not permit a minicomputer version—the low-end 360 Model 20 was about as small and cheap as IBM could go with the line, and it was already partly incompatible with the rest of the line. Some basic architectural features, especially the use of input/output channels and standardized interfaces to tape and disk drives, prevented going much lower in price.

So despite the name of the series, there were sectors missing from the full circle promised by System/360—sectors that would grow in the next two decades. Orders for the 360 strained IBM's resources, although IBM was still big enough and astute enough not to ignore the minicomputer market, and in 1969 it responded with System/3, an incompatible computer that could be rented for as low as $1000 a month.[43] It was "a candid concession that System/360 could not bridge the widening opportunities in the marketplace."[44] System/3 was a successful product, because it was easier to adapt for small business and accounting jobs than minicomputers, which often carried their laboratory workbench heritage with them. One interesting feature of the System/3 was its use of a new and smaller punched card that could encode 96 characters. Even as this product was introduced, IBM was developing the storage medium that would forever displace the punched card: the *floppy disk*. As soon as inexpensive disk storage was made available for the computer, customers abandoned the incompatible punched card. The venerable 80-column card, however, continued to be popular through the next decade.

Fortune magazine's quote that "you bet your company" on the 360, which disturbed IBM's senior management, turned out to be true, but for the wrong reason. It was success, not failure, that threatened the company's existence. Success meant the need to raise capital quickly, build new plants, hire new workers, and expand production. It meant that anyone who wanted the company to adopt an alternate style of computing, such as the ones that Project MAC or the minicomputer companies promoted, would have to swim against the rushing current of 360 orders.

The Period of Soaring Stocks

Spurred on by Defense Department spending for the Vietnam War, and by NASA's insatiable appetite for computing power to get a man on the Moon, the late 1960s was a time of growth and prosperity for the computer industry in the United States. For those who remember the personal computer explosion of the 1980s, it is easy to overlook earlier events. From about 1966 to 1968, almost any stock that had "-ex," "-tronics," or simply "-tron" in its name rode an upward trajectory that rivalled the Moon rockets.[45] John Brooks, a well-known business journalist and astute observer of Wall Street, labeled them the "go-go years," referring to the rabid chants of brokers watching their fortunes ascend with the daily stock ticker.[46]

Some of the go-go stocks were issued by companies that focused on the base technology of transistors and the newly invented integrated circuit. Others were brand-new minicomputer companies following in Digital Equipment Corporation's footsteps. Others were the time-sharing utilities. Others were software and service companies, which sprouted to help ease customers into the complexity of operating the complex and expensive new mainframes. Some were "plug-compatible manufacturers," which sold, at lower cost, pieces of a system that were compatible with IBM's product line. Finally, there were third-party leasing companies that lived under an "umbrella" of IBM's pricing and leasing policy. We shall begin with the latter group.

Leasing Companies Having entered into a consent decree with the U.S. government in 1956, IBM agreed to sell as well as lease its computers. Leasing continued to predominate, however, because IBM had the capital that few others had. And many customers liked leasing, which did not tie up their capital, and made it possible to cancel a lease and move to the better product if new technology came along. IBM, in turn, received a steady flow of cash from its leases, although it had to meet the challenge of competitors offering machines with newer technology.

The key to the emergence of leasing companies in the mid-1960s was the perception that IBM was charging artificially high rents on its equipment—specifically, that IBM was charging rent on the expectation that the computers would become obsolescent, and therefore returned from lease, in as little as five years. The leasing companies reckoned that customers would want to hang on to them longer—up to ten years. Beginning with a company called Leasco that started in a Brooklyn loft in 1961, these companies would buy mainframes from IBM (as permitted by the consent decree), and then rent them to customers at up to 20 percent less than what IBM charged.[47] These companies had little trouble raising the necessary capital from Wall Street, as they were able to convince financiers that they were sheltered under an umbrella of artificially high prices charged by IBM.

The question remained whether IBM's prices were, in fact, too high. Given the enormous cost of bringing out System/360, IBM naturally wanted to get that investment back. There were other factors, however. System/360 offered its own software-compatible path for a customer to migrate to a new computer. With this generation of mainframes, the cost of writing software was higher than ever before, which might encourage customers to hold onto a computer even if a competitor offered better hardware but no software compatibility. Financial models that predicted how long a customer might keep a mainframe had to take this into account, but by how much?

Even as the 360 was going to its first customers, IBM recognized that the pace of technology was not going to slacken. The company spent lavishly on research facilities, building up laboratories in Yorktown Heights, New York, San Jose, California, and in Europe. That was an expense the leasing companies did not have to bear, but it meant that IBM could ensure that at least some technical innovation was under its control. To the extent that IBM dominated the computer industry, it could mete out this innovation gradually, thus not making its installed base obsolete too quickly. But dominance was a fleeting thing: if IBM held back too much, another company was sure to enter in. And others did, as the examples of Control Data and RCA will reveal.

Just how much of the market IBM controlled became the subject of another federal antitrust action beginning in 1969, but no single company, no matter how big, could control the pace of the underlying technology. GE's sales to customers who wanted time-sharing facilities was only one example. In any event, at the end of the 1960s IBM announced a successor to System/360—the 370 line. This line was software-compatible with the 360, but it was better suited for time-sharing, and it used integrated circuits for its processor and memory. System/370 probably came sooner than IBM wanted, but by 1970 it had no choice.[48] Given the timing of System/370's announcement, perhaps IBM's prices were not too high after all.

In the late 1970s the company had to respond to other pressures as well. These came from the minicomputer companies whose products were evolving to handle mainframe applications. In 1978, IBM introduced a low-cost 4300-series, compatible with the 360-370 line, and a midsized AS/400, which was not compatible. It responded to pressures from the makers of large mainframes by bringing out a successor to the larger models of the 370—the 3030 series. None of these competitive pressures—from other mainframes, from the minicomputer, from time-sharing, and from low-cost workstations (discussed later)—were enough to do serious damage, but their combination, evolving by 1990 into a networked system of inexpensive workstations, would.

System/360's dominance of the market was certainly shorter than the ten years that the leasing companies had gambled on, although not every customer felt the need to upgrade immediately. In any event, the pace of technology, combined with the end of the bull market in 1971 and with IBM's careful manipulation of pricing and product announce-ments, served to fold up the pricing umbrella by the mid-1970s, leaving investors in the leasing companies with heavy financial losses.

Compatible Mainframes   The second consequence of the announcement of System/360 was a redefinition of the role of IBM's principal compe-titors, and led to the emergence of smaller companies aimed directly at the 360 line. During the initial SPREAD Committee discussions, some-one expressed the fear that by introducing a broad line of machines, "the competition would be out after each [specific model] with a rifle." Unconstrained by a need for compatibility, someone could bring out a machine with far better performance for the same cost.[49] The commit-tee had to argue that the advantages of having a path for upward migration for 360-customers would overwhelm any advantages of a competitor's shot at a particular model.

At the highest end there would be no higher model to migrate to anyway. Control Data Corporation, which introduced the small-scale 160A already mentioned, came out with its 6600 computer in 1964 (figure 5.4). Designed by Seymour Cray and soon dubbed a "super-computer," the 6600 offered what Seymour Cray wanted to deliver in a computer—the fastest performance possible, period.[50] In terms of absolute sales, the CDC 6600 was not much of a threat, but its customers were unusual: the weapons laboratories like Lawrence Livermore, the large aerodynamics research organizations, the National Security Agency, and others for whom performance was all that mattered. These customers might collectively buy only a few units, but other, less glamorous customers held them in high regard. Whatever systems they chose was therefore reported and discussed seriously in the trade press.

Figure 5.4
Console of the Control Data Corporation's CDC-6600, ca. 1964. The 6600 was
designed by Seymour Cray and popularized the term "supercomputer." (*Source*:
Control Data Corporation Archives.)

IBM countered with a System/360, Model 91, but it was late in delivering
it, and its performance never matched the CDC machines.

Here the compatibility issue took its toll. The 6600 owed a lot of its
performance to Seymour Cray's talent for design, and Cray had the
further advantage that he could "start with a clean sheet of paper"—
unconstrained by compatibility issues—whenever he designed a computer.[51] IBM lost the battle for the high end to CDC (and later on to Cray
Research, founded by Seymour Cray in 1972). Control Data Corporation
eventually sued IBM, alleging that the Model 91 was a "phantom,"
announced before it was ready, in order to kill the CDC 6600. Whatever
the merits of the lawsuit, except for a few exceptional customers, most
preferred the advantages of software compatibility. Even many weapons
laboratories, with their unlimited budgets, installed one or more
System/360s alongside their CDC 6600.

Another threat to the 360 came from RCA. RCA had tried and failed
to enter commercial computing in the 1950s with the BIZMAC, but in
1959 RCA had better luck with the model 501, a small commercial

computer that was best known for being one of the first to be supplied with a compiler for the COBOL programming language. Professor Saul Rosen of Purdue University once said, "It was quite slow, and...the COBOL compiler was also very slow, but for many users a slow COBOL was better than no COBOL."[52] Another transistorized computer, the Model 301, was capable of real-time operation. (One was used by NASA to sequence and control the Saturn rockets that took men to the Moon.) Through the late 1950s and early 1960s, the company focused on bringing color television to a consumer market, giving digital electronics a lower priority.

Late in 1964 RCA announced a bolder offensive, the Spectra 70 series. This was a line of four computers that would execute, without modification, software written for corresponding models of the IBM 360 line. And they would cost up to 40 percent less. This was the competition IBM had feared. RCA did not need to plant spies inside IBM's laboratories—they could rely on specifications supplied to customers and software developers. The System/360 project was so big that IBM had to share a lot of information about it, and it was powerless to stop someone from building what later on would be called a "clone." Building 360-compatible computers also became a quick way for the Soviet Union to construct powerful mainframes.

RCA's aggressive pricing came from several factors. Because it had escaped the cost of designing the architecture that IBM had borne, its development costs were less than one tenth of IBM's.[53] And by starting later, RCA could also take advantage of advances in component technology. Two models, the Spectra 70/45 and the 70/55, used true integrated circuits, and thus offered better performance for the dollar than the 360.

RCA's Spectra 70 Series was successful, but sales withered after IBM returned fire in 1970 with its System/370, which also used integrated circuits. After incurring massive losses with no end in sight, RCA announced in 1971 that it was leaving the computer business. For a bargain price, the installed customer base was bought by UNIVAC (now a division of Sperry), and Digital Equipment Corporation bought RCA's brand-new manufacturing plant in Marlboro, Massachusetts. UNIVAC continued to service the RCA machines and carefully cultivated the hundreds of companies that owned them, eventually easing them over to UNIVAC mainframes. This echoed the sale of GE's computer business to Honeywell in 1970: an electronics giant selling out to a company more knowledgeable about marketing business equipment. Sperry UNIVAC got a good deal. It paid only $70 million for RCA's business, compared to the $234 million that Honeywell had paid for GE's.[54] Digital Equipment Corporation got the "deal of a lifetime" in the Marlboro plant, finding its modern facilities just what it needed at a time when it was expanding rapidly.[55]

The Plug-Compatible Manufacturers

RCA's failure did not invalidate the basic economics of copying the 360 architecture. Other companies with far less capital than RCA proved successful, not by copying the entire line, but by targeting pieces of the 360 system: memory units, tape drives, and central processing units. These companies, also operating under the umbrella of IBM's pricing policies, established the "plug-compatible manufacturer," or PCM business, another defining segment of the go-go years.

In 1970, Gene Amdahl, one of the company's star computer designers, left IBM to found a company that would make a compatible processor. The Amdahl Corporation began installations in 1975 of its Model 470 V/6, a machine that, like the CDC-6600, competed with the top of the IBM line. More than that, it far outperformed it. Unlike the CDC 6600, the Amdahl processor could run IBM 360 software.[56] Well before that time, companies like Memorex, Telex, Ampex, Storage Technology, and CalComp were offering tape drives, disk drives, and even main memory units, that one could simply plug into an IBM 360 or 370 installation in place of IBM's own equipment (hence the name), giving equal or better performance at a lower price. Coupled with an Amdahl processor, one could thus build a complete computer system that ran all the 360 software almost without any "Big Blue" hardware. (IBM's 360 products were painted a distinctive blue.)

These were the true "rifle shots" that Fred Brooks had worried about, and they did a lot of damage. On Wall Street, stock in these companies soared. IBM responded in some cases by repackaging and repricing its products to make the cost difference less. Those actions spawned no fewer than *ten* lawsuits between 1969 and 1975, charging IBM with antitrust violations.[57] This was in addition to the U.S. Justice Department's own antitrust suit, launched in 1969, and the Control Data lawsuit over the Model 91 already mentioned. Most of these suits were settled by 1980, by which time the rush of technology had rendered their substance irrelevant.

Some of the plug-compatibles prospered throughout all this, although the bear market of the 1970s took its toll. Amdahl survived, mainly through financing from Fujitsu, a Japanese company that had been casting about for ways to enter the U.S. computer market. In 1975 Fujitsu began building Amdahl computers in Japan for sale in the United States, relying on Gene Amdahl's talent to offer a competitive design based on the latest integrated-circuit technology. Another Japanese company, Hitachi, began making and selling plug-compatible mainframes as well, sold in the United States under the name National Advanced Systems, a division of National Semiconductor Corporation.[58] This was the first serious Japanese competition ever faced by the U.S. computer industry. A by-product of the PCM phenomenon, it long outlasted most of the PCM companies themselves. IBM's counter punch, the faster 3030-series introduced in the late 1970s, slowed the defections of customers to Amdahl's machines, leaving a competitive environment with the Japanese firmly entrenched. Gene Amdahl left the company still bearing his name to form Trilogy in 1980, but Trilogy never achieved the success he had hoped for. Many of the tape and disk manufacturers folded or merged with one another; few survived into the workstation era.[59]

UNIVAC, SDS   Another consequence of the System/360 was that several competitors found small sectors in the full circle where the 360's coverage was spotty. Besides the supercomputers and minicomputers covering the high and low ends, in the middle some scientific and engineering customers found that the performance of mid-range 360s suffered in comparison to the elegant 7090s they were obliged to abandon. These customers were more willing than others to go with a competitor, primarily because they were used to writing much of their own software. Sperry UNIVAC developed a version of its 1100 series of mainframes that employed integrated circuits and offered fast processing speeds. The 1108 mainframe, announced in 1964, sold especially well.[60] The 1100-series of computers was a strong challenge to the 360 line and were the basis for most of UNIVAC's profits into the early 1970s. Sales were especially strong to government and military agencies such as the Federal Aviation Administration.[61]

More dramatic was the success of a company that started up at the same time as CDC and DEC, namely Scientific Data Systems of California. SDS was founded by Max Palevsky, a philosophy major in college who found that, because of the invention of the digital computer, the 1950s was the "first time in history that a philosopher could get a job!"[62] Palevsky had first worked on a computer project at Bendix, then joined

Packard-Bell, a small electronics firm that he claims he found in the Yellow Pages. He convinced the company that they ought to enter the computer business. His training as a philosopher helped him understand not only computer logic but also the art of argumentation. The result, the inexpensive PB-250, was a modest success when it was introduced in 1960.[63] Palevsky developed an especially good relationship with the German rocket engineers who had come after World War II to Huntsville, Alabama, where they were involved with the Army's ballistic missile program. The computer that Palevsky had worked on at Bendix, though of an unconventional design, had caught the attention of this group, who were looking for ways of controlling a missile in real time.

In 1961 Palevsky left Packard Bell and founded his own company, Scientific Data Systems, raising around $1 million in venture capital. Within a year SDS had introduced a computer, the Model 910, and the company was profitable. Palevsky attributes the 910's success to its superior Input/Output facilities. It also made effective use of the latest developments in component technology. The 910 was one of the first nonmilitary computers to use silicon transistors, and a model delivered in 1965 was one of the first (along with RCA's) to use integrated circuits.[64]

By 1964 SDS had revenues greater than those of DEC—of course, it had started out with over ten times the capital. In 1969, when SDS had sales of $100 million and after-tax revenues of $10 million, Palevsky sold the company to Xerox for $900 million worth of Xerox stock.[65] (A few years later Xerox wrote off the division, incurring a loss of $1.3 billion.) Because of its modest beginnings, SDS is often compared to DEC, and its computers are sometimes placed in the class of minicomputers. The 12-bit Model 92 was in some ways similar to the PDP-8, but SDS's main business was the 24-bit 910 and 920—large-scale scientific computers, not as powerful as Control Data's supercomputers, but much more capable than minicomputers. Another model introduced in 1965, the 940, was explicitly marketed for time-sharing use, and it was also a success.[66] Many West Coast time-sharing companies, including the pioneer Tymshare, were based on it. The 940 was used for many pioneering research projects in human-computer interaction and networking in the Palo Alto, California area (before it became known as "Silicon Valley"). In a sense, it was the West Coast counterpart of the DEC PDP-10, a well-engineered, time-sharing system that had its greatest impact in advanced computing research. By using the best components, and by tailoring the design to suit scientific applications, SDS computers, like the UNIVAC 1108, were the natural descendants of the IBM 7090. They were both sold and leased, and prices ranged up to $250,000.

In 1967 SDS announced a more powerful computer, the Sigma 7, which cost around $1 million.[67] Palevsky's Huntsville connections served his company well. By the early 1960s the facilities were transferred from the Army to NASA, where, under the leadership of Wernher von Braun, the "rocket team" was charged with developing the boosters that would take men to the Moon and back. IBM hardware handled the bulk of the Center's chores, but SDS computers were installed to do real-time simulations and tests of the rockets' guidance systems. Drawing on a relationship established when Palevsky was working for Bendix, Helmut Hoelzer and Charles Bradshaw chose to install SDS computers after becoming disillusioned with RCA machines they had initially ordered for that purpose.[68]

SDS's fortunes rose and fell with the Apollo program: even as men were walking on the Moon in 1969, NASA was cutting back and having to plan for operations on smaller budgets. Xerox bought Palevsky's company at a value ten times its earnings, expecting that SDS, now the XDS division, would grow. Some journalists claimed that Palevsky knew he was selling a company with no future, but Palevsky stated, under oath for the United States vs. IBM antitrust trial, that he believed otherwise.[69] The division did not grow, and Xerox closed XDS in 1975. SDS had no adequate plan for expanding its products beyond the narrow niche it occupied—again revealing the wisdom of IBM's System/360 philosophy. But Xerox must also shoulder the blame. The company had built up the finest research laboratory for computing in the world, in Palo Alto, California, but it failed to fit these two pieces of its organization together, much less fit both of them into its core business of selling copiers.

Software Houses

A final measure of how the System/360 redefined the computer industry was in its effect on software and "service bureaus."[70] The idea of forming a company that bought or rented a computer to deliver a solution to another company's problem was not new. The first may have been Computer Usage Company, founded in 1955, which developed programs for the IBM 701 and 704 for industrial clients.[71] The major computer companies had their own in-house service bureaus that performed the same services—IBM's went back to the era of tabulators, and Control Data Corporation's service business was as important financially to the company as its hardware sales.

One of the pioneering independent companies was Automatic Data Processing, founded as Automatic Payrolls in 1949 by Henry Taub in Paterson, New Jersey. ADP's core business was handling payroll calculations for small and medium-sized companies. It primarily used IBM tabulating machinery, even after it acquired its first computer in 1961. The following year ADP's revenues reached $1 million.[72] It took a conservative approach to technology, using the computer to process data in batches of punched cards just as it had with its tabulators. Its first salesman, Frank Lautenberg, continued Taub's conservative and profit-oriented approach when he took over as CEO in 1975. (Lautenberg later became a U.S. senator from New Jersey.)[73]

Computer Sciences Corporation was founded in 1959 by Fletcher Jones and Roy Nutt, who had worked in the southern California aerospace industry. As described in chapter 3, CSC's first contract was to write a compiler for a business programming language ("FACT") for Honeywell. That evolved into a company that concentrated more on scientific and engineering applications, for customers like the NASA-Goddard Space Flight Center and the Jet Propulsion Laboratory. CSC also did basic systems programming for the large mainframes being sold in the mid-1960s.[74] Another major company that had a similar mix of scientific and commercial work was Informatics, founded by Walter F. Bauer in 1963.

In contrast to the minicomputer companies, who let third party OEMs customize a system for specific customers, IBM had a policy of including that support, including systems analysis and programming, into the already substantial price of the hardware. In 1968 IBM agreed to charge for these services separately; still, the complexity of setting up any System/360 meant that IBM had to work closely with its customers to ensure that an installation went well. The decision to "unbundle" turned what had been a trickle into a flood of third-party mainframe software and systems houses.[75]

The complexity of systems like the IBM 360 and its competitors opened up new vistas. Manufacturers were hard-pressed to deliver all the software needed to make these computers useful, because these machines were designed to handle multiple tasks at the same time, support remote terminals, be connected to one another in networks, and deliver other features not present in the mainframes of the late 1950s. The introduction of commercial time-sharing systems opened up still another avenue for growth. Many new software companies, like American Management Systems (AMS), were formed with the specific goal of getting customers up to speed with this new and complex technology.

While mindful of the impact a company like AMS would have on revenues from its own software teams, IBM was probably relieved to have such a company around to step into the breach. IBM was at the time unable to deliver system and programming software that was as good as its System/360 hardware. The original operating system software intended for the 360 was delivered late, and when it was delivered it did not work very well. And the programming language PL/I, intended to be the main language for the System/360, was not well received. The question arose, how could IBM, which could carry off such an ambitious introduction of new hardware, fail so badly in delivering software for it? Fred Brooks wrote a book to answer that question, *The Mythical Man-Month*, which has become a classic statement of the difficulties of managing complex software projects.[76]

After its decision to unbundle software pricing from hardware in 1969, IBM became, in effect, a software house as well. That decision has been described as an attempt to forestall rumored antitrust action. (If so, it did not work, because the Justice Department filed suit the month after IBM's announcement.) It is more accurate to say that IBM acknowledged that the computer industry had irrevocably changed, that software and services were becoming a separate industry anyway.[77]

The spectrum of service and software providers not only ran from scientific to commercial, it also included an axis of government and military contractors. These provided what came to be known as "systems integration" for specialized applications. One example was Electronic Data Systems (EDS), founded by H. Ross Perot in 1962. Perot had been a star salesman for IBM, and he had proposed that IBM set up a division that would sell computer time, instead of the computers themselves, to customers. When IBM turned him down he started EDS. After a shaky start, the company prospered, growing rapidly in the mid-1960s after the passage of the Medicare Act by Congress in 1965. Much of EDS's business was to customers in the federal government.[78]

The Cold War, especially after Sputnik in 1957, led to work for a variety of companies to manage systems for defense agencies. This business had deep roots, going back to the founding of the RAND Corporation and its spin-off, the System Development Corporation (SDC), to develop air defense software.[79] What was new was that, for the first time, there appeared companies that hoped to make profits only by contracting for systems work, that were not, like SDC, federally funded extensions of a defense agency. Ramo-Woldridge, centered in southern California, was perhaps the most successful of these. It was founded in 1953, when Simon Ramo and Dean Woldridge left Hughes Aircraft to form a company that focused on classified missiles and space operations work. R-W was later acquired by Thompson, an automotive supplier based in Cleveland, Ohio. That marriage of a "rust belt" industry with "high tech" might have seemed a poor one, but the result, TRW, became one of the most profitable of these companies. A major reason was that Thompson supplied a manufacturing capability that the other systems houses lacked, which enabled TRW to win bids for complex (mostly classified) space projects as a prime supplier. In the mid-1960s, with net sales around $500 million, TRW began branching into nonmilitary commercial work, building a division that developed a database of credit information.[80] The company remained focused on military software and space systems, however. One of its employees, Barry Boehm, helped found the discipline of "software engineering." Another person TRW employed briefly, Bill Gates, helped develop software for a computer network that managed the flow of water through the series of dams on the Columbia River. (We shall return to Gates's experience with TRW and his subsequent career in a later chapter.)

Besides TRW and the federally funded companies like SDC or MITRE, there were dozens of smaller fry as well. Their common denominator was that they supplied software and support services for a profit. Most of these began in southern California, like TRW, often founded by aerospace engineers. Some of them, wanting to be closer to the Pentagon, moved to the Washington, D.C., area, more specifically, to the open farmland in northern Virginia just beyond the District's Beltway (completed in 1964). Here land was cheap, and the new highways made access to the Defense agencies easy. (These agencies, like the Pentagon itself, were mainly on the Virginia side of the Potomac.)[81] Most of them have done very well, especially by profiting from defense contracts during Ronald Reagan's first term as president. The major aerospace and defense companies also opened up divisions to serve this market. The end of the Cold War has thrown these companies into turmoil, but the systems analysis they pioneered has been of lasting value and is now an accepted practice in most modern industries.

A final consequence of the System/360 was, indirectly, the antitrust action filed by the U.S. Justice Department in January 1969, on the last business day of the Johnson Administration. The suit dragged on for twelve years, generating enormous amounts of paper and work for teams of lawyers from all sides. (The documents produced for the trial have been a windfall for historians.) IBM continued to be profitable and to introduce new and innovative products during this time; its revenues tripled and its market share stayed at about 70 percent. One must wonder what the company might have done otherwise. The premise of the action was that IBM's actions, and its dominance of the business, were detrimental to the "dwarfs." In January 1982, with a new administration in power, the Justice Department dismissed the case, stating that it was "without merit."[82] By 1982 the place of the mainframe was being threatened by the personal computer, which had already been on the market for a few years, and by local-area networking, just invented. These developments, not the Justice Department, restructured the industry, in spite of IBM's role as a successful marketer of personal computers. Whether IBM would have acted more aggressively in establishing its dominance of the PC market had there been no threat of litigation remains unanswered.

The Fate of the BUNCH

The Justice Department suit implied that the BUNCH's very existence was being threatened by IBM's policies. Ironically, each of the BUNCH faced a depressing fate that had little to do with IBM.

In 1986 Burroughs and UNIVAC merged into a company called Unisys, which briefly became the second-largest computer company. In its travels from Eckert and Mauchly, to Remington Rand, to Sperry, to Burroughs, the name UNIVAC was somewhere dropped. By 1986 few remembered that "UNIVAC" was once synonymous with "computer," like "Scotch" tape or "Thermos" bottle. The casual abandonment of this venerated name was perhaps symbolic of the troubles of Unisys; with a few years it began suffering losses and fell to the lower ranks. It cut employment drastically, and after some painful restructuring began to show some profits.

In the 1980s NCR made a brave attempt to adopt the new architectures based on cheap microprocessors and the nonproprietary UNIX operating system. It was one of the first large system companies to do so. NCR also pioneered in building systems that gave mainframe performance from clusters of smaller, microprocessor-based subunits—a Holy Grail that many others had sought with little success. But its innovative culture made the company a takeover target. In 1991, a now-deregulated

AT&T, seeking to vault into a competitive position in large commercial systems, bought NCR in a hostile takeover. Like the Burroughs-Univac combination, this was also a disappointment. AT&T promised NCR employees that it would preserve the computer company's management structure, culture, and even the initials (to mean "Networked Computing Resources" instead of "National Cash Register"). But a few years later AT&T broke all three promises when companies like SUN and Silicon Graphics beat them to market with these kinds of products. AT&T spun off NCR as an independent company in 1996.

Honeywell allied itself with the Nippon Electric Company (NEC) to build its mainframes, which were IBM compatible. It had also been allied since the 1970s with the French company Machines Bull and the Italian company Olivetti. Beginning in 1986, Honeywell began a retreat out of the mainframe business and the next year turned it completely over to Bull, with NEC a minor partner.[83] Honeywell continued supplying the U.S. military market with domestic products, and along with Sperry became a leading supplier of specialized aerospace computers, military and civilian—a growing field as new-generation aircraft adopted "fly-by-wire" controls. In the mid-1980s Honeywell developed, under military contract, a set of specialized chips called VHSIC (Very High Speed Integrated Circuits), which were resistant to radiation. But unlike the situation two decades earlier, military contracts for integrated circuits did not lead nicely to commercial products.[84]

Control Data had an unusual history. It developed a healthy business of manufacturing tape drives and printers for competitors' computers, and it entered the service business as well. In 1968, with its stock riding the crest of the go-go years, it used that stock to acquire the Baltimore finance company Commercial Credit—a company many times larger than CDC. The acquisition gave CDC a source of funds to finance its diversification. Some observers charge that CDC milked the assets of Commercial Credit and drained it of its vitality over the next two decades, a foreshadowing of the leveraged buyouts of the 1980s.[85] Unlike most of the companies that brought suit against IBM, Control Data achieved a favorable settlement in 1973. That resulted in IBM's transferring its own Service Bureau to CDC.[86]

These victories made Bill Norris, CDC's founder and chairman, look like a wily fox, but we now know that Norris made the unforgivable error of taking his eye off the advancing pace of technology.[87] CDC's success came from the superior performance of its products, especially super-computers—a class of machines that CDC pioneered. Norris's ability to win in the courtroom or play with inflated stock was no substitute. CDC never really survived Seymour Cray's leaving. In 1972 Cray founded Cray Research, with a laboratory next to his house in Chippewa Falls, Wisconsin, and set out to recreate the spirit of CDC's early days. The CRAY-1 was introduced in 1976 and inaugurated a series of successful supercomputers. CDC continued to introduce supercomputers, but none could match the products from Seymour Cray's laboratory.

Even more heartbreaking was the failure of CDC's PLATO, an interactive, graphics-based system intended for education and training at all levels, from kindergarten on up (figure 5.5). It promised, for the expert and lay-person alike, easy and direct access to information from libraries and archives worldwide. CDC spent millions developing PLATO and had a large pilot installation operating at the University of Illinois by the mid-1970s.[88] But ultimately it failed. The reasons are complex. PLATO required a central CDC mainframe to run on, the terminals were expensive, and PLATO may have been too far ahead of its time. In 1994 most of the predictions for PLATO came true, via the Internet and using a system called the World Wide Web. (Note that the federal government paid most of the R&D costs of these systems.) By then it was too late for CDC to reap any benefits from PLATO. The company began losing large amounts of money in the mid-1980s, and in 1986 Bill Norris resigned. CDC survived, but only as a supplier of specialized hardware and services, mainly to an ever-shrinking military market.

Conclusion

John Brooks's "go-go years" are now a distant memory. The stories of Apple, Microsoft, and other companies from the 1980s and 1990s make those of an earlier era seem tame by comparison. People remember the high-flying financial doings, but they forget that those were the years when the foundation was laid for later transformations of the computer industry. That foundation included building large systems using integrated circuits, large data stores using disk files, and above all complex software written in high-level languages. The rise of independent software and systems houses, as well as plug-compatible manufacturers, also foreshadowed a time when software companies would become equal if not dominant partners in computing, and when clones of computer architectures also became common. Finally, it was a time when Wall Street learned that computers, semiconductors, and software deserved as much attention as the Reading Railroad or United States Steel.

Figure 5.5
CDC's PLATO System. (*top*) One use for PLATO was to store and retrieve engineering drawings and data. (*middle*) Another use, one that was widely publicized, was for education. (*bottom*) A PLATO terminal being used by a handicapped person (note the brace leaning against the desk). William Norris, the head of Control Data, wrote and spoke extensively on the social benefits of computing when made available to lay persons. The photograph inadvertently reveals why PLATO ultimately failed. In the background is an early model of a personal computer from Radio Shack. It is very primitive in comparison to PLATO, but eventually personal computers became the basis for delivering computing and telecommunications to the home, at a fraction of the cost of PLATO. (*Source*: Charles Babbage Institute, University of Minnesota.)

6

The Chip and Its Impact, 1965–1975

Just as the IBM System/360 transformed mainframe computing, so did a series of new machines transform minicomputing in the late 1960s. At first these two computing segments operated independently, but during the 1970s they began to coalesce. Behind these changes was an invention called the integrated circuit, now known universally as "the chip."

Minicomputers such as the PDP-8 did not threaten mainframe business; they exploited an untapped market and lived in symbiosis with their large cousins. Some thought it might be possible to do a mainframe's work with an ensemble of minis, at far lower cost. Mainframe salesmen, citing "Grosch's Law," argued that this tempting idea went against a fundamental characteristic of computers that favored large systems. Named for Herb Grosch (figure 6.1), a colorful figure in the computer business, this law stated that a computer system that was twice as big (i.e., that cost you twice as much money) got you not twice but four times as much computing power. If you bought two small computers, giving you two times the power of a single one, you would not do as well as you would if you used the money to buy a single larger computer.[1]

Believers in that law cited several reasons for it. Computers of that era used magnetic cores for storage. The cores themselves were cheap, but the support circuitry needed to read, write, and erase information on them was expensive. And a certain amount of that circuitry was required whether a memory capacity was large or small. That made the cost per bit higher for small memories than for large, so it was more economical to choose the latter, with an accompanying large processor system to take advantage of it. The most compelling reason was that no one really knew how to link small computers to one another and get coordinated performance out of the ensemble. It would have been like trying to fly

Figure 6.1
Herbert Grosch, ca. 1955. (*Source*: Herbert Grosch.)

passengers across the Atlantic with an armada of biplanes instead of a single jumbo jet. Eventually both barriers would fall, with the advent of semiconductor memory and new network architectures. By the time that happened—around the mid 1980s—the minicomputer itself had been replaced by a microprocessor-based workstation.[2] But as minicomputers had grown more and more capable through the late 1960s, they had slowly begun a penetration into mainframe territory while opening up new areas of application. Grosch's Law held, but it no longer ruled.

The force that drove the minicomputer was an improvement in its basic circuits, which began with the integrated circuit (IC) in 1959. The IC, or chip, replaced transistors, resistors, and other discrete circuits in the processing units of computers; it also replaced cores for the memory units. The chip's impact on society has been the subject of endless discussion and analysis. This chapter, too, will offer an analysis, recognizing that the chip was an evolutionary development whose origins go back to the circuit designs of the first electronic digital computers, and perhaps before that.

The von Neumann architecture described a computer in terms of its four basic functional units—memory, processor, input, and output. Below that level were the functional building blocks, which carried out the logical operations "AND," "OR," "NOT," "EXCLUSIVE OR," and a few others. Below *that* were circuits that each required a few—up to about a dozen—components that electrical engineers were familiar with: tubes (later transistors), resistors, capacitors, inductors, and wire. In the 1940s anyone who built a computer had to design from that level. But as computer design emerged as a discipline of its own, it did so at a higher level, the level of the logical functions operating on sets of binary digits. Thus arose the idea of assembling components into modules whose electrical properties were standardized, and which carried out a logical function. Using standardized modules simplified not only computer design but also testing and maintenance, both crucial activities in the era of fragile vacuum tubes.

J. Presper Eckert pioneered in using modules in the ENIAC to handle a group of decimal digits, and in the UNIVAC to handle digits coded in binary, a key and often overlooked invention that ensured the long-term usefulness of those two computers, at a time when other computers seldom worked more than an hour at a time.[3] When IBM entered the business with its Model 701, it also developed circuit modules—over two thousand different ones were required. For its transistorized machines it developed a compact and versatile "Standard Modular System" that reduced the number of different types.[4] Digital Equipment Corporation's first, and only, products for its first year of existence were logic modules, and the success of its PDP-8 depended on "flip-chip" modules that consisted of discrete devices mounted on small circuit boards.

Patents for devices that combined more than one operation on a single circuit were filed in 1959 by Jack Kilby of Texas Instruments and Robert Noyce of Fairchild Semiconductor. Their invention, dubbed at first "Micrologic," then the "Integrated Circuit" by Fairchild, was simply another step along this path.[5] Both Kilby and Noyce were aware of the prevailing opinion that existing methods of miniaturization and of interconnecting devices, including those described above, were inadequate. A substantial push for something new had come from the U.S. Air Force, which needed ever more sophisticated electronic equipment on-board ballistic missiles and airplanes, both of which had stringent weight, power consumption, and space requirements. (A closer look at the Air Force's needs reveals that reliability, more than size, was foremost on its mind.[6]) The civilian electronics market, which wanted something as well, was primarily concerned with the costs and errors that accompanied the wiring of computer circuits by hand. For the PDP-8's production, automatic wire-wrap machines connected the flip-chip modules. That eliminated, in Gordon Bell's words, "a whole floor full of little ladies wiring computers," although building a computer was still labor-intensive.[7] In short, "[a] large segment of the technical community was on the lookout for a solution of the problem because it was clear that a ready market awaited the successful inventor."[8]

Modern integrated circuits, when examined under a microscope, look like the plan of a large, futuristic metropolis. The analogy with architectural design or city planning is appropriate when describing chip design and layout. Chips manage the flow of power, signals, and heat just as cities handle the flow of people, goods, and energy. A more illuminating analogy is with printing, especially printing by photographic methods. Modern integrated circuits are inexpensive for the same reasons that a paperback book is inexpensive—the material is cheap and they can be mass produced. They store a lot of information in a small volume just as microfilm does. Historically, the relationship between printing, photography, and microelectronics has been a close one.

Modules like Digital Equipment Corporation's flip chips interconnected components by etching a pattern on a plastic board covered with copper or some other conductor; the board was then dipped into a solvent that removed all the conductor except what was protected by the etched pattern. This technique was pioneered during the Second World War in several places, including the Centrallab Division of the Globe-Union Company in Milwaukee, Wisconsin, where circuits were produced for an artillery fuze used by allied forces. Other work was done at the National Bureau of Standards in Washington, D.C.[9] Some of this work was based on patents taken out by Paul Eisler, an Austrian refugee who worked in England during the war, Eisler claims his printed circuits were used in the war's most famous example of miniaturized electronics, the Proximity Fuze, although others dispute that claim.[10] In his patent granted in 1948, Eisler describes his invention as "a process based on the printing of a representation of the conductive metal."[11] After the war the "printed circuit," as it became known, was adopted by the U.S. Army's Signal Corps for further development. The Army called it "Auto-Sembly" to emphasize production rather than its miniaturization.[12] It was the ancestor of printed circuits, familiar to both the consumer and military markets, and still in use.[13]

Throughout the 1950s, the U.S. armed services pressed for a solution to the interconnection problem, seeing it as a possible way to increase reliability. Reliability was of special concern to the U.S. Air Force, which had found itself embarrassed by failures of multimillion dollar rocket launches, failures later found to have been caused by a faulty component that cost at most a few dollars. The Air Force mounted a direct attack on this problem for the Minuteman ballistic missile program, setting up a formal procedure that penetrated deep into the production lines of the components' manufacturers.

At the same time it inaugurated an ambitious program it called "molecular electronics," whose goal was to develop new devices made of substances whose individual molecules did the switching. Just how that would be done was unspecified, but the Air Force awarded a $2 million development contract to Westinghouse in April 1959—within months of the invention of the IC—to try.[14] Later on Westinghouse received another $2.6 million. The idea never really went anywhere. Two years after awarding the contract, the Air Force and Westinghouse reported substantial progress, but the press, reporting that the "USAF Hedges Molectronics Bets," called the use of ICs an "interim step" needed to reduce the size and complexity of airborne electronics.[15] The term "molecular electronics" quietly vanished from subsequent reports.

The Air Force's push for higher reliability of parts for the Minuteman ballistic missile had a greater impact on the electronics industry because it did achieve a breakthrough in reliability. Suppliers introduced "clean rooms," where workers wore gowns to keep dust away from the materials they were working with. Invented at the Sandia National Laboratories in the early 1960s for atomic weapons assembly, such rooms were washed by a constant flow of ultra-filtered air flowing from the ceiling to the floor.[16] Eventually the industry would build fabrication rooms, or "fabs," that were many times cleaner than a hospital. They would control the impurities of materials almost to an atom-by-atom level, at temperatures and pressures regulated precisely. The electronics industry developed these techniques to make transistors for Minuteman. The culture took root.

At every step of the production of every electronic component used in Minuteman, a log was kept that spelled out exactly what was done to the part, and by whom. If a part failed a subsequent test, even a test performed months later, one could go back and find out where it had been. If the failure was due to a faulty production run, then every system that used parts from that run could be identified and removed from service. Suppliers who could not or would not follow these procedures were dropped.[17] Those who passed the test found an additional benefit: they could market their components elsewhere as meeting the "Minuteman Hi-Rel" standard, charging a premium over components produced by their competitors. Eventually the estimated hundred-fold reduction of failure rates demanded by the Air Force came to be accepted as the norm for the commercial world as well.[18] In a reverse of Gresham's Law, high-quality drove low-quality goods from the market.

This program came at a steep price. Each Minuteman in a silo cost between $3 and $10 million, of which up to 40 percent was for the electronics.[19] And the Hi-Rel program's emphasis remained on discrete components, although the clean-room production techniques were later transferred to IC production. However successful it was for the Minuteman, the Hi-Rel program did not automatically lead to advances in commercial, much less consumer, markets.[20]

The Invention of the Integrated Circuit

In the early 1960s the Air Force initiated the development of an improved Minuteman, one whose guidance requirements were far greater than the existing missile's computer could handle. For mainly political reasons, "those who wished other capabilities from ICBMs [intercontinental ballistic missiles] were unable to start afresh with an entirely new missile. Instead, they had to seek to build what they wanted into successive generations of Minuteman."[21] The reengineering of Minuteman's guidance system led, by the mid-1960s, to massive Air Force purchases for the newly invented IC, and it was those purchases that helped propel the IC into the commercial marketplace.

Before discussing those events, it is worth looking at the circumstances surrounding the IC's invention. As important as the military and NASA were as customers for the IC, they had little to do with shaping its invention.

After graduating from the University of Illinois with a degree in Electrical Engineering in 1947, Jack Kilby took a job at Centrallab in Milwaukee—the industrial leader in printed circuits and miniaturization. At first he worked on printed circuit design; later he became involved in getting the company to make products using germanium transistors. "By 1957 ... it was clear that major expenditures would soon be required. The military market represented a major opportunity, but required silicon devices.... The advantages of the diffused transistor were becoming apparent, and its development would also have required expenditures beyond the capabilities of Centrallab.... I decided to leave the company."[22] The following year he joined Texas Instruments in Dallas, already known in the industry for having pioneered the shift from germanium to silicon transistors. "My duties were not precisely defined, but it was understood that I would work in the general area of microminiaturization."[23] Texas Instruments (TI) was one among many companies that recognized the potential market, both military and civilian, for such devices. But how to build them?

Jack Kilby is a tall, modest man whose quiet manner reflects the practical approach to problems people often associate with Midwesterners. He was born in Jefferson City, Missouri, and grew up in the farming and oil-well supply town of Great Bend, Kansas, named after the southern turn that the Arkansas River takes after coming out of the Rockies. His father was an engineer for a local electrical utility.[24] He recalls learning from his father that the cost of something was as important a variable in an engineering solution as any other.[25]

As others at TI and elsewhere were doing in 1958, Kilby looked at microminiaturization and made an assessment of the various government-funded projects then underway. Among those projects was one that TI was already involved with, called Micro-Module, which involved depositing components on a ceramic wafer.[26] Kilby did not find this approach cost effective (although IBM chose a variation of it for its System/360). In the summer of 1958 he came up with a fresh approach—to make all the individual components, not just the transistors, out of germanium or silicon. That swam against the tide of prevailing economics in the electronics business, where resistors sold for pennies, and profits came from shaving a few tenths of a cent from their production cost. A resistor made of silicon had to cost a lot more than one made of carbon. But Kilby reasoned that if resistors and other components were made of the same material as the transistors, an entire circuit could be fashioned out of a single block of semiconductor material. Whatever increased costs there were for the individual components would be more than offset by not having to set up separate production, packaging, and wiring processes for each.

Jack Kilby built an ordinary circuit with all components, including its resistors and capacitor, made of silicon instead of the usual materials, in August, 1958. In September he built another circuit, only this time all the components were made from a single piece of material—a thin 1/16-inch × 7/16-inch wafer of germanium. (The company's abilities to work with silicon for this demonstration were not quite up to the task.) He and two technicians laboriously laid out and constructed the few components on the wafer and connected them to one another by fine gold wires. The result, an oscillator, worked. In early 1959 he applied for a patent, which was granted in 1964 (figure 6.2).[27] Texas Instruments christened it the "solid circuit." It was a genuine innovation, a radical departure from the military-sponsored micromodule, molecular electronics, and other miniaturization schemes then being pursued.[28]

Robert Noyce also grew up in the Midwest, in Grinell, Iowa, where his father was a Congregational minister. Some ascribe Noyce's inventiveness to Protestant values of dissent and finding one's own road to salvation,[29] but not all Protestant faiths shared that, and one would not describe Noyce or the other Midwestern inventors as religious. A more likely explanation is the culture of self-sufficiency characteristic of Midwestern farming communities, even though only one or two of the inventors in this group actually grew up on farms. In any event, the Corn Belt in the 1930s and 1940s was fertile ground for digital electronics.

(a)   **June 23, 1964**   J. S. KILBY   **3,138,743**

MINIATURIZED ELECTRONIC CIRCUITS

Filed Feb. 6, 1959   4 Sheets—Sheet 2

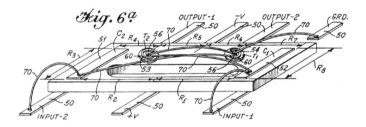

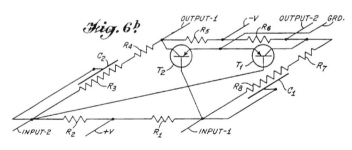

INVENTOR

Jack S. Kilby

BY

Stevens, Davis, Miller & Mosher
ATTORNEYS

(b)

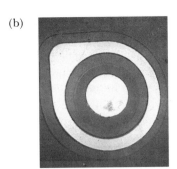

(c)

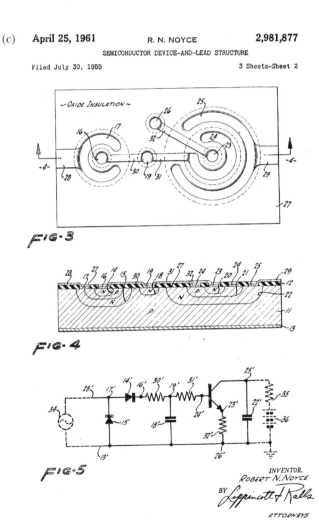

April 25, 1961   R. N. NOYCE   2,981,877
             SEMICONDUCTOR DEVICE-AND-LEAD STRUCTURE

Filed July 30, 1959   3 Sheets-Sheet 2

FIG·3

FIG·4

FIG·5

INVENTOR.
ROBERT N. NOYCE
BY
Lippincott & Ralls
ATTORNEYS

Figure 6.2
The chip. (*a*) Patent for integrated circuit by Jack Kilby. (*b*) Planar transistor.
(*Source*: Fairchild Semiconductor.) (*c*) Patent for integrated circuit by Robert
Noyce.

Robert Noyce was working at Fairchild Semiconductor in Mountain View, California, when he heard of Kilby's invention. He had been thinking along the same lines, and in January 1959 he described in his lab notebook a scheme for doing essentially the same thing Kilby had done, only with a piece of silicon.[30] One of his coworkers at Fairchild, Swiss-born Jean Hoerni, had paved the way by developing a process for making silicon transistors that was well-suited for photo-etching production techniques, making it possible to mass-produce ICs cheaply.[31] It was called the "planar process," and as the name implies, it produced transistors that were flat. (Other techniques required raised metal lines or even wires somehow attached to the surface to connect a transistor.) The process was best suited to silicon, where layers of silicon oxide—"one of the best insulators known to man," Noyce recalled—could be built up and used to isolate one device from another.[32] For Noyce the invention of the IC was less the result of a sudden flash of insight as of a gradual build-up of engineering knowledge about materials, fabrication, and circuits, most of which had occurred at Fairchild since the company's founding in 1957. (By coincidence, the money used to start Fairchild Semiconductor came from a camera company, Fairchild Camera and Instrument. Sherman Fairchild, after whom the company was named, was the largest individual stockholder in IBM—his father helped set up IBM in the early part of the century.)[33]

Noyce applied for a patent, too, in July 1959, a few months after Kilby. Years later the courts would sort out the dispute over who the "real" inventor was, giving each person and his respective company a share of the credit. But most acknowledge that Noyce's idea to incorporate Hoerni's planar process, which allowed one to make the electrical connections in the same process as making the devices themselves, was the key to the dramatic progress in integrated electronics that followed.

Hoerni did not share in the patents for the integrated circuit, but his contribution is well known. "I can go into any semiconductor factory in the world and see something I developed being used. That's very satisfying."[34] His and Noyce's contributions illustrate how inventors cultivate a solution to a problem first of all visually, in what historian Eugene Ferguson calls the "mind's eye."[35] Although the invention required a thorough knowledge of the physics and chemistry of silicon and the minute quantities of other materials added to it, a nonverbal, visual process lay behind it.[36]

These steps toward the IC's invention had nothing to do with Air Force or military support. Neither Fairchild nor Texas Instruments were among the first companies awarded Air Force contracts for miniaturization. The shift from germanium to silicon was pioneered at Texas Instruments well before it was adopted for military work. Kilby's insight of using a single piece of material to build traditional devices went against the Air Force's molecular electronics and the Army's micromodule concepts. And the planar process was an internal Fairchild innovation.[37]

But once the IC was invented, the U.S. aerospace community played a crucial role by providing a market. The "advanced" Minuteman was a brand-new missile wrapped around an existing airframe. Autonetics, the division of North American Aviation that had the contract for the guidance system, chose integrated circuits as the best way to meet its requirements. The computer they designed for it used about 2,000 integrated and 4,000 discrete circuits, compared to the 15,000 discrete circuits used in Minuteman I, which had a simpler guidance requirement.[38] Autonetics published comparisons of the two types of circuits to help bolster their decision. According to Kilby, "In the early 1960s these comparisons seemed very dramatic, and probably did more than anything else to establish the acceptability of integrated circuits to the military."[39] Minuteman II first flew in September 1964; a year later the trade press reported that "Minuteman is top Semiconductor User," with a production rate of six to seven missiles a week.[40] The industry had a history of boom and bust cycles caused by overcapacity in its transistor plants. Were it not for Minuteman II they would not have established volume production lines for ICs: "Minuteman's schedule called for over 4,000 circuits a week from Texas Instruments, Westinghouse, and RCA."[41]

Fairchild was not among the three major suppliers for Minuteman. Noyce believed that military contracts stifled innovation—he cited the Air Force's molecular electronics as an example of approaching innovation from the wrong direction. He was especially bothered by the perception that with military funding,

the direction of the research was being determined by people less competent in seeing where it ought to go, and a lot of time of the researchers themselves was spent communicating with military people through progress reports or visits or whatever.[42]

However, before long, the company recognized the value of a military market: "Military and space applications accounted for essentially the entire integrated circuits market last year [1963], and will use over 95 percent of the circuits produced this year."[43]

Although reluctant to get involved in military contracts, Fairchild did pursue an opportunity to sell integrated circuits to NASA for its Apollo Guidance Computer (figure 6.3).[44] Apollo, whose goal was to put a man on the Moon by the end of the 1960s, was not a military program. Its guidance system was the product of the MIT Instrumentation Laboratory, which under the leadership of Charles Stark Draper was also responsible for the design of guidance computers for the Polaris and Poseidon missiles. Like Minuteman, Apollo's designers started out with modest on-board guidance requirements. Initially most guidance was to be handled from the ground; as late as 1964 it was to use an *analog* computer.[45] However, as the magnitude of the Lunar mission manifested itself the computer was redesigned and asked to do a lot more. The lab had been among the first to purchase integrated circuits from TI in 1959. After NASA selected the Instrumentation lab to be responsible for the Apollo guidance system in August 1961, Eldon Hall of the lab opened discussions with TI and Fairchild (figure 6.4). The IC's small size and weight were attractive, although Hall was concerned about the lack of data on manufacturing reliable numbers of them in quantity. In a decision that looks inevitable with hindsight, he decided to use ICs in the computer, adopting Fairchild's "micrologic" design with production chips from Philco-Ford, Texas Instruments, and Fairchild. His selection of Fairchild's design may have been due to Noyce's personal interest in the MIT representatives who visited him several times in 1961 and 1962. (Noyce was a graduate of MIT.)[46] NASA approved Hall's decision in November 1962, and his team completed a prototype that first operated in February 1965, about a year after the Minuteman II was first flown.[47]

In contrast to the Minuteman computer, which used over twenty types of ICs, the Apollo computer used only one type, employing simple logic.[48] Each Apollo Guidance Computer contained about 5,000 of these chips.[49] The current "revolution" in microelectronics thus owes a lot to both the Minuteman and the Apollo programs. The Minuteman was first: it used integrated circuits in a critical application only a few years after they were invented. Apollo took the next and equally critical step: it was designed from the start to exploit the advantages of integrated logic.

Around 75 Apollo Guidance Computers were built, of which about 25 actually flew in space. During that time, from the initial purchase of prototype chips to their installation in production models of the Apollo

Figure 6.3
Launch of the Saturn V/Apollo 11 spacecraft, July 1969. The relationship between the U.S. space program and the advance of computing technology was a complex one. The demands of programs like Apollo and Minuteman advanced the state of the art of microelectronics and computer circuits. Advances in computing, on the other hand, shaped the way programs like Apollo were designed and operated. (*Source*: NASA.)

computer, the price dropped from $1,000 a chip to between $20 and $30.[50] The Apollo contract, like the earlier one for Minuteman, gave semiconductor companies a market for integrated circuits, which in turn they could now sell to a civilian market. By the time of the last Apollo flight in 1975 (the Apollo-Soyuz mission), one astronaut carried a pocket calculator (an HP-65) whose capabilities were greater than the on-board computer's. Such was the pace of innovation set in motion by the aerospace community.

Figure 6.4
Eldon Hall, head of the Apollo Computer Division at the MIT Instrumentation
Laboratory, ca. 1968. (*Source*: Charles Stark Draper Laboratory.)

Commercial Impact of the Chip

The biggest impact of this invention on commercial computing was in
the minicomputer, not the mainframe, industry. For its System/360 line,
IBM developed "solid logic technology," a scheme similar to micro-
module, in which circuits were deposited on a ceramic substrate about
half an inch thick, with metallic conducting channels printed on it.[51] By
the time of the 360's announcement in April 1964 the integrated circuit
was rapidly proving itself, and some in IBM worried that it would be left
behind with obsolete technology. An internal IBM memorandum written
in September 1963 stated that ICs "do not constitute a competitive
threat either now or in the next five years," while another internal
report written in September 1964 argued that rapid progress in "monolithics" (IBM's term for ICs) had been made, and that IBM had a "2–4

year lag in practical experience" and needed "6 months to a year to catch up" in IC expertise.[52]

Both memorandums were right: IBM had learned to produce solid logic technology circuits reliably and in large quantities, which had served the System/360 well. But to remain competitive, it adopted ICs for the System/370, announced at the end of the decade. But as early as the first deliveries of the System/360 in the spring of 1965, Scientific Data Systems (SDS) had announced a computer that used ICs. When RCA decided to compete against IBM with a family of 360-compatible machines, it also decided to go with integrated circuits. By 1966, both SDS and RCA computers were being delivered and IBM's "five year" lead was now one year. Integrated circuits went from Kilby's crude laboratory device to practical use in commercial computers faster than anticipated. Part of the reason was the eagerness of military and aerospace customers; credit is also due to Noyce's observation that the basic techniques of IC fabrication were an evolutionary advance over planar transistor production.

Second-Generation Minicomputers Digital Equipment Corporation showed how one could enter the computer business with a modest amount of capital and a modest physical plant. With inexpensive but powerful ICs on the market, the road opened up for others to follow DEC's example. DEC did not dominate in minicomputers in the same way IBM dominated mainframes. Unlike the BUNCH, DEC's competitors did not feel they had to answer every product announcement, or offer software-compatible products. Technical innovation, at low cost and in a compact package, mattered more. The result was that the performance of minis increased at a phenomenal rate from 1965 through the 1970s. Prices remained low and even dropped. To enter this market, one needed to have a grasp of the latest technology, but banks and venture capital firms did not fear—as they did with those who wanted to compete with IBM—that Digital would crush the newcomer by virtue of its dominant market share.

Between 1968 and 1972, around one hundred new companies or divisions of established companies offered minicomputers on the commercial market, an average of one every three weeks for that five-year period. Table 6.1 lists some of the more well-known among them.[53]

There were a few barriers that one had to surmount to enter the business, but most barriers were low.[54] Semiconductor firms were offering a ready supply of inexpensive chips, which provided basic logic in a simple package. By 1970 the IC makers had adopted a standard that laid a solid foundation on which the computer industry would grow for the next two decades. That was to supply a set of chips that used transistors to implement the logic, called "transistor-transistor logic" (TTL). TTL chips were inexpensive and easy to design with.[55] They were packaged in a black plastic or ceramic case, with leads arranged along either side like a flattened caterpillar. Engineers at Fairchild introduced this "dual in-line package" (DIP) in 1964. Rex Rice led this effort; his work was based in part on work patented by Nathan Pritikin (a self-made

Table 6.1
Minicomputers, 1965–1974

Manufacturer	Computer	Year
California Data Processors	XI/35	1974
Cincinnati Milacron	CIP/2200	1970
Computer Automation	LSI "Naked Mini" series	1972
Computer Terminal Corporation	Datapoint 2200	
Data General	Nova	1969
	Supernova	1971
Digital Computer Controls	DCC-116, 112	1972
Digital Equipment Corp.	PDP-11 series	1970
General Automation	SPC-16	1971
General Electric	GEPAC 4010, 4020	
GRI Computer Corp.	GRI-99 series	1972
GTE Information Systems	IS/1000	1970
Hewlett-Packard	2100 Series	1971
Honeywell	H-316	ca. 1970
	DDP-516	ca. 1971
IBM	System 3	1969
Interdata	Model 70	
Lockheed Electronics	MAC-16	1968
	SUE	1972
Modular Computer Systems	MODCOMP line	1971
Motorola	MDP-1000	1968
Prime Computer	300 series	1973
Raytheon	500 series	1974
Scientific Data Systems	SDS-910	1962
	SDS-920	1965
Systems Engineering Labs	SEL-810	
Texas Instruments	960, 980	1974
Varian Associates	520	
	620	1972
Westinghouse	W-2500	1971

inventor who later became famous for advocating a low-cholesterol diet). The package was rugged and easy to handle.[56]

By 1970 a way of connecting the chips to one another had also standardized. The printed circuit board, pioneered by Globe-Union, had evolved to handle integrated circuits as well. A minicomputer designer could now lay out a single large printed circuit board, with places for all the ICs necessary for the circuits of a small computer. On an assembly line (possibly located in Asia and staffed by women to save labor costs), a person (or machine) would "stuff" chips into holes on one side of the board. She would then place the board in a chamber, where a wave of molten solder would slide across the pins protruding through to the other side, attaching them securely to the printed connections. The process was fast, reliable, and yielded a rugged product. It was Fairchild engineers who first explored this method of packaging and wiring, on an experimental computer called SYMBOL, in the late 1960s. SYMBOL was built to explore a new computer architecture; its packaging and wiring proved far more influential than its architecture.[57] This combination—TTL logic, dual in-line packaging, and large, wave-soldered printed circuit boards—remained a standard into the 1990s.[58]

The Founding of Intel   Digital Equipment Corporation made its mark by architectural advances like Direct Memory Access. The second wave of minicomputer products showed an equally-remarkable set of architectural innovations. The first was to settle on sixteen bits for the mini's word length. This move followed the lead of IBM, after its announcement of the System/360 in 1964. System/360 used a 32-bit word, but it also set a standard of 8 bits as the "byte," the basic group of bits to encode a letter of the alphabet or other character. An 8-bit byte allowed 256 different combinations—far more than needed for numbers, punctuation and the upper and lower case alphabet—but eight was a power of two, which simplified certain aspects of a computer's design. And the "extra" combinations left room for future growth.[59] IBM's choice of an 8-bit byte became a standard, so much so that few remember that 6 bits had been the standard among minicomputer manufacturers, for whom the "extra" bits made up a proportionally greater cost of the overall system.

There was one difference: minicomputers did not use IBM's EBCDIC code for a character but developed an 8-bit extension of ASCII instead. This decision had some long-term consequences. It widened the gap between IBM and the rest of the computer industry. Because the eighth bit of ASCII was not standardized, it led also to a proliferation of different standard codes for mini, and later personal, computers. The ultimate consequences of this split would play out in the 1980s, after IBM entered the personal computer market with an ASCII machine.

DEC's PDP-8, with its 12-bit word length and a 6-bit code for each character, was shipped after the 360's announcement. For Gardner Hendrie, an engineer working on a 14-bit mini at the rival Computer Controls Corporation, IBM's announcement of the System/360 was a bombshell. Over the summer of 1964 CCC redesigned its machine, and in October it announced the DDP-116, the first 16-bit minicomputer.[60] The machine was profitable for CCC. The company grew rapidly and for a while it looked like it would overtake Digital Equipment Corporation, but in 1966 Honeywell bought CCC, and it lost its independence. Honeywell continued the 16-bit line, but perhaps because it was devoting more attention to mainframes, the product line withered.

Honeywell's foray into the minicomputer market ended up as a minor diversion. No one would remember Honeywell minicomputers were it not for one exceptional event associated with them. Beginning in 1967 the Advanced Research Projects Agency (ARPA) of the Department of Defense began a series of meetings to discuss how to link computers across the country in a network.[61] ARPA had several reasons for wanting such a network, the main one being a desire to share resources among places receiving its funding at a time when computers were large and expensive. Early in the process the ARPA researchers recognized that the diverse nature of the computers being connected was a major problem. They proposed to solve it by using a minicomputer at each node, between the main computer being networked and the network itself. For this interface message processor (IMP), they chose a Honeywell DDP-516, a descendent of the 16-bit CCC mini.[62] Four nodes (i.e., four IMPS) were operating west of the Rockies in December 1969; a year later there were ten, spanning the country. By 1971 ARPANET consisted of fifteen nodes connecting twenty-three host computers. Of those, nine were PDP-10s, five were IBM System/360s, one was the Illiac-IV, and the rest were assorted minicomputers and mainframes.[63] ARPANET was demonstrated publicly at a hotel in Washington, D.C., in October 1972; by that year there were thirty nodes.[64] ARPANET was dismantled in 1988, but it will always be remembered as the precursor of the Internet, which burst into the public's consciousness in the 1990s. By that time IMPS were no longer needed to connect the nodes; they were replaced by a set of software and hardware developments that made them unnecessary. A few DDP-516 computers were saved and have been carefully preserved in museums and private collections as the modern equivalent of Samuel Morse's telegraph apparatus of the 1840s.[65]

Another company that quickly recognized the advantages of a 16-bit word length was Data General. Edson DeCastro, an engineer at Digital Equipment Corporation, grew frustrated with DEC's plans for a 16-bit computer, which DEC was calling the PDP-X. Developing a 16-bit machine was a major assignment, as DEC was well aware of Computer Controls Corporation's success. DeCastro proposed a design for the PDP-X, but DEC's management turned it down. In the spring of 1968 he and two other DEC engineers resigned and founded Data General in the neighboring town of Hudson. Their goal was to produce a 16-bit mini.[66]

Data General's founding as a descendent of Digital Equipment echoed a phenomenon that came to define the computer culture, especially in Silicon Valley on the West Coast. One observer remarked: "In this business, . . . capital assets in the traditional sense of plant, equipment, and raw materials counted for nothing. . . . Brainpower was the entire franchise."[67] Ken Olsen felt otherwise. He believed that DeCastro had developed the PDP-X while a DEC employee and now was using that design to launch Data General. The Nova, the 16-bit machine that Data General announced at the end of 1968, was not the rejected PDP-X, however. It was simpler and more elegant. Those who used one still talk of it as a "clean machine," a rare example of a truly optimal design—complex, but no more complex than necessary.[68]

The Nova also incorporated advances in circuits and packaging not available the previous year. It used chips that contained more transistors than earlier computers had used. These were later called medium scale integration (MSI) to distinguish them from their predecessors. (Later on came large scale integration [LSI] and finally very large scale integration [VLSI], based on the number of transistors on a chip.) These, in turn, were mounted on a single printed circuit board about 15 inches square. This board was larger than what was standard at the time and made for a very compact package.

In mid-1971 the company introduced an advanced "Super Nova," which incorporated still another technical innovation, integrated circuits instead of magnetic cores for random access memory (RAM). Although it had always been possible to make ICs that would store information, core memories had been about ten times cheaper because of the experience in producing them, going back to SAGE. A breakthrough came in 1970, when a supercomputer project at the University of Illinois chose 256-bit memory chips from Fairchild for its central memory (figure 6.5). The Illiac-IV was an ambitious attempt to break out of the von Neumann architecture of sequential processing. The Illiac-IV was beset with many difficulties, including student unrest on the Urbana campus, as well as problems implementing the ambitious design. Its greatest legacy was probably its use of semiconductor memory, which paved the way for Data General's commercial use.[69]

The Super Nova established semiconductor RAM's viability for commercial computers. It was only one of several indicators that a revolution in silicon was about to happen. We have already mentioned IBM's decision to go to monolithic memory for its System/370. Before

(a)

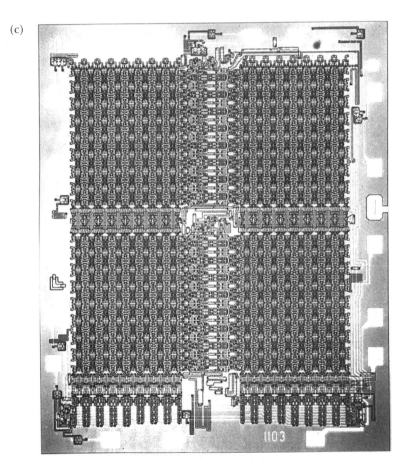

Figure 6.5
(*a*) Fairchild memory chip, used in Illiac-IV, and the first practical alternative to magnetic core. The chip stored 256 bits. (*Source*: Fairchild Semiconductor.) (*b*) Data General Super Nova. (*Source*: Data General Corporation.) (*c*) Intel 1103 memory chip. Capacity 1,024 bits. (*Source*: Charles Babbage Institute, University of Minnesota.)

the Illiac-IV, System/370, or the Super Nova appeared, an event took place that would be even more significant. In July 1968 Robert Noyce and Gordon Moore left Fairchild and founded a new company, whose initial focus would be on semiconductor memory. Once again, "Brainpower was the entire franchise": Noyce and Moore had been among the original founders of Fairchild's semiconductor division in 1957, having left a company founded by William Shockley that began transistor production in the lower San Francisco peninsula. The two men, soon joined by Andrew Grove, also of Fairchild, chose a name that would be a contraction of the words "integrated electronics": Intel.[70] In 1970 it introduced a 1,024-bit dynamic RAM chip, the 1103. That, plus deliveries of Super Novas and IBM System/370s, marked the beginning of the end for magnetic core.[71] From that modest beginning, Intel would become a dominant force in computing in the 1980s and 1990s. (Its subsequent path will be discussed in following chapters).

The Super Nova was packaged in a compact rectangular box. On the front panel were rows of lights indicating the status of its internal registers, with parallel rows of switches that allowed one to set or reset individual bits in those registers. Later generations of computers would do away with the front panel. With that exception, the hardware configuration of the Nova—a rectangular box, large printed circuit boards, and chips for both the processor and memory—has persisted to the present day.

The PDP-11   Digital Equipment Corporation found itself at a disadvantage by 1969. It had introduced a model of the PDP-8 that used integrated circuits (the PDP 8/I), but its 12-bit word length and limited instruction repertoire did not compare well to the machines from CCC/Honeywell or Data General. Other companies, large and small, were also entering the field.[72] Stung by DeCastro's defection, DEC made another try at designing a 16-bit mini. In March 1969 four DEC engineers flew to Pittsburgh to consult with Gordon Bell, who had taken a temporary leave from DEC to teach at Carnegie Mellon University, and with William Wulff, also a professor there. The Carnegie professors were not enthusiastic about the proposed design. An alternate design proposed by one of the DEC engineers, Harold McFarland, showed more promise. The group "decided to discard about a year's worth of work" and redesign the computer around McFarland's ideas.[73] DEC called the new machine the PDP-11, announced it in January 1970, and began deliveries a few months later.

In the course of the redesign, the team came up with an innovation that allowed it to regain the ground it had lost to Data General. That was a redefinition of the concept of a *bus*, a set of wires that served all major sections of the machine in a common and standard way. The notion was not new. The electromechanical Mark I, installed at Harvard in the 1940s, used one (spelled "buss").[74] The Whirlwind also had a bus, as did the Nova, and models of the PDP-8. But the PDP-11's Unibus carried this concept further: nearly all major units of the machine, including memory and I/O devices, were connected to a common, 56-line bus. That made the machine especially easy for DEC as well as its customers to configure for specialized applications or to expand.[75]

The bus architecture, like the packaging pioneered by the Nova, has prevailed in computer design ever since.[76] The 16-bit wordlength of this generation would double to 32, and then 64 bits, but it would remain a power of two.

The PDP-11 quickly surpassed its rivals and continued to fuel Digital Equipment Corporation's growth. Sales of the PDP-8 had enlarged the company from about 900 employees in 1965 to 5,800 in 1970. With the help of the PDP-11, it grew to 36,000 employees by 1977.[77] Over 170,000 PDP-11's were sold in the 1970s.[78] A recession hit the computer industry around 1970, but DEC, with its PDP-11, and Data General, with the Nova, survived and prospered. Competitors fell by the wayside or found only small niche markets. The go-go years were over (they would return), but they left in their wake a redefinition of the computer industry and its technology.

The Nova's success came from its elegant design and innovative packaging; the PDP-11's from its innovative architecture, which opened up minicomputers to a host of applications that had previously been the domain of mainframe computers. In addition to the Unibus, the machine employed a number of addressing "modes," which allowed one access to data in a variety of flexible and powerful ways. "Digital's traditional business is to sell architecture," said one engineer who worked for DEC in those years. In his view, the PDP-11's architecture was "wonderful," although the way the company implemented it was overly complex compared to the way Data General did things.[79] It was a much more complex machine than the PDP-8. According to Gordon Bell, "The PDP-11 was initially a hard machine to understand and was marketable only to those with extensive computer experience."[80]

The PDP-11's power meant that those who *did* understand it could develop software to make it easy for others to use, which suggests another way to distinguish this generation of minicomputers: users programmed them in familiar languages like FORTRAN rather than in machine code, an activity that seemed close to black magic for users of the PDP-8. The new minicomputers also came with tools that allowed easy editing and simplified finding and correcting bugs in programs. For many customers it was the best of both worlds—the flexibility and ease of use of a time-sharing system, with none of the overhead and expense of time-sharing a mainframe.

Direct-Access Computing Triumphant

While minicomputer systems were maturing, mainframe systems were evolving to offer similar, interactive capabilities. Time-sharing a main-frame was difficult in the late 1960s; by the mid-1970s, time-sharing packages were robust and stable parts of many System/370 installations. In addition to its Conversational Monitoring System (CMS), IBM offered a time sharing option (TSO) for its 370 computers beginning in 1971, while General Electric/Honeywell offered a successful commercial system based on its work at Dartmouth College (discussed later). For large and small systems, time-sharing became an acceptable and economical way of using a computer.

A key factor was the development of disk storage that offered rapid and direct access to large amounts of data. IBM had pioneered the use of disk storage with RAMAC in the late 1950s, but for the next ten years sequentially accessed tape, not disks, remained the mainstay of mass storage on mainframes. With the System/370, IBM introduced new models of disk storage that offered dramatically increased performance. During the late 1960s and early 1970s, the cost of storing data on disks dropped twentyfold, while the capacity of a typical disk storage system increased fortyfold.[81]

At many installations, the tape reels were joined by disk packs—they looked like cake boxes—as mass storage media. By 1980 it became common to use drives with disks that were not removable, or if they were, they were sealed with their read-write heads, to maintain tight tolerances. IBM came up with a drive that initially had two spindles holding 30 megabytes of data each: people called them "Winchester" drives, after the 30-30 rifle. IBM called disks "Direct Access Storage Devices," shortened to the acronym DASD. The commercial success of these products led other companies to rush in with disk drives that were plug-compatible with IBM's, leading once again to a flurry of activity on

Wall Street and the inevitable lawsuits against IBM for alleged antitrust activity.[82]

A little-recognized factor in the triumph of direct access computing was the IBM software product CICS, developed in the late 1960s for the electric utility industry but soon found throughout the commercial world.[83] CICS was not tailored to any specific application but rather allowed customers to write their own programs to permit direct query and retrieval of data from large databases. Among the first to exploit it were gas and electric utilities, who could use it to answer customer queries over the telephone. It has also transformed retail sales in the United States. Consider, for example, a typical transaction that occurs day and night, year round: a person calls a mail order house in Maine, asks about price, color, size, and availability of a pair of shoes, confirms the order, pays by credit card, and orders it to be shipped that afternoon for overnight delivery. By giving out a customer number printed on the catalog, the company determines the person's correct address including ZIP code, as well credit history and recent buying patterns. The person's credit card account is checked and debited; the telephone company bills the mail-order house for the telephone call, the inventory record is updated, and perhaps an order is sent to a factory in Asia to make more shoes. The overnight carrier is notified of the shipment, and an invoice for that is generated as well. The whole transaction takes a few minutes. Most of the inventory and billing information is transferred directly from one computer to another. There are only a few paper records. CICS, or software similar to it, is used in many of these operations. This kind of activity has become so common that traditional retail buying at a downtown department store is increasingly seen as exceptional. The mainframe's ability to handle this kind of data kept it viable in spite of the increasing competition from minicomputers and workstations.

Computer Science Education

Another place affected by these trends was the academic world, where the technological advances of the 1970s transformed both research and the teaching of computer programming.

Although universities had been teaching courses on computing since the 1950s, a batch environment was hardly optimum for that. Running batches of programs worked best when the programs themselves were error-free. In an installation like the NASA-Ames Research Center, where the computer-processed wind tunnel data, it was important that the program plow through each day's data as fast as possible. After an initial shakedown period, whatever errors the programs had were found and corrected. The compiler was a sophisticated program that used a lot of the computer's memory and time, but that time was well spent since the compiler generated machine code that executed quickly.

In a teaching environment the opposite was the case: the programs were bound to contain errors. Because the programs were short and typically did not handle large quantities of data, it mattered less that the compiled program execute quickly. Since many students would be submitting different programs, it did matter that the compiler work quickly, and that it produce, besides machine code, a detailed diagnosis of the errors it found, so that the student would know how to correct them. In fact, many commercial installations found a need to tinker with a program and thus recompile more frequently than NASA's wind tunnel experience would suggest. These, too, found a need to telescope the operations of batch computing that had grown up around mainframe installations.

The batch method of computer use remained at center stage at many universities, if for no other reason than that universities wanted their students to become familiar with the kind of systems that they would find in the industrial world upon graduation. Several university departments developed systems that would compile a student's program, and immediately direct the machine to execute it. These "load-and-go" compilers allowed the student to get a much quicker response.[84] (The University of Michigan MAD system, described in chapter 3, was an early version of this.)

The most innovative work in adapting batch computing to teaching was done at the University of Waterloo Canada. Waterloo's computer science department was among the first to be founded (in 1962), and before long, under the leadership of J. Wesley Graham, it was teaching computing to more undergraduates than almost any other school in Canada. In 1967 it apparently also owned the most powerful computer in all of Canada (a System/360 Model 75). By that year the department already had developed a FORTRAN compiler for another computer. It was called WATFOR (Waterloo FORTRAN) and was based on a similar compiler developed at the University of Wisconsin in the early 1960s:

WATFOR was written by four third-year math students in the summer of 1965. It was a fast in-core compiler with good error diagnostics, which proved especially useful to students for debugging their programs, as well as speeding up execution.[85]

The compiler allowed the university computer to run 6,000 jobs an hour, and by Graham's estimate reduced the cost of running a student's program from ten dollars to ten cents.[86] WATFOR was rewritten for the 360 Model 75 when it arrived; it was also upgraded (and given the whimsical name WATFIV). Graham's textbook *Fortran IV with WATFOR and WATFIV* influenced a generation of computer science students, who learned programming from it.[87] The university developed a similar compiler for the COBOL language, called WATBOL, for the 360/75 in 1972; it also developed Waterloo SCRIPT, a text-processing program widely used at a time when stand-alone word processors were rare. Waterloo distributed this software to academic computing centers world-wide, earning the university a steady stream of revenue from modest service fees.

BASIC at Dartmouth

Time-sharing offered another avenue for university instruction. A time-sharing environment could be set up to handle small programs and data sets from many users, giving each a rapid diagnosis of any errors he or she might have made. It was not enough simply to have time-sharing; one also had to design the system so that users could write programs easily and receive a quick and intelligible diagnosis of any errors. Under the leadership of John G. Kemeny, chairman of the mathematics department (and later president of the college), Dartmouth began building such a system in 1963. Kemeny had done calculations using punched-card equipment for the design of atomic weapons at Los Alamos. That experience led him to believe that "next to the original development of general-purpose high-speed computers the most important event was the coming of man-machine interaction."[88] He wanted a system that would teach interactive computing to all of Dartmouth's students—not just those studying computer science or engineering or physics. He was aware of work being done in the Cambridge area, including the IBM-based CTSS and a system running on a DEC PDP-1.[89] Whereas MIT went on from these modest beginnings to the more ambitious Project MAC, Kemeny and Thomas E. Kurtz (also of the Dartmouth mathematics department) decided to build a modest system around a programming language especially tailored to the needs of Dartmouth students. They called that language BASIC.

Bell Laboratories and Project MAC had chosen General Electric computers for their ambitious time-sharing systems, and now Dartmouth, too, chose GE. Dartmouth used a General Electric 235 computer connected to a smaller GE Datanet computer, which collected and managed the signals from the Teletype terminals scattered across the campus. Dartmouth's system was not a general-purpose time-sharing system, as defined by proponents like John McCarthy. It was tightly focused on undergraduate education, at a school where only 25 percent of students majored in science or engineering. MIT wanted a variety of languages available on its systems; at Dartmouth, the students programmed only in BASIC.[90]

The Dartmouth experience was a success for both Dartmouth and GE. But Dartmouth's model of open, free access, like the college library, did not prevail. General Electric offered a commercial version, and a few other universities adopted the model, and these continued to be supported after General Electric sold its computer business to Honeywell. Mainframe computers were still expensive, and what worked for a small, private college like Dartmouth did not necessarily work at a state university with a large, diverse graduate and undergraduate student body. Most universities felt a need to charge a fee based on the time a student was connected to the computer system, with higher fees charged for the time his or her program used the central processor. Who actually paid this fee varied from one university to the next, although often the National Science Foundation or some other government agency was involved.[91] In many cases, little real money was paid. The computer manufacturer gave the university a discount on the hardware; it may also have claimed a tax deduction; the university found ways to bury the remaining charges into some other, federally funded (sometimes military) research project. Many universities continued to teach computing using punched cards, Fortran, COBOL, and batch processing.

Although the Dartmouth model had only a modest influence, the programming language they developed, BASIC, became one of the most widely used computer programming languages in the world with an influence that extended well beyond the time-sharing system for which it was written. We saw how the IBM System/360 was able to evolve for decades while retaining its essential structure; so too did BASIC evolve to serve new markets while preserving its ease of use. It eventually became a language that propelled the personal computer into the mainstream, along with the company that provided the best BASIC for personal computers, the Microsoft Corporation.

The crucial step in the evolution of BASIC was taken in 1971 at the Digital Equipment Corporation. For the just-announced PDP-11, DEC developed a system called "Resource Sharing Time Sharing" (RSTS-11) that allowed time-sharing on the PDP-11. Initially it was offered for the Model 20, the simplest PDP-11; later versions ran on bigger models. The PDP-11/20 had a fraction of the power of the GE mainframe at Dartmouth, and with its 56K core memory would be considered a toy by 1990s standards. It had no facilities in hardware for protecting memory locations from alteration, either deliberate or accidental—something most thought was absolutely necessary for time-sharing.[92]

A team of engineers led by Mark Bramhall implemented RSTS-11 entirely in BASIC, but a version of BASIC with some interesting extensions. RSTS-11 needed to make system calls to log on or off and the like, which was implemented by a command called "SYS." A user could recall the individual bytes stored at a specific location in memory by another command, called PEEK. A PDP-11 user with special privileges could POKE bytes directly into memory—the reverse of PEEK, although unlike PEEK, this was a very dangerous command that could destroy the viability of time-sharing. None of these commands would have been feasible on the General Electric system used at Dartmouth, but for the PDP-11 they worked, and worked well.

Besides adding these commands, DEC engineers further modified BASIC so that it also could be implemented without taking up much memory. These implementations severely compromised some of Kemeny and Kurtz's principles of what the language ought to look like, something that upset the Dartmouth professors.[93] But as a result, one no longer needed machine language for even a simple minicomputer installation. This combination of features of DEC's BASIC—its ability to do low-level system calls or byte transfers, and its ability to fit on machines with limited memory—would be adopted by the Microsoft Corporation a few years later for its version of BASIC for the first personal computers.

Time-sharing systems based on more advanced models of the new minicomputers, like the PDP-11/45 and the Hewlett-Packard HP-2000, were very popular. These systems provided an alternative to the mainframe for computer science departments. Through the 1970s, as that discipline emerged, many universities and colleges found they could build a respectable curriculum around a time-shared minicomputer at a modest cost. For the beginning student there was BASIC; for those more advanced, there were more advanced languages, such as Pascal, then in favor as a better teaching language. This variety compensated for the fact that these students did not experience the flavor of the world of data processing using the "big iron" of the IBM System/370 and its giant

OS/MVS operating system. In some cases, these students learned something far in advance of what students using mainframes learned—the UNIX operating system, which had been developed on DEC minicomputers. This generation of computers thus created a new generation of students; students who took for granted the small size, low cost, and interactive use of the minicomputer, and the power of the UNIX operating system.

By the mid-1970s, the minicomputer had established strong positions in several markets and had moved out of its niche as an embedded processor for the OEM market. What held it back from the business data-processing market was the mainframe's ability to move enormous quantities of data through its channels, back and forth to rows of tape drives and "disk farms." But the minicomputer took better advantage than mainframes of advances in integrated circuits, packaging, and processor architecture. Its future seemed bright indeed. What happened next, however, was not what its creators intended. The mini generated the seeds of its own destruction, by preparing the way for personal computers that came from an entirely different source.

7

The Personal Computer, 1972–1977

Ready or not, computers are coming to the people.
That's good news, maybe the best since psychedelics.

Those words introduced a story in the fifth anniversary issue of *Rolling Stone*[1] (December 7, 1972). "Spacewar: Fanatic Life and Symbolic Death Among the Computer Bums" was written by Stewart Brand, a lanky Californian who had already made a name for himself as the publisher of the *Whole Earth Catalog*. Brand's resumé was unique, even for an acknowledged hero of the counterculture. At Stanford in the 1960s, he had participated in Defense Department–sponsored experiments with hallucinogenic drugs. In 1968 he had helped Doug Engelbart demonstrate his work on interactive computing at a now-legendary session of the Fall Joint Computer Conference in San Francisco.[2] Brand was no stranger to computers or to the novel ways one might employ them as interactive tools.

Brand was right. Computers did come to the people. The spread of computing to a mass market probably had a greater effect on society than the spread of mind-altering drugs. Personal computing, however, did not arrive in the way that Brand—or almost anyone else—thought it would. The development of personal computing followed a trajectory that is difficult to explain as rational. When trying to describe those years, from 1972 through 1977, one is reminded of Mark Twain's words: "Very few things happen at the right time, and the rest do not happen at all. The conscientious historian will correct these defects."[3] This chapter will examine how computers came "to the people," not as Twain's historian would have written it, but as it really occurred.

What triggered Brand's insight was watching people at the Stanford Artificial Intelligence Laboratory playing a computer game, Spacewar. Spacewar revealed computing as far from the do-not-fold-spindle-ormutilate punched-card environment as one could possibly find. The hardware they were using was not "personal," but the way it was being used was personal: for fun, interactively, with no concern for how many ticks of the processor one was using. That was what people wanted when, two years later, personal computers burst into the market.

Spacewar was running on a PDP-10. In terms of its hardware, a PDP-10 had nothing in common with the personal computers of the next decades.[4] It was large—even DEC's own literature called it a mainframe.[5] It had a 36-bit word length. A full system cost around a half million dollars and easily took up a room of its own. It used discrete transistors and magnetic cores, not integrated circuits, for logic and memory.[6] Still, one can think of the PDP-10 as an ancestor of the personal computer. It was designed from the start to support interactive use. Although its time-sharing abilities were not as ambitious as those of MIT's Project MAC, it worked well. Of all the early time-sharing systems, the PDP-10 best created an illusion that each user was being given the full attention and resources of the computer. That illusion, in turn, created a mental model of what computing could be—a mental model that would later be realized in genuine personal computers.[7]

Chapter 5 discussed the early development of time-sharing and the selection of a General Electric computer for Project MAC at MIT. While that was going on, the MIT Artificial Intelligence Laboratory obtained a DEC PDP-6, the PDP-10's immediate predecessor, for its research (figure 7.1). According to the folklore, MIT students, especially members of the Tech Model Railroad Club, worked closely with DEC on the PDP-6, especially in developing an operating system for it, which would later have an influence on the PDP-10's system software.[8] As a pun on the Compatible Time Sharing System that was running on an IBM mainframe nearby, the students called their PDP-6 system ITS—Incompatible Time Sharing System.[9] The PDP-6 did not have the disk storage necessary to make it a viable time-sharing system and only about twenty were sold. The PDP-10 did have a random-access disk system, which allowed its users direct access to their own personal files.[10] Like other DEC computers, the PDP-10 also allowed users to load personal files and programs onto inexpensive reels of DECtape, which fitted easily into a briefcase.

The feeling that a PDP-10 was one's own personal computer came from its operating system—especially from the way it managed the flow of information to and from the disks or tapes. With MIT's help, DEC supplied a system called "TOPS-10," beginning in 1972. In the

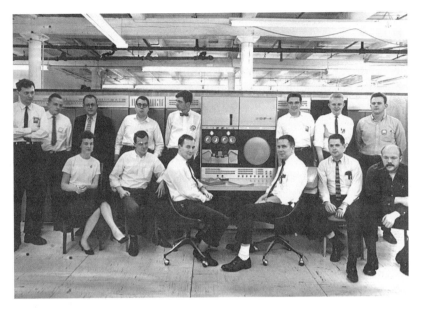

Figure 7.1
One of the most influential computers of all time, the DEC PDP-6, flanked by its creators at the Mill, 1964. C. Gordon Bell is at the left, wearing the sports jacket. The PDP-6 did not sell well but was the prototype for the more successful PDP-10 and DEC System-20. It would have as much of an impact on the course of computing as the much more celebrated PDP-8, also introduced at that time. (*Source:* Digital Equipment Corporation.)

introduction to the TOPS-10 manual, the authors stated, "Our goal has always been that in a properly configured system, each user has the feeling that he owns his portion of the machine for the time he needs to use it."[11] Users could easily create, modify, store, and recall blocks of data from a terminal. The system called these blocks by the already-familiar term, "files." Files were named by one to six characters, followed by a period, then a three-character extension (which typically told what type of file it was, e.g.: xxxxxx.BAS for a program written in BASIC). By typing DIR at a terminal users could obtain a directory of all the files residing on a disk. They could easily send the contents of a file to a desired output device, which typically consisted of a three-letter code, for example, LPT for line printer, or TTY for Teletype.[12]

A small portion of TOPS-10 was always present in core memory. Other programs were stored on the disk and could be called up as necessary. One, called PIP (Peripheral Interchange Program), allowed users to move files in a variety of ways to and from input/output equipment. Another program, TECO (Text Editor and Corrector), allowed users to edit and manipulate text from a terminal. DDT (Dynamic Debugging Tool) allowed users to analyze programs and correct errors without going through the long turnaround times that plagued batch processing.

For PDP-10 users, TOPS-10 was a marvel of simplicity and elegance and gave them the illusion that they were in personal control. TOPS-10 was like a Volkswagen Beetle: basic, simple, and easy to understand and work with.[13] Using a PDP-10 was not only fun but addictive. It was no accident that Brand saw people playing Spacewar on one, or that it was also the computer on which Adventure—perhaps the most long-lasting of all computer games—was written.[14]

On the West Coast another system appeared with similar capabilities, the SDS-940, offered by Scientific Data Systems (SDS) of southern California. The 940 was an extension of a conventional computer, the SDS- 930, modified by researchers at Berkeley with support from the Defense Department's Advanced Research Projects Agency. The 940 was more polished than the PDP-10, and it performed well. Still, the PDP-10 seemed to be preferred. At the Xerox Palo Alto Research Center, the legendary lab where so much of personal computing would be created, the staff was encouraged to use SDS machines, since Xerox had just purchased SDS. But the researchers there resisted and instead built a clone of a PDP-10, which they called MAXC—Multiple Access Xerox Computer—the name a pun on Max Palevsky, the founder of SDS.[15] (Palevsky, after becoming very wealthy from the sale of SDS to Xerox, dabbled in Hollywood movies, politics, and culture—and joined the board of *Rolling Stone*. Palevsky also became a venture capitalist with that money, helping to fund Intel, among other companies.)[16]

For a while, when Wall Street was enamored of anything connected with computers, it was easy to raise money to buy or lease a PDP-10 or SDS-940, and then sell computer time to engineering companies or other customers. Most of these firms were undercapitalized and did not understand the complexities of what they were selling. Like their counterparts in the electric utility industry, they had to have enough capacity to handle peak loads, in order not to discourage customers. But that meant that during off-peak times they would be wasting unused and expensive computing equipment. The capital requirements necessary to manage the cycles of the business were as large as they were in the electric power business, which had gone through decades of chaos and turmoil before settling down. Only a few survived,[17] and even fewer, like

Tymshare of Cupertino, California, did well (although it was sold to McDonnell-Douglas in the early 1980s).[18] Among those many companies, one is worth mentioning, Computer Center Corporation, or C-Cubed, which installed one of the first PDP-10s in the Seattle area in 1968. While it was getting started, it offered a local teenager, Bill Gates, free time on the computer in exchange for helping find and rid the system of bugs. C-Cubed folded in 1970, having given Gates a taste of the potential of interactive computing.[19]

Many of those who had access to these systems saw the future of computing. But the financial troubles of time-sharing companies also showed that it would be difficult to make personal, interactive use widely available. There were attempts to make terminals accessible to the public for free or at low cost—the most famous being the Resource One project in the San Francisco Bay area (partially funded by the *Whole Earth Catalog*). But it did not last, either.[20]

Calculators and Corporate Personal Computer Projects

Economics prevented the spread of computing to the public from the top down—from large mainframes through time-shared terminals. But while those attempts were underway, the underlying technology was advancing rapidly. Could personal computing arrive from the bottom up—from advances in semiconductor electronics?

Many engineers believe that a mental model of the personal computer was irrelevant. They believe that no one invented the personal computer, it simply flowed from advances in semiconductors. Chuck House, an engineer involved with the early Hewlett-Packard calculators, said, "One could uncharitably say that we invented essentially nothing; we simply took all the ideas that were out there and figured out how to implement them cost-effectively." Gordon Bell stated, "The semiconductor density has really been the driving force, and as you reach different density levels, different machines pop out of that in time."[21] To them, inventions are like a piece of fruit that falls to the ground when it is ripe, and the inventor is given credit for doing little more than picking it up. If that were true, one would find a steady progression of machines offering personal, interactive use, as advances in semiconductors made them viable. And these would have come from established firms who had the engineering and manufacturing resources to translate those advances into products.

Products that took advantage of advances in semiconductors did appear on the market. It is worth looking at them to see whether they validate or refute the bottom-up explanation of the PC's invention.

The first electronic computers were of course, operated, as if they were personal computers. Once a person was granted access to a machine (after literally waiting in a queue), he or she had the whole computer to use, for whatever purpose. That gave way to more restricted access, but those at MIT and Lincoln Labs who used the Whirlwind, TX-0, and TX-2 that way never forgot its advantages. In 1962 some of them developed a computer called the LINC, made of Digital Equipment Corporation logic modules and intended for use by a researcher as a personal tool. A demonstration project, funded by the NIH, made sixteen LINCs available to biomedical researchers. DEC produced commercial versions, and by the late 1960s, about 1,200 were in use as personal computers. A key feature of the LINC was its compact tape drive and tapes that one could easily carry around: the forerunner of DECtape. The ease of getting at data on the tape was radically different from the clumsy access of tape in mainframes, and this ease would be repeated with the introduction of floppy-disk systems on personal computers.[22] DEC also marketed a computer that was a combination of a LINC and a PDP-8, for $43,000. Although DECtape soon was offered on nearly all DEC's products, the LINC did not achieve the same kind of commercial success as the PDP-8 and PDP-11 lines of minicomputers.[23]

Advances in chip density first made an impact on personal devices in calculators.[24] For decades there had been a small market for machines that could perform the four functions of arithmetic, plus square root. In the 1950s and 1960s the calculator industry was dominated by firms such as Friden and Marchant in the United States, and Odhner in Europe. Their products were complex, heavy, and expensive.[25] In 1964 Wang Laboratories, a company founded by An Wang, a Chinese immigrant who had worked with Howard Aiken at Harvard, came out with an electronic calculator. The Wang LOCI offered more functions, at a lower cost, than the best mechanical machines. Its successor, the Wang 300, was even easier to use and cheaper, partly because Wang deliberately set the price of the 300 to undercut the competitive mechanical calculators from Friden and others.[26] (Only one or two of the mechanical calculator firms survived the transition to electronics.) A few years later Hewlett-Packard, known for its oscilloscopes and electronic test equipment, came out with the HP-9100A, a calculator selling for just under $5,000. And the Italian firm Olivetti came out with the Programma 101, a $3,500 calculator intended primarily for accounting and statistical work. Besides direct calculation, these machines could also execute a short sequence of steps recorded on magnetic cards.[27] Like the LINC, these calculators used discrete circuits. To display digits, the Wang used "Nixie" tubes, an ingenious tube invented by Burroughs in 1957. HP used a small cathode-ray tube, as might be expected from a company that made oscilloscopes.

By 1970 the first of a line of dramatically cheaper and smaller calculators appeared that used integrated circuits.[28] They were about the size of a paperback book and cost as little as $400. A number of wealthy consumers bought them immediately, but it wasn't until Bowmar advertised a Bowmar Brain for less than $250 for the 1971 Christmas season that the calculator burst into public consciousness.[29] Prices plummeted: under $150 in 1972; under $100 by 1973; under $50 by 1976; finally they became cheap enough to be given away as promotional trinkets.[30] Meanwhile Hewlett-Packard stunned the market in early 1972 with the HP-35, a $400 pocket calculator that performed all the logarithmic and trigonometric functions required by engineers and scientists. Within a few years the slide rule joined the mechanical calculator on the shelves of museums.[31]

Like processed foods, whose cost is mostly in the packaging and marketing, so with calculators: technology no longer determined commercial success. Two Japanese firms with consumer marketing skills, Casio and Sharp, soon dominated. Thirty years after the completion of the half-million dollar ENIAC, digital devices became throw-away commodities. The pioneering calculator companies either stopped making calculators, as did Wang, or went bankrupt, as did Bowmar. Hewlett-Packard survived by concentrating on more advanced and expensive models; Texas Instruments survived by cutting costs.

The commodity prices make it easy to forget that these calculators were ingenious pieces of engineering. Some of them could store sequences of keystrokes in their memory and thus execute short programs. The first of the programmable pocket calculators was Hewlett-Packard's HP-65, introduced in early 1974 for $795 (figure 7.2). Texas Instruments and others soon followed. As powerful as they were, the trade press was hesitant to call them computers, even if Hewlett-Packard introduced the HP-65 as a "personal computer" (possibly the first use of that term in print).[32] Their limited programming was offset by their built-in ability to compute logarithms and trigonometric functions, and to use floating-point arithmetic to ten

Figure 7.2
HP-65. (*Source:* Smithsonian Institution.)

decimal digits of precision. Few mainframes could do that without custom-written software.

The introduction of pocket programmable calculators had several profound effects on the direction of computing technology. The first was that the calculator, like the Minuteman and Apollo programs of the 1960s, created a market where suppliers could count on a long production run, and thereby gain economies of scale and a low price. As chip density, and therefore capability, increased, chip manufacturers faced the same problem that Henry Ford had faced with his Model T: only long production runs of the same product led to low prices, but markets did not stay static. That was especially true of integrated circuits, which by nature became ever more specialized in their function as the levels of integration increased. (The only exception was in memory chips, which is one reason why Intel was founded to focus on memories.) The calculator offered the first consumer market for logic chips that allowed companies to amortize the high costs of designing complex integrated circuits. The dramatic drop in prices of calculators between 1971 and 1976 showed just how potent this force was.[33]

The second effect was just as important. Pocket calculators, especially those that were programmable, unleashed the force of personal creativity and energy of masses of individuals. This force had already created the hacker culture at MIT and Stanford (observed with trepidation by at least one MIT professor).[34] Their story is one of the more colorful among the dry technical narratives of hardware and software design. They and their accomplishments, suitably embellished, have become favorite topics of the popular press. Of course their strange personal habits made a good story, but were they true? Developing system software was hard work, not likely to be done well by a salaried employee, working normal hours and with a family to go home to in the evening. Time-sharing freed all users from the tyranny of submitting decks of cards and waiting for a printout, but it forced some users to work late at night, when the time-shared systems were lightly loaded and thus more responsive.

The assertion that hackers created modern interactive computing is about half-right. In sheer numbers there may never have been more than a few hundred people fortunate enough to be allowed to "hack" (that is, not do a programming job specified by one's employer) on a computer like the PDP-10. By 1975, there were over 25,000 HP-65 programmable calculators in use, each one owned by an individual who could do whatever he or she wished to with it.[35] Who were these people? HP-65 users were not "strange". Nearly all were adult professional men, including civil and electrical engineers, lawyers, financial people, pilots, and so on. Only a few were students (or professors), because they cost $795. Most purchased the HP-65 because they had a practical need for calculation in their jobs. But this was a *personal* machine—one could take it home at night. These users—perhaps 5 or 10 percent of those who owned machines—did not fit the popular notion of hackers as kids with "[t]heir rumpled clothes, their unwashed and unshaven faces, and their uncombed hair."[36] But their passion for programming made them the intellectual cousins of the students in the Tech Model Railroad Club. And their numbers—only to increase as the prices of calculators dropped—were the first indication that personal computing was truly a mass phenomenon.

Hewlett-Packard and Texas Instruments were unprepared for these events. They sold the machines as commodities; they could ill-afford a sales force that could walk a customer through the complex learning process needed to get the most out of one. That was what IBM salesmen were known for—but they sold multimillion dollar mainframes.

Calculators were designed to be easy enough to use to make that unnecessary, at least for basic tasks. What was unexpected was how much more some of those customers wanted to do. Finding little help from the supplier, they turned to one another. Users groups, clubs, newsletters, and publications proliferated.

This supporting infrastructure was critical to the success of personal computing; in the following decade it would become an industry all its own. Many histories of the personal computer emphasize this point; they often cite the role of the Homebrew Computer Club, which met near the Stanford campus in the mid-1970s, as especially important.[37] The calculator users groups were also important, though for different reasons. As the primitive first personal computers like the Altair gave way to more complete systems, a number of calculator owners purchased one of them as well. In the club newsletters there were continuous discussions of the advantages and drawbacks of each—the one machine having the ability to evaluate complex mathematical expressions with ease, the other more primitive but *potentially* capable of doing all that and more.[38] There was no such thing as a typical member of the Homebrew Computer Club, although calculator owners tended to be professionals whose jobs required calculation during the day, and who thought of other uses at night. Many of them were bitten by the PC bug; at the same time they took a show-me attitude toward the computer. Could you rely on one? Could you use one to design a radar antenna? Could it handle a medium-sized mailing list? Was the personal computer a serious machine? At first the answers were, "not yet," but gradually, with some firm prodding by this community, the balance shifted. Groups like the Homebrew Computer Club emphasized the "personal" in personal computer; calculator users emphasized the word computer.

Ever since time-sharing and minicomputers revealed an alternative to mainframe computing, there have been prophets and evangelists who raged against the world of punched cards and computer rooms, promising a digital paradise of truly interactive tools. The most famous was Ted Nelson, whose self-published book *Computer Lib* proclaimed (with a raised fist on the cover): "You can and must understand computers *now*."[39] By 1974 enough of these dreams had become real that the specific abilities—and limits—of actual "dream machines" (the alternate title to Nelson's book) had to be faced. Some of the dreamers, including Nelson, were unable to make the transition. They dismissed the pocket calculator. They thought it was puny, too cheap, couldn't do graphics, wasn't a "von Neumann machine," and so on.[40] For them, the dream machine was better, even if (or because) it was unbuilt.[41] By 1985 there would be millions of IBM Personal Computers and their copies in the offices and homes of ordinary people. These computers would use a processor that was developed for other purposes, and adapted for the personal computer almost by accident. But they would be real and a constant source of inspiration and creativity to many who used them, as well as an equal source of frustration for those who knew how much better they could be.

The Microprocessor

Calculators showed what integrated circuits could do, but they did not open up a direct avenue to personal interactive computing. The chips used in them were too specialized for numerical calculation to form a basis for a general-purpose computer. Their architecture was ad-hoc and closely guarded by each manufacturer. What was needed was a set of integrated circuits—or even a single integrated circuit—that incorporated the basic architecture of a general-purpose, stored-program computer.[42] Such a chip, called a "microprocessor," did appear.

In 1964 Gordon Moore, then of Fairchild and soon a cofounder of Intel, noted that from the time of the invention of integrated circuits in 1958, the number of circuits that one could place on a single integrated circuit was doubling every year.[43] By simply plotting this rate on a piece of semi-log graph paper, "Moore's Law" predicted that by the mid 1970s one could buy a chip containing logic circuits equivalent to those used in a 1950s-era mainframe. (Recall that the UNIVAC I had about 3,000 tubes, about the same number of active elements contained in the first microprocessor discussed below.) By the late 1960s transistor-transistor logic (TTL) was well established, but a new type of semiconductor called metal-oxide semiconductor (MOS), emerged as a way to place even more logic elements on a chip.[44] MOS was used by Intel to produce its pioneering 1103 memory chip, and it was a key to the success of pocket calculators. The chip density permitted by MOS brought the concept of a computer-on-a-chip into focus among engineers at Intel, Texas Instruments, and other semiconductor firms. That did not mean that such a device was perceived as useful. If it was generally known that enough transistors could be placed on a chip to make a computer, it was also generally believed that the market for such a chip was so low that its sales would never recoup the large development costs required.[45]

By 1971 the idea was realized in silicon. Several engineers deserve credit for the invention. Ted Hoff, an engineer at Intel, was responsible for the initial concept, Federico Faggin of Intel deserves credit for its realization in silicon, and Gary Boone of Texas Instruments designed similar circuits around that time. In 1990, years after the microprocessor became a household commodity and after years of litigation, Gil Hyatt, an independent inventor from La Palma, California, received a patent on it. Outside the courts he has few supporters, and recent court rulings may have invalidated his claim entirely.[46]

The story of the microprocessor's invention at Intel has been told many times.[47] In essence, it is a story encountered before: Intel was asked to design a special-purpose system for a customer. It found that by designing a general-purpose computer and using software to tailor it to the customer's needs, the product would have a larger market.

Intel's customer for this circuit was Busicom, a Japanese company that was a top seller of hand-held calculators. Busicom sought to produce a line of products with different capabilities, each aimed at a different market segment. It envisioned a set of custom-designed chips that incorporated the logic for the advanced mathematical functions. Intel's management assigned Marcian E. Hoff, who had joined the company in 1968 (Intel's twelfth employee), to work with Busicom.

Intel's focus had always been on semiconductor memory chips. It had shied away from logic chips like those suggested by Busicom, since it felt that markets for them were limited. Hoff's insight was to recognize that by designing fewer logic chips with more general capabilities, one could satisfy Busicom's needs elegantly. Hoff was inspired by the PDP-8, which had a very small set of instructions, but which its thousands of users had programmed to do a variety of things. He also recalled using an IBM 1620, a small scientific computer with an extremely limited instruction set that nevertheless could be programmed to do a lot of useful work.

Hoff proposed a logic chip that incorporated more of the concepts of a general-purpose computer (figure 7.3). A critical feature was the ability to call up a subroutine, execute it, and return to the main program as needed.[48] He proposed to do that with a register that kept track of where a program was in its execution and saved that status when interrupted to perform a subroutine. Subroutines themselves could be interrupted, with return addresses stored on a "stack": an arrangement of registers that automatically retrieved data on a last-in-first-out basis.[49]

With this ability, the chip could carry out complex operations stored as subroutines in memory, and avoid having those functions perma-

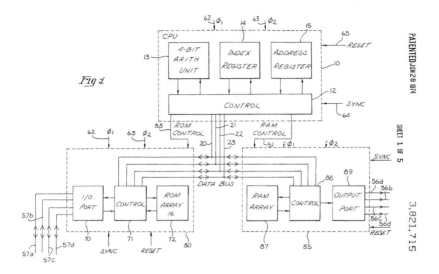

Figure 7.3
(*top*) Patent for a "Memory System for a Multi-Chip Digital Computer," by M. E. Hoff, Stanley Mazor, and Federico Faggin of Intel. The patent was not specifically for a "computer on a chip," but note that all the functional blocks found in the processor of a stored-program computer are shown in this drawing. (*bottom*) Intel 8080. (*Source:* Smithsonian Institution.)

nently wired onto the chip. Doing it Hoff's way would be slower, but in a calculator that did not matter, since a person could not press keys that fast anyway. The complexity of the logic would now reside in software stored in the memory chips, so one was not getting something for nothing. But Intel was a memory company, and it knew that it could provide memory chips with enough capacity. As an added inducement, sales of the logic chips would mean more sales of its bread-and-butter memories.

That flexibility meant that the set of chips could be used for many other applications besides calculators. Busicom was in a highly competitive and volatile market, and Intel recognized that. (Busicom eventually went bankrupt.) Robert Noyce negotiated a deal with Busicom to provide it with chips at a lower cost, giving Intel in return the right to market the chips to other customers for noncalculator applications. From these unsophisticated negotiations with Busicom, in Noyce's words, came a pivotal moment in the history of computing.[50]

The result was a set of four chips, first advertised in a trade journal in late 1971, which included "a microprogrammable computer on a chip!"[51] That was the 4004, on which one found all the basic registers and control functions of a tiny, general-purpose stored-program computer. The other chips contained a read-only memory (ROM), random-access memory (RAM), and a chip to handle output functions. The 4004 became the historical milestone, but the other chips were important as well, especially the ROM chip that supplied the code that turned a general-purpose processor into something that could meet a customer's needs. (Also at Intel, a team led by Dov Frohman developed a ROM chip that could be easily reprogrammed and erased by exposure to ultraviolet light. Called an EPROM (erasable programmable read-only memory) and introduced in 1971, it made the concept of system design using a microprocessor practical.)[52]

The detailed design of the 4004 was done by Stan Mazor. Federico Faggin was also crucial in making the concept practical. Masatoshi Shima, a representative from Busicom, also contributed. Many histories of the invention give Hoff sole credit; all players, including Hoff, now agree that that is not accurate. Faggin left Intel in 1974 to found a rival company, Zilog. Intel, in competition with Zilog, felt no need to advertise Faggin's talents in its promotional literature, although Intel never showed any outward hostility to its ex-employee.[53] The issue of whom to credit reveals the way many people think of invention: Hoff had the idea of putting a general-purpose computer on a chip, Faggin and the others "merely" implemented that idea in silicon. At the time, Intel was not sure what it had invented either: Intel's patent attorney resisted Hoff's desire at the time to patent the work as a "computer."[54] Intel obtained two patents on the 4004, covering its architecture and implementation; Hoff's name appears on only one of them. (That opened the door to rival claims for patent royalties from TI, and eventually Gil Hyatt.)

The 4004 worked with groups of four bits at a time—enough to code decimal digits but no more. At almost the same time as the work with

Busicom, Intel entered into a similar agreement with Computer Terminal Corporation (later called Datapoint) of San Antonio, Texas, to produce a set of chips for a terminal to be attached to mainframe computers. Again, Mazor and Hoff proposed a microprocessor to handle the terminal's logic. Their proposed chip would handle data in 8-bit chunks, enough to process a full byte at a time. By the time Intel had completed its design, Datapoint had decided to go with conventional TTL chips. Intel offered the chip, which they called the 8008, as a commercial product in April 1972.[55]

In late 1972, a 4-bit microprocessor was offered by Rockwell, an automotive company that had merged with North American Aviation, maker of the Minuteman Guidance System. In 1973 a half dozen other companies began offering microprocessors as well. Intel responded to the competition in April 1974 by announcing the 8080, an 8-bit chip that could address much more memory and required fewer support chips than the 8008. The company set the price at $360—a somewhat arbitrary figure, as Intel had no experience selling chips like these one at a time. (Folklore has it that the $360 price was set to suggest a comparison with the IBM System/360.)[56] A significant advance over the 8008, the 8080 could execute programs written for the other chip, a compatibility that would prove crucial to Intel's dominance of the market. The 8080 was the first of the microprocessors whose instruction set and memory addressing capability approached those of the minicomputers of the day.[57]

From Microprocessor to Personal Computer

There were now, in early 1974, two converging forces at work. From one direction were the semiconductor engineers with their ever-more-powerful microprocessors and ever-more-capacious memory chips. From the other direction were users of time-sharing systems, who saw a PDP-10 or XDS 940 as a basis for public access to computing. When these forces met in the middle, they would bring about a revolution in personal computing.

They almost did not meet. For the two years between Brand's observation and the appearance of the Altair, the two forces were rushing past one another. The time-sharing systems had trouble making money even from industrial clients, and the public systems like Community Memory were also struggling. At the other end, semiconductor companies did not think of their products as a possible basis for a personal computer.

A general-purpose computer based on a microprocessor did appear in 1973. In May of that year Thi T. Truong, an immigrant to France from Viet Nam, had his electronics company design and build a computer based on the Intel 8008 microprocessor. The MICRAL was a rugged and well-designed computer, with a bus architecture and internal slots on its circuit board for expansion. A base model cost under $2,000, and it found a market replacing minicomputers for simple control operations. Around two thousand were sold in the next two years, none of them beyond an industrial market.[58] It is regarded as the first microprocessor-based computer to be sold in the commercial marketplace. Because of the limitations of the 8008, its location in France, and above all, the failure by its creators to see what it "really" was, it never broke out of its niche as a replacement for minicomputers in limited industrial locations.

The perception of the MICRAL as something to replace the mini was echoed at Intel as well. Intel's mental model of its product was this: an *industrial* customer bought an 8080 and wrote specialized software for it, which was then burned into a read-only-memory to give a system with the desired functions. The resulting inexpensive product (no longer programmable) was then put on the market as an embedded controller in an industrial system. A major reason for that mental model was the understanding of how hard it was to program a microprocessor. It seemed absurd to ask untrained consumers to program when Intel's traditional customers, hardware designers, were themselves uncomfortable with programming.

With these embedded uses in mind, microprocessor suppliers developed educational packages intended to ease customers into system design. These kits included the microprocessor, some RAM and ROM chips, and some other chips that handled timing and control, all mounted on a printed circuit board. They also included written material that gave a tutorial on how to program the system. This effort took Intel far from its core business of making chips, but the company hoped to recoup the current losses later on with volume sales of components.[59] These kits were sold for around $200 or given away to engineers who might later generate volume sales.

Intel and the others also built more sophisticated "Development Systems," on which a customer could actually test the software for an application (figure 7.4). These were fully assembled products that sold

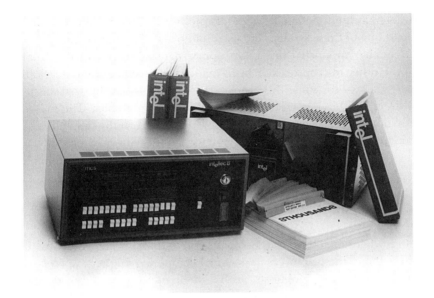

Figure 7.4
Intellec-8 Development System. This was, in fact, a general-purpose computer, but Intel did not market it as such. Intel intended that customers buy them to assist in writing and debugging microprocessor software that would go into embedded systems. A few were purchased and used as alternatives to minicomputers. (*Source:* Intel.)

for around $10,000. To use these systems, customers also needed specialized software that would allow them to write programs using a language like FORTRAN, and then "cross-compile" it for the microprocessor—that is, from the FORTRAN program generate machine code, not for the computer on which it was written, but for the microprocessor. The company hired Gary Kildall, an instructor at the Naval Postgraduate School in Monterey, California, to develop a language based on IBM's PL/I.[60] He called it PL/M, and in 1973 Intel offered it to customers. Initially this software was intended to be run on a large mainframe, but it was soon available for minicomputers, and finally to microprocessor-based systems. In 1974 Intel offered a development system, the Intellec 4, which included its own resident PL/M compiler (i.e., one did not need a mainframe or a mini to compile the code).[61] A similar Intellec-8 introduced the 8-bit microprocessors.

With these development systems, Intel had in fact invented a personal computer. But the company did not realize it. These kits were not marketed as the functional computers they were. Occasionally someone bought one of these systems and used it in place of a minicomputer, but Intel neither supported that effort nor recognized its potential.[62] Intel and the other microprocessor firms made money selling these development systems—for some they were very profitable—but the goal was to use them as a lever to open up volume purchases of chips. The public could not buy one. The chip suppliers were focused on the difficulties in getting embedded systems to do useful work; they did not think that the public would be willing to put up with the difficulties of programming just to own their own computer.

Role of Hobbyists

Here is where the electronics hobbyists and enthusiasts come in. Were it not for them, the two forces in personal computing might have crossed without converging. Hobbyists, at that moment, were willing to do the work needed to make microprocessor-based systems practical.

This community had a long history of technical innovation—it was radio amateurs, for example, who opened up the high-frequency radio spectrum for long-distance radio communications after World War I. After World War II, the hobby expanded beyond amateur radio to include high-fidelity music reproduction, automatic controls, and simple robotics. A cornucopia of war surplus equipment from the U.S. Army Signal Corps found its way into individual hands, further fueling the phenomenon. (A block in lower Manhattan known as "Radio Row," where the World Trade Center now stands, was a famous source of surplus electronic gear.)[63] The shift from vacuum tubes to integrated circuits made it harder for an individual to build a circuit on a breadboard at home, but inexpensive TTL chips now contained whole circuits themselves.[64] As the hobby evolved rapidly from analog to digital applications, this group supplied a key component in creating the personal computer: It provided an infrastructure of support that neither the computer companies nor the chip makers could.

This infrastructure included a variety of electronics magazines. Some were aimed at particular segments, for example, *QST* for radio amateurs. Two of them, *Popular Electronics* and *Radio-Electronics*, were of general interest and sold at newsstands; they covered high-fidelity audio, shortwave radio, television, and assorted gadgets for the home and car. Each issue typically had at least one construction project. For these projects the magazine would make arrangements with small electronics companies to supply a printed circuit board, already etched and drilled, as well as specialized components that readers might have difficulty finding locally. By scanning the back issues of these magazines we can trace how hobbyists moved from analog to digital designs.

A machine called the Kenbak-1, made of medium and small-scale integrated circuits, was advertised in the September 1971 issue of *Scientific American.* The advertisement called it suitable for "private individuals," but it was really intended for schools. The Kenbak may be the first personal computer, but it did not use a microprocessor, and its capabilities were quite limited.

The Scelbi-8H was announced in a tiny advertisement in the back of the March 1974 issue of *QST.* It used an Intel 8008, and thus may be the first microprocessor-based computer marketed to the public. According to the advertisement, "Kit prices for the new Scelbi-8H mini-computer start as low as $440!"[65] It is not known how many machines Scelbi sold, but the company went on to play an important part in the early personal computer phenomenon.[66]

In July 1974, *Radio-Electronics* announced a kit based on the Intel 8008, under the headline "Build the Mark-8: Your Personal Minicomputer."[67] The project was much more ambitious than what typically appeared in that magazine. The article gave only a simple description and asked readers to order a separate, $5.00 booklet for detailed instructions. The Mark-8 was designed by Jonathan Titus of Virginia Polytechnic University in Blacksburg. The number of machines actually built may range in the hundreds, although the magazine reportedly sold "thousands" of booklets. At least one Mark-8 users club sprang up, in Denver, whose members designed an ingenious method of storing programs on an audio cassette recorder.[68] Readers were directed to a company in Englewood, New Jersey, that supplied a set of circuit boards for $47.50, and to Intel for the 8008 chip (for $120.00). The Mark-8's appearance in *Radio-Electronics* was a strong factor in the decision by its rival *Popular Electronics* to introduce the Altair kit six months later.[69]

These kits were just a few of many projects described in the hobbyist magazines. They reflected a conscious effort by the community to bring digital electronics, with all its promise and complexity, to amateurs who were familiar only with simpler radio or audio equipment. It was not an easy transition: construction of both the Mark-8 and the TV-typewriter (described next) was too complex to be described in a magazine article; readers had to order a separate booklet to get complete plans. *Radio-Electronics* explained to its readers that "[w]e do not intend to do an article this way as a regular practice."[70] Although digital circuits were more complex than what the magazine had been handling, it recognized that the electronics world was moving in that direction and that its readers wanted such projects.

Other articles described simpler digital devices—timers, games, clocks, keyboards, and measuring instruments—that used inexpensive TTL chips. One influential project was the TV-Typewriter, designed by Don Lancaster and published in *Radio-Electronics* in September 1973. This device allowed readers to display alphanumeric characters, encoded in ASCII, on an ordinary television set. It presaged the advent of CRT terminals as the primary input-output device for personal computers—one major distinction between the PC culture and that of the minicomputer, which relied on the Teletype. Lee Felsenstein called the TV-Typewriter "the opening shot of the computer revolution."[71]

Altair

1974 was the *annus mirabilis* of personal computing. In January, Hewlett-Packard introduced its HP-65 programmable calculator. That summer Intel announced the 8080 microprocessor. In July, *Radio-Electronics* described the Mark-8. In late December, subscribers to *Popular Electronics* received their January 1975 issue in the mail, with a prototype of the "Altair" minicomputer on the cover (figure 7.5), and an article describing how readers could obtain one for less than $400. This announcement ranks with IBM's announcement of the System/360 a decade earlier as one of the most significant in the history of computing. But what a difference a decade made: the Altair was a genuine personal computer.

H. Edward Roberts, the Altair's designer, deserves credit as the inventor of the personal computer. The Altair was a capable, inexpensive computer designed around the Intel 8080 microprocessor. Although calling Roberts the inventor makes sense only in the context of all that came before him, including the crucial steps described above, he does deserve the credit. Mark Twain said that historians have to rearrange past events so they make more sense. If so, the invention of the personal computer at a small model-rocket hobby shop in Albuquerque cries out for some creative rearrangement. Its utter improbability and unpredictability have led some to credit many other places with the invention, places that are more sensible, such as the Xerox Palo Alto Research Center, or Digital Equipment Corporation, or even IBM. But Albuquer-

Figure 7.5
MITS Altair 8800 Computer. The front panel was copied from the Data General Nova. The machine shown in this photograph was one of the first produced and was owned by Forrest Mims, an electronics hobbyist and frequent contributor to *Popular Electronics*, who had briefly worked at MITS. (*Source:* Smithsonian Institution.)

que it was, for it was only at MITS that the technical and social components of personal computing converged.

Consider first the technical. None of the other hobbyist projects had the impact of the Altair's announcement. Why? One reason was that it was designed and promoted as a capable minicomputer, as powerful as those offered by DEC or Data General. The magazine article, written by Ed Roberts and William Yates, makes this point over and over: "a full-blown computer than can hold its own against sophisticated minicomputers"; "not a 'demonstrator' or a souped-up calculator"; or "performance competes with current commercial minicomputers."[72] The physical appearance of the Altair computer suggested its minicomputer lineage. It looked like the Data General Nova: it had a rectangular metal case, a front panel of switches that controlled the contents of internal registers, and small lights indicating the presence of a binary one or zero. Inside the Altair's case, there was a machine built mainly of TTL integrated circuits (except for the microprocessor, which was a MOS device), packaged in dual-in-line packages, soldered onto circuit boards. Signals and power traveled from one part of the machine to another on a bus. The Altair used integrated circuits, not magnetic cores, for its primary memory. The *Popular Electronics* cover called the Altair the "world's first minicomputer kit"; except for its use of a microprocessor, that accurately described its physical construction and design.[73]

But the Altair as advertised was ten times cheaper than minicomputers were in 1975. The magazine offered an Altair for under $400 as a kit, and a few hundred more already assembled. The magazine cover said that readers could "save over $1,000." In fact, the cheapest PDP-8 cost several thousand dollars. Of course, a PDP-8 was a fully assembled, operating computer that was considerably more capable than the basic Altair, but that did not really matter in this case. (Just what one got for $400 will be discussed later.) The low cost resulted mainly from its use of the Intel 8080 microprocessor, just introduced. Intel had quoted a price of $360 for small quantities of 8080s, but Intel's quote was not based on a careful analysis of how to sell the 8080 to this market. MITS bought them for only $75 each.[74]

The 8080 had more instructions and was faster and more capable than the 8008 that the Mark-8 and Scelbi-8 used. It also permitted a simpler design since it required only six instead of twenty supporting chips to make a functional system. Other improvements over the 8008 were its ability to address up to 64 thousand bytes of memory (vs. the 8008's 16 thousand), and its use of main memory for the stack, which permitted essentially unlimited levels of subroutines instead of the 8008's seven levels.

The 8080 processor was only one architectural advantage the Altair had over its predecessors. Just as important was its use of an open bus.[75] According to folklore, the bus architecture almost did not happen. After building the prototype Altair, Roberts photographed it and shipped it via Railway Express to the offices of *Popular Electronics* in New York. Railway Express, a vestige of an earlier American industrial revolution, was about to go bankrupt; it lost the package. The magazine cover issue showed the prototype, with its light-colored front panel and the words "Altair 8800" on the upper left. That machine had a set of four large circuit boards stacked on top of one another, with a wide ribbon cable carrying 100 lines from one board to another. After that machine was lost, Robert redesigned the Altair. He switched to a larger deep blue cabinet and discarded the 100-wire ribbon cable. In the new design, wires connected to a rigid backplane carried the signals from one board to another. That allowed hobbyists to add a set of connectors that could accept other cards besides the initial four.[76]

The $400 kit came with only two cards to plug into the bus: those two, plus a circuit board to control the front panel and the power supply, made up the whole computer. The inside looked quite bare. But laboriously soldering a set of wires to an expansion chassis created a full set of slots into which a lot of cards could be plugged. MITS was already designing cards for more memory, I/O and other functions.

Following the tradition established by Digital Equipment Corporation, Roberts did not hold specifications of the bus as a company secret. That allowed others to design and market cards for the Altair. That decision was as important to the Altair's success as its choice of an 8080 processor. It also explains one of the great ironies of the Altair, that it inaugurated the PC era although it was neither reliable nor very well-designed. Had it not been possible for other companies to offer plug-in cards that improved on the original MITS design, the Altair might have made no greater impact than the Mark-8 had. The bus architecture also led to the company's demise a few years later, since it allowed other companies to market compatible cards and, later, compatible computers. But by then the floodgates had opened. If MITS was unable to deliver on its promises of making the Altair a serious machine (though it tried), other companies would step in. MITS continued developing plug-in cards and peripheral equipment, but the flood of orders was too much for the small company.

So while it was true that for $400 hobbyists got very little, they could get the rest—or design and build the rest. Marketing the computer as a bare-bones kit offered a way for thousands of people to bootstrap their way into the computer age, at a pace that they, not a computer company, could control.

Assembling the Altair was much more difficult than assembling other electronics kits, such as those sold by the Heath Company or Dynaco. MITS offered to sell "completely assembled and tested" computers for $498, but with such a backlog of orders, readers were faced with the choice of ordering the kit and getting something in a couple of months, or ordering the assembled computer and perhaps waiting a year or more.[77] Most ordered the kit and looked to one another for support in finding the inevitable wiring errors and poorly soldered connections that they would make. The audience of electronics hobbyists, at whom the magazine article was aimed, compared the Altair not to the simple Heathkits, but to building a computer from scratch, which was almost impossible: not only was it hard to design a computer, it was impossible to obtain the necessary chips. Chips were inexpensive, but only if they were purchased in large quantities, and anyway, most semiconductor firms had no distribution channels set up for single unit or retail sales. Partly because of this, customers felt, rightly, that they were getting an incredible bargain.

The limited capabilities of the basic Altair, plus the loss of the only existing Altair by the time the *Popular Electronics* article appeared, led to the notion that it was a sham, a "humbug," not a serious product at all.[78] The creators of the Altair fully intended to deliver a serious computer whose capabilities were on a par with minicomputers then on the market. Making those deliveries proved to be a lot harder than they anticipated. Fortunately, hobbyists understood that. But there should be no mistake about it: the Altair was real.

MITS and the editors of *Popular Electronics* had found a way to bring the dramatic advances in integrated circuits to individuals. The first customers were hobbyists, and the first thing they did with these machines, once they got them running, was play games.[79] Roberts was trying to sell it as a machine for serious work, however. In the *Popular Electronics* article he proposed a list of twenty-three applications, none of them games.[80] Because it was several years before anyone could supply peripheral equipment, memory, and software, serious applications were rare at first. That, combined with the primitive capabilities of other machines like the Mark-8, led again to an assumption that the Altair was not a serious computer. Many of the proposed applications hinted at in the 1975 article were eventually implemented. Years later one could still find an occasional Altair (or more frequently, an Altair clone) embedded into a system just like its minicomputer cousins.

The next three years, from January 1975 through the end of 1977, saw a burst of energy and creativity in computing that had almost no equal in its history. The Altair had opened the floodgates, even though its shortcomings were clear to everyone. One could do little more than get it to blink a pattern of lights on the front panel. And even that was not easy: one had to flick the toggle switches for each program step, then deposit that number into a memory location, then repeat that for the next step, and so on—hopefully the power did not go off while this was going on—until the whole program (less than 256 bytes long!) was in memory. Bruised fingers from flipping the small toggle switches were the least of the frustrations. In spite of all that, the bus architecture meant that other companies could design boards to remedy each of these shortcomings, or even design a copy of the Altair itself, as IMSAI and others did.[81]

But the people at MITS and their hangers-on created more than just a computer. This $400 computer inspired the extensive support of user groups, informal newsletters, commercial magazines, local clubs, conventions, and even retail stores. This social activity went far beyond traditional computer user groups, like SHARE for IBM or DECUS for Digital. Like the calculator users groups, these were open and informal, and offered more to the neophyte. All of this sprang up with the Altair, and many of the publications and groups lived long after the last Altair computer itself was sold.

Other companies, beginning with Processor Technology, soon began offering plug-in boards that gave the machine more memory. Another board provided a way of connecting the machine to a Teletype, which allowed fingers to heal. But Teletypes were not easy to come by—an individual not affiliated with a corporation or university could only buy one secondhand, and even then they were expensive. Before long, hobbyists-led small companies began offering ways of hooking up a television set and a keyboard (although Don Lancaster's TV Typewriter was not the design these followed). The board that connected to the Teletype sent data serially—one bit at a time; another board was designed that sent out data in parallel, for connection to a line printer that minicomputers used, although like the Teletype these were expensive and hard to come by.[82]

The Altair lost its data when the power was shut off, but before long MITS designed an interface that put out data as audio tones, to store programs on cheap audio cassettes. A group of hobbyists met in Kansas City in late 1975 and established a "Kansas City Standard" for the audio tones stored on cassettes, so that programs could be exchanged from one computer to another.[83] Some companies brought out inexpensive paper tape readers that did not require the purchase of a Teletype. Others developed a tape cartridge like the old 8-track audio systems, which looped a piece of tape around and around. Cassette storage was slow and cumbersome—users usually had to record several copies of a program and make several tries before successfully loading it into the computer. Inadequate mass storage limited the spread of PCs until the "floppy" disk was adapted.

The floppy was invented by David L. Noble at IBM for a completely different purpose. When IBM introduced the System/370, which used semiconductor memory, it needed a way to store the computer's initial control program, as well as to hold the machine's microprogram. That had not been a problem for the System/360, which used magnetic cores that held their contents when the power was switched off. From this need came the 8-inch diameter flexible diskette, which IBM announced in 1971.[84] Before long, people recognized that it could be used for other purposes besides the somewhat limited one for which it had been invented. In particular, Alan Shugart, who had once worked for IBM, recognized that the floppy's simplicity and low cost made it the ideal storage medium for low-cost computer systems.[85] Nevertheless, floppy drives were rare in the first few years of personal computing. IBM's hardware innovation was not enough; there had to be an equivalent innovation in system software to make the floppy practical. Before that story is told, we shall look first at the more immediate issue of developing a high-level language for the PC.

Software: BASIC

The lack of a practical mass storage device was one of two barriers that blocked the spread of personal, interactive computing. The other was a way to write applications software. By 1977 two remarkable and influential pieces of software—Microsoft BASIC and the CP/M Operating System—overcame those barriers.

In creating the Altair, Ed Roberts had to make a number of choices: what processor to use, the design of the bus (even whether to use a bus at all), the packaging, and so on. One such decision was the choice of a programming language. Given the wide acceptance of BASIC it is hard to imagine that there ever was a choice, but there was. BASIC was not invented for small computers. The creators of BASIC abhorred the changes others made to shoehorn the language onto systems smaller than a mainframe. Even in its mainframe version, BASIC had severe limitations—on the numbers and types of variables it allowed, for example. In the view of academic computer scientists, the versions of BASIC developed for minicomputers were even worse—full of ad hoc patches and modifications. Many professors disparaged BASIC as a toy language that fostered poor programming habits, and they refused to teach it. Serious programming was done in FORTRAN—an old and venerable but still capable language.

If, in 1974, one asked for a modern, concise, well-designed language to replace FORTRAN, the answer might have been APL, an interactive language invented at IBM by Kenneth Iverson in the early 1960s. A team within IBM designed a personal computer in 1973 that supported APL, the "SCAMP," although a commercial version of that computer sold poorly.[86] Or PL/I: IBM had thrown its resources into this language, which it hoped would replace both FORTRAN and COBOL. Gary Kildall chose a subset of PL/I for the Intel microprocessor development kit.

BASIC's strength was that it was easy to learn. More significant, it already had a track record of running on computers with limited memory. Roberts stated that he had considered FORTRAN and APL, before he decided the Altair was to have BASIC.[87]

William Gates III was born in 1955, at a time when work on FORTRAN was just underway. He was a student at Harvard when the famous cover of *Popular Electronics* appeared describing the Altair. According to one biographer, his friend Paul Allen saw the magazine and showed it to Gates, and the two immediately decided that they would write a BASIC compiler for the machine.[88] Whether it was Gates's or Roberts's decision to go with BASIC for the Altair, BASIC it was (figure 7.6).

In a newsletter sent out to Altair customers, Gates and Allen stated that a version of BASIC that required only 4K bytes of memory would be available in June 1975, and that more powerful versions would be available soon after. The cost, for those who also purchased Altair memory boards, was $60 for 4K BASIC, $75 for 8K, and $150 for "extended" BASIC (requiring disk or other mass storage). Those who wanted the language to run on another 8080-based system had to pay $500.[89]

In a burst of energy, Gates and Allen, with the help of Monte Davidoff, wrote not only a BASIC that fit into very little memory; they wrote a BASIC with a lot of features and impressive performance. The language was true to its Dartmouth roots in that it was easy to learn. It broke with those roots by providing a way to move from BASIC commands to instructions written in machine language. That was primarily through a USR command, which was borrowed from software written for DEC minicomputers (where the acronym stood for user service routine).[90] A programmer could even directly put bytes into or pull data out of specific memory locations, through the PEEK and POKE commands— which would have caused havoc on the time-shared Dartmouth system. Like USR, these commands were also derived from prior work done by DEC programmers, who came up with them for a time-sharing system they wrote in BASIC for the PDP-11. Those commands allowed users to

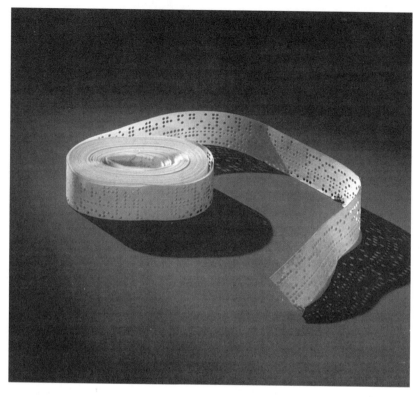

Figure 7.6
Paper tape containing BASIC, version 1.1, from the Smithsonian Collections. According to a letter by Bill Gates in the December 1975 issue of the Altair Users Group newsletter, *Computer Notes*: "If anyone is using BASIC version 1.1, you have a copy of a tape that was stolen back in March. No customers were ever shipped 1.1, as it was experimental and is full of bugs!" (*Source:* Smithsonian Institution.)

pass from BASIC to machine language easily—a crucial feature for getting a small system to do useful work.

These extensions kept their BASIC within its memory constraints while giving it the performance of a more sophisticated language. Yet it remained an interactive, conversational language that novices could learn and use. The BASIC they wrote for the Altair, with its skillful combination of features taken from Dartmouth and from the Digital Equipment Corporation, was the key to Gates's and Allen's success in establishing a personal computer software industry.

The developers of this language were not formally trained in computer science or mathematics as were Kemeny and Kurtz. They were introduced to computing in a somewhat different way. Bill Gates's private school in Seattle had a General Electric time-sharing system available for its pupils in 1968, a time when few students even in universities had such access. Later on he had access to an even better time-shared system: a PDP-10 owned by the Computer Center Corporation. Later still, he worked with a system of PDP-10s and PDP-11s used to control hydroelectric power for the Bonneville Power Administration. One of his mentors at Bonneville Power was John Norton, a TRW employee who had worked on the Apollo Program and who was a legend among programmers for the quality of his work.[91]

When he was writing BASIC for the Altair, Gates was at Harvard. He did not have access to an 8080-based system, but he did have access to a PDP-10 at Harvard's computing center (named after Howard Aiken). He and fellow student Monte Davidoff used the PDP-10 to write the language, based on the written specifications of the Intel 8080. In early 1975 Paul Allen flew to Albuquerque and demonstrated it to Roberts and Yates. It worked. Soon after, MITS advertised its availability for the Altair. Others were also writing BASIC interpreters for the Altair and for the other small computers now flooding the market, but none was as good as Gates's and Allen's, and it was not long before word of that got around.

It seemed that Roberts and his company had made one brilliant decision after another: the 8080 processor, the bus architecture, and now BASIC. However, by late 1975 Gates and Allen were not seeing it that way. Gates insists that he never became a MITS employee (although Allen was until 1976), and that under the name "Micro Soft," later "Micro-Soft," he and Allen retained the rights to their BASIC.[92] In a now-legendary "Open Letter to Hobbyists," distributed in early 1976, Gates complained about people making illicit copies of his BASIC by duplicating the paper tape. Gates claimed "the value of the computer time we have used [to develop the language] exceeds $40,000." He said that if he and his programmers were not paid, they would have little incentive to develop more software for personal computers, such as an APL language for the 8080 processor. He argued that illicit copying put all personal computing at risk: "Nothing would please me more than to hire ten programmers and deluge the hobby market with good software."[93]

Gates did his initial work on the PDP-10 while still an undergraduate at Harvard. Students were not to use that computer for commercial purposes, although these distinctions were not as clear then as they would be later. The language itself was the invention of Kemeney and Kurtz of Dartmouth; the extensions that were crucial to its success came from programmers at the Digital Equipment Corporation, especially Mark Bramhall, who led the effort to develop a time-sharing system (RSTS-11) for the PDP-11. Digital, the only commercial entity among the above group, did not think of its software as a commodity to sell; it was what the company did to get people to buy hardware.[94]

Bill Gates had recognized what Roberts and all the others had not: that with the advent of cheap, personal computers, software could and should come to the fore as the principal driving agent in computing. And only by charging money for it—even though it had originally been free—could that happen. By 1978 his company, now called "Microsoft," had severed its relationship with MITS and was moving from Albuquerque to the Seattle suburb of Bellvue. (MITS itself had lost its identity, having been bought by Pertec in 1977.) Computers were indeed coming to "the people," as Stewart Brand had predicted in 1972. But the driving force was not the counterculture vision of a Utopia of shared and free information; it was the force of the marketplace. Gates made good on his promise to "hire ten programmers and deluge the...market" (figure 7.7).

System Software: The Final Piece of the Puzzle

Gary Kildall's entree into personal computing software was as a consultant for Intel, where he developed languages for system development. While doing that he recognized that the floppy disk would make a good mass storage device for small systems, if it could be properly adapted. To do that he wrote a small program that managed the flow of information to and from a floppy disk drive. As with the selection of BASIC, it appears in hindsight to be obvious and inevitable that the floppy disk would be the personal computer's mass storage medium. That ignores the fact that it was never intended for that use. As with the adaptation of BASIC, the floppy had to be recast into a new role. As with BASIC, doing that took the work of a number of individuals, but the primary effort came from one man, Gary Kildall.

A disk had several advantages over magnetic or paper tape. For one, it was faster. For another, users could both read and write data on it. Its primary advantage was that a disk had "random" access: Users did not have to run through the entire spool of tape to get at a specific piece of data. To accomplish this, however, required tricky programming—some-

Figure 7.7
Microsoft Team, ca. 1978. This photograph shows Microsoft as it was moving from Albuquerque, where the Altair was built, to the Seattle area, where Bill Gates (lower left) and Paul Allen (lower right) were from. It was still a small company that focused mainly on supplying programming languages for personal computers. (*Source:* Microsoft.)

thing IBM had called, for one of its mainframe systems, a Disk Operating System, or DOS.[95]

A personal computer DOS had little to do with mainframe operating systems. There was no need to schedule and coordinate the jobs of many users: an Altair had one user. There was no need to "spool" or otherwise direct data to a roomful of chain printers, card punches, and tape drives: a personal computer had only a couple of ports to worry about. What *was* needed was rapid and accurate storage and retrieval of files from a floppy disk. A typical file would, in fact, be stored as a set of fragments, inserted at whatever free spaces were available on the disk. It was the job of the operating system to find those free spaces, put data there, retrieve it later on, and reassemble the fragments. All that gave the user an illusion that the disk was just like a traditional file cabinet filled with folders containing paper files.

Once again, Digital Equipment Corporation was the pioneer, in part because of its culture; because of the experience many of its employees had had with the TX-0 at MIT, one of the first computers to have a conversational, interactive feel to it. For its early systems DEC introduced DECtape, which although a tape, allowed programmers rapid access to data written in the middle, as well as at the ends, of the reel.[96] The PDP-10s had powerful DECtape as well as disk storage abilities; its operating systems were crucial in creating the illusion of personal computing that had so impressed observers like Stewart Brand.

In the late 1960s DEC produced OS/8 for the PDP-8, which had the feel of the PDP-10 but ran on a machine with very limited memory. OS-8 opened everyone's eyes at DEC; it showed that small computers could have capabilities as sophisticated as mainframes, without the bloat that characterized mainframe system software. Advanced versions of the PDP-11 had an operating system called RT-11 (offered in 1974), which was similar to OS/8, and which further refined the concept of managing data on disks.[97] These were the roots of personal computer operating systems. DEC's role in creating this software ranks with its invention of the minicomputer as major contributions to the creation of personal computing.

Gary Kildall developed PL/M for the Intel 8080. He used an IBM System/360, and PL/M was similar to IBM's PL/I. While working on that project Kildall wrote a small control program for the mainframe's disk drive. "It turned out that the operating system, which was called CP/M for Control Program for Micros, was useful, too, fortunately."[98] Kildall said that PL/M was "the base for CP/M," even though the commands were clearly derived from Digital's, not IBM's software.[99] For example, specifying the drive in use by a letter; giving file names a period and three-character extension; and using the DIR (Directory) command, PIP, and DDT were DEC features carried over without change.[100] CP/M was announced to hobbyists as "similar to DECSYSTEM 10" in an article by Jim Warren in *Dr. Dobb's Journal of Computer Calisthenics and Orthodontia* [sic] in April 1976. Warren was excited by CP/M, stating that it was "well designed, based on an easy-to-use operating system that has been around for a DECade.[sic]"[101] Suggested prices were well under $100, with a complete floppy system that included a drive and a controller for around $800—not cheap, but clearly superior to the alternatives of cassette, paper tape, or any other form of tape. CP/M was the final piece of the puzzle that, when made available, made personal computers a practical reality.

Gary Kildall and his wife, Dorothy McEwen, eased themselves into the commercial software business while he also worked as an instructor at the Naval Postgraduate School in Monterey, California (figure 7.8). As interest in CP/M picked up, he found himself writing variations of it for other customers. The publicity in *Dr. Dobb's Journal* led to enough sales to convince him of the potential market for CP/M. In 1976 he quit his job and with Dorothy founded a company, Digital Research (initially Intergalactic Digital Research), whose main product was CP/M.[102]

The next year, 1977, he designed a version with an important difference. IMSAI, the company that had built a "clone" of the Altair (figure 7.9), wanted a license to use CP/M for its products. Working with IMSAI employee Glen Ewing, Kildall rewrote CP/M so that only a small portion of it needed to be customized for the specifics of the IMSAI. The rest would be common code that would not have to be rewritten each time a new computer or disk drive came along. He called the specialized code the BIOS—Basic Input/Output System.[103] This change standardized the system software in the same way that the 100-pin Altair bus had

Figure 7.8
Gary Kildall. A DEC VT-100 terminal is visible in the background. (*Source:* Kristen Kildall.)

Figure 7.9
IMSAI 8080, one of the most successful copies of the Altair, with a video monitor and a disk storage system supplied by Micropolis. (*Source:* Smithsonian Institution.)

standardized hardware. IMSAI's computer system became a standard, with its rugged power supply, room for expansion with plenty of internal slots, external floppy drive, and CP/M.

End of the Pioneering Phase, 1977

By 1977 the pieces were all in place. The Altair's design shortcomings were corrected, if not by MITS then by other companies. Microsoft BASIC allowed programmers to write interesting and, for the first time, serious software for these machines. The ethic of charging money for this software gave an incentive to such programmers, although software piracy also became established. Computers were also being offered with BASIC supplied on a read-only-memory (ROM), the manufacturer paying Microsoft a simple royalty fee. (With the start-up codes also in ROM, there was no longer a need for the front panel, with its array of lights and switches.) Eight-inch floppy disk drives, controlled by CP/M, provided a way to develop and exchange software that was independent of particular models. Machines came with standardized serial and parallel ports, and connections for printers, keyboards, and video monitors. Finally, by 1977 there was a strong and healthy industry of publications, software companies, and support groups to bring the novice on board. The personal computer had arrived.

8

Augmenting Human Intellect, 1975–1985

In the mid-1970s, amid the grassroots energy and creativity of the small systems world, what else was happening? When the established computer companies saw personal computers appear, they, too, entered a period of creativity and technical advance. At first there was little overlap. By 1985, though, there was overlap and more: the paradigm of personal computing based on inexpensive microprocessors forced itself onto the industry. This chapter looks at how that happened.

Digital Equipment Corporation

Digital Equipment Corporation built the foundation for interactive personal computing with its minicomputers and its software. What were they doing when Intel announced its 8080, a device with the essentials of a minicomputer on one chip? "We were just in the throes of building the VAX."[1] The VAX was an extension of the PDP-11 that reached toward mainframe performance. It was a major undertaking for DEC and strained the company's resources. As IBM had done with its System/360, Digital "bet the company" on the VAX—a move toward higher performance and larger systems.

Many within DEC felt that the company was not so much a mini-computer builder as it was a company that sold architecture.[2] Beginning with the TX-0, DEC's founders had taken pride in their ability to build high-performance computers—large or small—through innovative design. That may explain why DEC failed to counter the threat that companies like Intel posed to its business. To build a computer around the Intel 8080 meant surrendering decisions about architecture to a semiconductor house—how could they allow themselves to do that? The other alternative, licensing the PDP-11 instruction set to chip makers, who would produce microprocessors based on it, was likewise rejected.

The company thought that would be giving the "corporate jewels" away. Digital did produce the LSI-11, a single-board PDP-11, in 1974, but that did not lead to inexpensive systems as did the Intel 8080. A single-chip PDP-11, called T-11, was developed but never marketed. The microprocessor phenomenon passed the PDP-11 by, even though elements of its architecture turned up in microprocessor designs (especially the Motorola 6800).[3]

Planning for an extension to the PDP-11 began in 1974 or 1975. DEC announced the VAX, Model 11/780, in October 1977 (figure 8.1). The full name was VAX-11, which stood for Virtual Address eXtension [of the] PDP-11. The implication was that the VAX was simply a PDP-11 with a 32-bit instead of a 16-bit address space. In fact, the VAX was really a new machine. It could, however, execute existing PDP-11 software by setting a "mode bit" that called forth the PDP-11 instruction set. (Eventually the compatibility mode was dropped.)

DEC continued to market small computers at successively lower prices and in smaller packages, for example, the PDP-8/A, introduced in 1975 for under $3,000.[4] But the company preferred to develop and market higher performance. One reason it gave was that for a given application, the cost of the computer was only part of the total cost; there was also

Figure 8.1
VAX 11/780, ca. 1978. (*Source:* Digital Equipment Corporation.)

"the high fixed overhead costs associated with the [existing] application."[5] Apparently DEC did not feel it could achieve truly drastic price reductions, as MITS had done with the Altair. That argument, coupled with DEC's reluctance to turn over its skill in computer architecture to semiconductor companies, kept the company out of the personal computer market during the crucial years, 1974 to 1977, when it could most easily have entered it.

Just as DEC was not the first to market a 16-bit mini, it was not the first to extend address space beyond 16 bits. In 1973, Prime, also located off Route 128 in Massachusetts, shipped a 32-bit minicomputer. Prime grew rapidly until merging with Computervision in the late 1980s. Another company, Interdata, described a "mega-mini" in 1974. Their design was also commercially successful, and that year the company was bought by Perkin-Elmer, the Connecticut optics company.[6] Systems Engineering Laboratory of Fort Lauderdale, Florida, also introduced a 32-bit mini, which was popular with NASA and aerospace customers. S.E.L. was sold to Gould in 1980 and became the basis for that venerable company's entree into the computer business.[7] The impetus for these developments was the growing availability of relatively cheap semiconductor memory to replace magnetic core. These memory chips made it more practical to design machines with large main memories, which in turn demanded more address space.

If the VAX was only nominally an extension of the PDP-11, it was genuinely a "virtual" memory computer. An informal definition of this term is that it is a way to make a computer's small but fast main memory seem to be bigger than it is, by swapping data to and from a slower but larger memory on a disk. A more precise definition concerns the way this is done: first of all, overall performance must not be seriously degraded by this process, and second, the user should not have to know that this swapping is going on (hence the term: the memory is "virtually" large but in reality small).[8]

The need for a hierarchy of memories, each slower but larger than the one below it, was discussed in the Institute for Advanced Study reports by Burks, Goldstine, and von Neumann in the late 1940s. The Atlas, designed at Manchester University in England and built by Ferranti in 1962, was probably the first to use a design that gave the user the illusion of a single-level fast memory of large capacity.[9] It was one of the fastest computers in the world at the time and also one of the most influential on successive generations. A user of the Atlas saw a machine with a virtual memory of one million 48-bit words. The computer automatically swapped data between the core and the drum, based on the contents of a set of registers (a technique called associative memory-addressing).[10] Though influential, commercial versions of the Atlas were only a modest success for Ferranti. In the United States, Burroughs offered virtual memory, with some important architectural advances, in the mid-1960s. IBM offered it with System/370 models announced in 1972. (It is probably from marketing the 370 that the term virtual memory came into wide use.)[11] The SDS 940 time-sharing system also followed the Atlas design.

C. Gordon Bell led the initial design effort for the VAX, and Bill Strecker was its chief architect. Breaking through the limits of the PDP-11's 65 Kbyte address space was their primary goal. The VAX provided 2^{32} or 4.3 gigabytes (equivalent to one billion 32-bit words) of virtual address space. Its addressing scheme divided memory into blocks, called pages, and used an associative comparison to determine whether the desired page was in core or not. The VAX processor used sixteen 32-bit general registers, like the IBM 360. It also had a rich set of over 250 instructions with nine different addressing modes, which allowed a single instruction to carry out complex operations.[12]

The VAX was a commercial success, selling around 100,000 over the next decade and leaping over the other 32-bit minis even though it appeared later. The 11/780's performance, roughly calculated at one million instructions per second (MIPS), became a benchmark against which competitors would compare their machines into the 1990s. A whole family of "Vaxen" followed: the less-powerful 11/750 in 1980, the higher-performance 8600 in 1984, and the compact MicroVax II in 1985, among others.[13] These machines kept DEC profitable and dominant along Route 128. Even Data General, whose Nova had been such a strong competitor for the PDP-11, had trouble competing with the VAX, although it did introduce a 32-bit Eclipse in 1980, as chronicled in Tracy Kidder's bestseller *The Soul of a New Machine*.[14]

The VAX was a general-purpose computer that came with the standard languages and software. It sold to a wide market, but its biggest impact was on engineering and science. Prices started at $120,000, which was too expensive for a single engineer, but just cheap enough to serve a division at an aerospace, automotive, or chemical firm. For them the standard practice had been either to get in line to use the company's mainframe, or to sign up for time on a commercial time-sharing service. The VAX gave them computing power at hand. It had a solid, engineering-oriented operating system (VMS), and sophisticated I/O facilities for data collection.

Finally, the VAX came with a powerful and easy-to-use terminal, the VT-100. It had an impressive number of features, yet one felt that none was superfluous. It somehow managed to retain the comfortable feel of the old Teletype. One feature that many users loved was its ability to scroll a pixel at a time, rather than a line at a time. There was no practical reason to have this feature, and it failed to catch on with other terminal displays, but it had a great appeal. The VT-100's codes, using ASCII, did become the standard for terminals for the next twenty years.

A Word about UNIX

The impact of UNIX on commercial computing will be discussed more fully in the next chapter, and here we will just briefly describe its place with regard to the VAX. In addition to VMS, the VAX's PDP-11 ancestry meant that users could also run UNIX on it. UNIX was developed on DEC minicomputers, and for the first few years of UNIX's existence it ran only on DEC computers, mainly PDP-11s. The University of California at Berkeley's version of UNIX, which had an enormous influence on computing and on the Internet, was developed on a VAX. Still, DEC was ambivalent about UNIX for the VAX. Ken Olsen allegedly stated at one point that "UNIX is snake oil!" (The context in which he made that statement has been disputed.[15]) At any rate, the VAX could and did run Berkeley UNIX, and for at least the formative years, VAX computers were the most common nodes on the Internet.[16]

IBM and the Classic Mainframe Culture

In the mid-1970s, while the personal computer was being invented and while Digital was building the VAX, what was IBM doing? Like Digital, IBM was busy extending its existing line, with the high-end 3033 announced in early 1977, and the low-priced 4300-series announced in 1979. This latter series offered a dramatic increase in performance per dollar over the mid-range 370 systems then being marketed, an improvement that came mainly from using large-scale integrated circuits. These LSI chips were developed and designed by IBM in-house and did not resemble the ones being marketed by companies like Intel or Fairchild.[17] As System/370 installations grew in number and complexity, the issue of interconnecting them also arose. Bob Evans of IBM remarked that, in the early 1970s, the plethora of incompatible and ad hoc networking schemes resembled the chaos of computer architectures that IBM had sought to reduce a decade before.[18] The result was Systems

Network Architecture (SNA), first shipped in 1974. SNA was a layer cake of standards, spelled out in detail. It formed the basis for networking large computer systems into the 1990s.

In 1975 IBM introduced a product that might have seemed at odds with its mainframe orientation: a "personal" computer, Model 5100. This machine could fit on a desk and contained a processor, keyboard, cassette tape drive, and small video terminal in a single package. Prices began at $9,000 for a machine with 16 Kbytes of memory.[19] It supported both BASIC and APL (the developer of APL, Kenneth Iverson, had joined IBM in 1960), which the user could select by flipping a switch on the front panel. But little or no applications software was available; the third-party support community that grew up around the Altair failed to materialize for the 5100. Sales were modest but steady. (The "other" IBM personal computer will be discussed shortly.)

Another answer to the question of what IBM was doing is that it was in court. For IBM the 1970s was the decade of the lawsuit: *U.S. vs. IBM,* filed January 17, 1969, and dismissed in 1981. The charge was that IBM was in violation of antitrust laws by virtue of its dominance of the U.S. market for general-purpose electronic digital computers. The Justice Department based this charge on a definition of "market" that covered the business-oriented electronic data-processing activities served by main-frame computers, of which IBM held about 70 percent of the market and the "BUNCH" nearly all the rest. IBM countered by arguing that its competition was not just Burroughs, Univac, NCR, CDC, and Honeywell, but rather thousands of companies, large and small, that made and sold computers, peripherals, software, services, and the like. After a long discovery process, during which depositions were taken from represen-tatives of most of these companies, the case finally went to trial in May, 1975—that is, around the time that Bill Gates and Paul Allen were talking about developing BASIC for the Altair.

The discovery process and the testimony were thorough and detailed. Transcripts of the depositions and testimony run into thousands of pages.[20] But none of the gathering storm of personal systems made it into the trial. Neither Bill Gates nor Ed Roberts was called to testify or give depositions. The court focused its attention on the former "Dwarfs," especially RCA and GE, who had left the business. Occasion-ally firms that competed with IBM's mainframes at one or two places were examined. These included SDS (a subsidiary of Xerox by then), whose computers competed with the System/360 Model 44 for scientific applications, and Digital Equipment Corporation, not for its minicomputers but for its PDP-10. The court even looked closely at Singer, the venerable sewing machine company, which had purchased Friden in 1963 and built up a business in point-of-sale retail terminals. (The British company ICL bought Singer's computing business in 1976.)

Reading through the volumes of transcripts, one feels a sense of tragedy and unreality about the whole trial. The judge, David Edelstein, was often baffled by the avalanche of jargon that spewed forth from the expert witnesses each day; this typically resulted in his losing his temper by mid-afternoon. (The courtroom had a defective air-conditioning system, which did not help matters in the summer, either.) The money spent on hiring and retaining a team of top-notch attorneys (led by Nicholas Katzenbach for IBM) and their research staffs was money that did not go into the research and development of new computer technology. And yet both sides, with all their highly paid legal and research staffs, utterly and completely missed what everyone has since recognized as the obvious way that computing would evolve: toward microprocessor-based, networked desktop computing. There is no record of someone bringing an Apple II into the courthouse building in lower Manhattan; if someone had, would anyone have recognized it for what it was? By coincidence, just as the Apple II was being introduced at a computer fair in California in 1977, one expert witness testified, "I will be a little stronger than that ... it is most unlikely that any major new venture into the general purpose [sic] computer industry can be expected."[21] As late as 1986 one Justice Department economist, still fuming over the dismissal of the case, complained that "IBM faces no significant domestic or foreign competition that could threaten its dominance."[22] That statement was made the year that Microsoft offered its shares to the public. A few years later IBM began suffering unprecedented losses and began laying off employees for the first time in its history. A new crop of books soon appeared, these telling the story of how IBM had been outsmarted by Bill Gates. Other than writing tell-all books about IBM, everything else about the computer industry had fundamentally and irrevocably changed.[23]

In the end the combatants ran out of energy. The 1981 inauguration of Ronald Reagan, who had campaigned against an excessive exercise of federal power, was enough to end it. But what really killed the government's case was that, even neglecting the personal computer, there was vigorous and healthy competition throughout the decade. The failures of GE and RCA were more than offset by the successes of Digital Equipment, SDS, Amdahl, and software companies like EDS. The industry was too healthy: the personal fortunes amassed by Gene Amdahl and Max Palevsky made it hard to take the charges seriously.[24] In one of the rare instances of levity, IBM's lawyers were able to elicit more than a few chuckles in the courtroom when they described the enormous wealth that Palevsky—a philosophy student—made in a few years with machines aimed right at IBM's middle range of mainframes.

IBM continued to develop new products. In addition to the 4300 and 3030 mainframes, IBM went after the minicomputer companies with its System/38 in 1978, following that with its AS/400 in 1988. The AS/400 was aimed more at business than engineering customers, but otherwise it was a strong competitor to the VAX. It used advanced architectural features that IBM had planned for its follow-on to the System/370 but had not implemented. As such, the AS/400 represented IBM's most advanced technology, and it generated strong revenues for IBM into the 1990s, when its mainframe sales suffered from technological obsolescence.[25] IBM failed to bring other products to market at this time, however, a failure that ultimately hurt the company. It is not clear how much the antitrust suit had to do with that.

From "POTS" to "OLTP"

The concept of a computer utility, naively envisioned in the late 1960s as being like the electric power utilities, evolved in several directions in the 1970s. General Electric built a large international network from its association with Dartmouth. Using machines like the PDP-10 and SDS 940, other utilities offered unstructured computer time. By 1975 TYMNET comprised a network of twenty-six host computers, routed through eighty communications processors. The simple hub-and-spoke topology of time-sharing evolved into a web of multiple rings, so that the failure of one host would not bring the system down.[26]

At the same time, a more tightly structured and disciplined use of terminals for on-line access also appeared. This was tailored for a specific application, such as processing insurance claims, where the programs and types of data handled by a terminal were restricted. Many were private, though some were semipublic, such as the effort by the U.S. National Library of Medicine to put its century-old *Index Medicus* on-line. (By the end of the 1970s its *MEDLINE* system provided on-line searches of medical literature from research libraries worldwide.) These systems were more like the SAGE air-defense and SABRE airline reservations systems of the late 1950s than they were like the Dartmouth College model. A new acronym appeared to describe it, "OLTP" for "On-line Transaction Processing," to differentiate it from the less-structured connotations of "POTS" (Plain Old Time-Sharing). Thus although computer usage was no longer in batches of cards, some of the basic structure of a punched-card installation remained.

A number of companies introduced terminals to serve this market. Some were descended from the Teletype and combined a typewriter keyboard and a printing mechanism (e.g., the DECwriter II or Teletype Model 37, both ca. 1975). Others replicated the Teletype, only with a video screen instead of a printer. These, like the Lear-Siegler ADM-3, were sometimes called "dumb terminals," "glass teletypes," or "glass TTY": they offered little beyond simple data entry and viewing. In contrast to them were "smart" terminals that allowed users to see and edit a full screen of text, and which contained a measure of computing power. Besides the VT-100, DEC had produced several designed around a PDP-8 processor; another company that had some success was Datapoint of San Antonio. Recall that it was Datapoint's contract with Intel that led to the 8080 microprocessor; however, the Datapoint 2200 terminal did not use a microprocessor. Some of these terminals, especially the Datapoint, came close to becoming personal computers without the vendor realizing it.[27]

The VT-100 became the standard ASCII terminal, while a terminal introduced by IBM became the EBCDIC standard by 1980. That was the model 3270, announced in 1971.[28] The 3270 was the philosophical opposite of the DEC VT-100: it operated on the assumption that the user would be keying structured information into selected fields, with other fields (e.g., for a person's name or date of birth) replicated over and over for each record. Therefore, the terminal did not transmit information as it was keyed in but waited until a full screen was keyed in; then it sent only whatever was new to the computer (in compressed form). IBM mainframe installations now routinely included terminals and time-shared access through the time sharing option (TSO) software. Typically these terminals were segregated in special rooms near the mainframe installation. They were seldom found in a private office.

By 1980, as the lawsuit was coming to an end, IBM still dominated the industry. But more and more, IBM was floating in a slower channel of the river. That began in 1963 with the development of ASCII, when IBM adopted EBCDIC. In 1964 IBM chose a hybrid semiconductor technology over ICs. In 1970 it adopted integrated circuits of its own design, slightly different from the standard TTL chips then flooding the market.

In the mid-70s, IBM's Systems Network Architecture established a standard for networking large systems, but SNA was different from the networking schemes then being developed by the Defense Department. Finally, there were the different approaches to terminal design represented by the 3270 and VT-100. Only with hindsight can we discern a pattern.

IBM's introduction of the personal computer in 1981 brought the issue to a head. The IBM PC used ASCII, not EBCDIC. It used standard TTL and MOS chips from outside suppliers. And its connections with its keyboard and monitor were closer to the minicomputer than to the 3270. The PC's internal design reveals how the pressures of the marketplace were able to accomplish what the courts and the U.S. Justice Department could not do.

The mainframe, batch model of computing, whose high-water mark was the 7090, was giving way, not only to interactive but also to decentralized processing. The increasingly fuzzy line that distinguished "smart" terminals from stand-alone personal computers was one indication. New design questions came to the fore: how to apportion processing functions and memory among the terminals and a central system, and how to send data efficiently and reliably through such a network. IBM had embarked on the design of a "Future System" (FS) that attacked some of these issues head-on. Planning for FS began in the early 1970s, and IBM hoped to announce products to replace its System/370 line by 1975. But FS was abandoned in 1975, in part because its designers were unable to solve the architectural problems, and in part because the success of the System/370 architecture meant that IBM would put itself at an unacceptable risk to abandon that market to third-party vendors.[29] Some of the concepts found their way into the mid-range AS/400, but canceling FS was "the most expensive development-effort failure in IBM's history."[30]

Viatron

A start-up company from Route 128 had an idea with similar promise but equally dismal results. The John the Baptist of distributed computing was Viatron Computer Systems of Bedford, Massachusetts. It was the outgrowth of an Air Force Project from the mid-1960s called AESOP (Advanced Experimental System for On-line Planning). Prepared by the MITRE Corporation, AESOP envisioned a network of terminals that provided visual as well as text information to middle and high-level managers, including those without any sophistication in computing, to help them do their work with the same level of acceptance as the telephone:

The core of the management system . . . will be not so much the central processor or central memory. The real basis . . . will be the unique program of instructions which makes the central processor, the central memory, and the organization's store of data and formal quantitative models easily available to the manager through the window of his desk top display, thus making it possible for him to exert the full power of his intentions through the use of his simple lightgun pointer. As AESOP-type management systems are developed, managers will learn to converse and interact with the processor with ease and naturalness. They will also learn to communicate through the processor with other members of the organization.[31]

Two of the report's authors, Joseph Spiegel and Dr. Edward Bennett, left MITRE and cofounded Viatron in 1967. Bennett was successful in raising venture capital—these were the go-go years—and announced that by 1969 Viatron would be renting interactive terminals that would move processing onto the desktop. He also predicted that his company would surpass IBM in numbers of installed computers. System 21 terminals were to rent for the unbelievably low price of $40 a month.[32] The system included a keyboard, a 9-inch video display, and two cassette tape drives for storage of data and formatting information (figure 8.2). An optional attachment allowed users to disconnect the keyboard and tape unit and connect it to any standard television set for remote computing, say, in a hotel room. The terminal contained within it a "micro-processor" [sic] with 512 characters of memory. Other options included an optical character-recognition device, a "communications adapter," and an ingenious, Rube Goldberg–inspired "printing robot" which one placed over a standard IBM Selectric typewriter. Activating a set of solenoids, mechanical fingers pressed the Selectric keys to type clean output at 12 characters/second.[33]

The key to Viatron's impressive specifications was its use of MOS integrated circuits. This technique of integrated-circuit fabrication was the technical foundation for the microprocessor revolution of the 1970s, but it was immature in 1969. Viatron had to invest its start-up capital in perfecting MOS, and then it needed more money to gear up for volume production. That was too ambitious. By 1970, production lines were just starting, but the volume was small, and Viatron's sales and marketing were in disarray. At a meeting of the board held in Bennett's home in the summer of 1970, he found himself ousted from Viatron just as his wife was about to serve everyone dinner (they never ate the meal). The

Figure 8.2
Office automation: Viatron 21. (*Source:* Charles Babbage Institute, University of Minnesota.)

company delivered a few systems by 1971, but in April of that year it declared bankruptcy. Losses to venture capitalists ran upwards of $30 million in fiscal year 1969–1970 alone.[34] Viatron became just another of many companies to fail while attempting to topple IBM from the top of the industry.

Wang

Even by the metric of the go-go years, Viatron's trajectory was bizarre, which should not obscure the truth of Bennett's observation. Advances in MOS integrated circuits were making IBM's way of doing computing obsolete, at least in the office environment. The company that succeeded where Viatron failed was Wang Labs, which in an earlier era had pioneered in electronic calculators. By 1971 Wang recognized that calculators were becoming a commodity, with razor-thin profit margins dependent on packaging more than on technical innovation. Wang Labs began a transition to a minicomputer company, and by 1972 made a complete crossover with its Model 2200 "computing calculator"—a general-purpose computer, although Wang was careful not to market it as such. Like Digital Equipment in the late 1950s, Wang was reluctant to use the word "computer" because of the word's connotations. Wang had an astute sense of knowing when to get out of one market and into a new one about to open up. As Wang's profits soared, Wall Street analysts concocted elaborate theories to explain this, some based on an alleged innate sense that his Chinese ancestors gave him. Dr. Wang was, in fact, a conservative engineer who understood the technology of his company's products and who valued his company's independence. In this regard he was closer to his Yankee counterpart and neighbor, Ken Olsen of DEC, than he was to any Western stereotype of Oriental mind-set.[35]

An Wang chose next to direct the company toward what would later be known as "Office Automation." In the mid-1970s that meant word processing. Word processing has become so commonplace that it is hard to recall how absurd the concept was at a time when even small computers cost thousands and skilled typists were paid $1.25 an hour. An old story tells of how graduate students at MIT programmed the $120,000 PDP-1 to serve as an "expensive typewriter" in the early 1960s. IBM developed a program called TEXT-90 for the 7090, but that was used only for special applications and never penetrated the office environment. In the early 1960s, some members of the committee working on the ASCII standard argued that codes for lowercase letters were unnecessary and a waste of space.[36]

The term "word processing" came into use after 1964, when IBM announced the MTST—a version of its Selectric typewriter that could store and recall sequences of keystrokes on a magnetic tape cartridge.[37] An early Wang product, the Model 1200, was similar, but customers found its complexity daunting. Other companies that entered the field at this time included NBI ("Nothing But Initials") outside of Denver, Lanier in Atlanta, and CPT in Minneapolis.

The second time around Wang got it right. Wang engineers found out first of all what office people wanted. They realized that many users of word-processing equipment were terrified of losing a day's work by the inadvertent pressing of the wrong key. And it wasn't just secretaries who were prone to such actions: in 1981 ex-President Jimmy Carter lost a few pages of his memoirs—"I had labored over them for a couple of days"—

by pressing the wrong key on his $12,000 Lanier "No Problem" [sic] word-processing system. An anxious phone call to Lanier produced a utilities disk that allowed him to recover the data from the original diskette.[38] After this, Wang's engineers came up with a design that would make such a loss nearly impossible. They also decided on a terminal that used a cathode ray tube, which displayed half a page of text instead of the one or few lines that other systems used. Commands were accessed by a simple screen of menus. In a later era Wang's design might have been known by the cliché "user friendly"; it was also a "distributed" system. But the company used neither term in its marketing. Unlike other minicomputer companies, Wang did little OEM business; it sold machines to the people who were going to use it. Wang spared its customers—Wall Street brokerage houses, large banks, and oil companies at first—the technical jargon. (A decade later office workers were not so lucky, everything would get plastered with the term "user friendly" no matter how obtuse it was.)[39]

A major requirement was that the system have a speedy response. Time-sharing relieved users of the need to wait in a queue with a deck of punched cards, but on a busy day users faced an equally onerous wait at their terminals while the mainframe got around to each job. Unlike MIT hackers, office employees could not be expected to come in at midnight to do their work. The answer was to put some of the processing power into the terminal itself, with the central computer serving primarily for data storage and retrieval—commonplace after 1985, but a radical departure from time-sharing in 1975. The WPS (Wang Word Processing System) was unveiled at a trade show in New York in June 1976, and according to some accounts nearly caused a riot (figure 8.3).[40] A basic system, including hard disk storage, cost $30,000. Wang Labs, ranked forty-fifth in data-processing revenues in 1976, moved up to eighth place by 1983, just below IBM, DEC, and the remnants of the BUNCH. Some analysts thought Wang was in the best position of any company to become number two in the industry. (No Wall Street person would risk his career by predicting a new number one.) Others put the company's success into the pigeonhole of "office automation" rather than general-purpose computing, but what Wang was selling was at heart a general-purpose, distributed computer system. Wang's success was a vindication of Viatron's vision. However, Wang was unable to reinvent itself once again in the 1990s, when it faced competition from commodity personal computers running cheap word-processing software, and it too went bankrupt.[41]

Figure 8.3
Office automation: WANG Word Processing System. (*Source:* Charles Babbage
Institute, University of Minnesota.)

Xerox PARC

One of the ironies of the story of Wang is that despite its innovations, few
stories written about the 1970s talk about Wang. To read the literature
on these subjects, one would conclude that the Xerox Corporation was

the true pioneer in distributed, user-friendly computing; that the Xerox Palo Alto Research Center, which Stewart Brand so glowingly described in his 1972 *Rolling Stone* article, was the place where the future of computing was invented. Why was that so?

The Xerox Corporation set up a research laboratory in the Palo Alto foothills in 1970. Its goal was to anticipate the profound changes that technology would bring to the handling of information in the business world. As a company famous for its copiers, Xerox was understandably nervous about talk of a "paperless office." Xerox did not know if that would in fact happen, but it hoped that its Palo Alto Research Center (PARC) would help the company prosper through the storms.[42]

Two things made PARC's founding significant for computing. The first was the choice of Palo Alto: Jacob Goldman, director of corporate research at Xerox, had favored New Haven, Connecticut, but the person he hired to set up the lab, George Pake, favored Palo Alto and prevailed, even though it was far from Xerox's upstate New York base of operations and its Connecticut headquarters. The lab opened just as "Silicon Valley," led by Robert Noyce of the newly founded Intel, was taking form.

The second reason for PARC's significance took place in the halls of Congress. As protests mounted on college campuses over the U.S. involvement in Viet Nam, a parallel debate raged in Congress that included the role of universities as places where war-related research was being funded by the Defense Department. Senator J. William Fulbright was especially critical of the way he felt science research was losing its independence in the face of the "monolith" of the "military-industrial complex" (a term coined by President Eisenhower in 1961). In an amendment to the 1970 Military Procurement Authorization Bill, a committee chaired by Senator Mike Mansfield inserted language that "none of the funds authorized . . . may be used to carry out any research project or study unless such a study has a direct and apparent relationship to a specific military function or operation."[43] The committee did not intend to cripple basic research at universities, only to separate basic from applied research. Some members assumed that the National Science Foundation would take the DoD's place in funding basic research. Even before the passage of this "Mansfield Amendment," the DoD had moved to reduce spending on research not related to specific weapons systems; thus this movement had support among hawks as well as doves.

The NSF was never given the resources to take up the slack. At a few select universities, those doing advanced basic research on computing felt that they were at risk, because their work was almost entirely funded by the Defense Department's Advanced Research Projects Agency (ARPA).[44] At that precise moment, George Pake was scouring the country's universities for people to staff Xerox PARC. He found a crop of talented and ambitious people willing to move to Palo Alto. ARPA funding had not been indiscriminate but was heavily concentrated at a few universities—MIT, Carnegie-Mellon, Stanford, UC-Berkeley, UCLA, and the University of Utah—and researchers from nearly every one of them ended up at PARC, including Alan Kay and Robert Taylor from Utah, and Jerome Elkind and Robert Metcalfe from MIT.[45] There were also key transfers from other corporations, in particular from the Berkeley Computer Corporation (BCC), a struggling time-sharing company that was an outgrowth of an ARPA-funded project to adapt an SDS computer for time-sharing. Chuck Thacker and Butler Lampson were among the Berkeley Computer alumni who moved to PARC. All those cited above had had ARPA funding at some point in their careers, and Taylor had been head of ARPA's Information Processing Techniques Office.

Two ARPA researchers who did not move to PARC were the inspiration for what would transpire at Xerox's new lab. They were J.C.R. Licklider, a psychologist who initiated ARPA's foray into advanced computer research beginning in 1962, and Douglas Engelbart, an electrical engineer who had been at the Stanford Research Institute and then moved to Tymshare. In 1960, while employed at the Cambridge firm Bolt Beranek and Newman, Licklider published a paper titled "Man-Computer Symbiosis" in which he forecast a future of computing that "will involve a very close coupling between the human and electronic members of the partnership." In a following paper, "The Computer as a Communication Device," he spelled out his plan in detail.[46] He was writing at the heyday of batch processing, but in his paper Licklider identified several technical hurdles that he felt would be overcome. Some involved hardware limits, which existing trends in computer circuits would soon overcome. He argued that it was critical to develop efficient time-sharing operations. Other hurdles were more refractory: redefining the notions of programming and data storage as they were then practiced. In 1962 "Lick" joined ARPA, where he was given control over a fund that he could use to realize this vision of creating a "mechanically extended man."[47]

Douglas Engelbart was one of the first persons to apply for funding from ARPA's Information Processing Techniques Office in late 1962; he was seeking support for a "conceptual framework" for "augmenting human intellect."[48] Engelbart says that a chance encounter with Vannevar Bush's *Atlantic Monthly* article "As We May Think" (published in July 1945) inspired him to work on such a plan. Licklider directed him to work with the time-shared Q-32 experimental computer located in Santa Monica, through a leased line to Stanford; later Engelbart's group used a CDC 160A, the proto-minicomputer. The group spent its time studying and experimenting with ways to improve communication between human beings and computers. His most famous invention, first described in 1967, was the "mouse," which exhaustive tests showed was more efficient and effective than the light pen (used in the SAGE), the joystick, or other input devices.[49] Engelbart recalled that he was inspired by a device called a planimeter, which an engineer slid over a graph to calculate the area under a curve. Among many engineers this compact device was a common as a slide rule; it is now found only among antique dealers and museums.

In December 1968 Engelbart and a crew of over a dozen helpers (among them Stewart Brand) staged an ambitious presentation of his "Augmented Knowledge Workshop" at the Fall Joint Computer Conference in San Francisco. Interactive computer programs, controlled by a mouse, were presented to the audience through a system of projected video screens and a computer link to Palo Alto. Amazingly, everything worked. Although Engelbart stated later that he was disappointed in the audience's response, the presentation has since become legendary in the annals of interactive computing. Engelbart did not join Xerox-PARC, but many of his coworkers, including Bill English (who did the detail design of the mouse), did.[50]

What was so special about the mouse? The mouse provided a practical and superior method of interacting with a computer that did not strain a user's symbolic reasoning abilities. From the earliest days of the machine's existence, the difficulties of programming it were recognized. Most people can learn how to drive a car—a complex device and lethal if not used properly—with only minimal instruction and infrequent reference to an owner's manual tossed into the glove box. An automobile's control system presents its driver with a clear, direct connection between turning the steering wheel and changing direction, pressing on the gas pedal and accelerating, pressing on the brake pedal and slowing down. Compare that to, say, UNIX, with its two- or three-letter commands, in which the command to delete a file might differ from one to print a file only by adjacent keys. Automobiles—and the mouse—use eye-hand coordination, a skill human beings have learned over thousands of years of evolution, but a keyboard uses a mode of human thought that humans acquired comparatively recently. Researchers at PARC refined the mouse and integrated it into a system of visual displays and iconic symbols (another underutilized dimension of human cognition) on a video screen.

For the U.S. computing industry, the shift of research from ARPA to Xerox was a good thing; it forced the parameters of cost and marketing onto their products. It is said that Xerox failed to make the transition to commercial products successfully; it "fumbled the future," as one writer described it. Apple, not Xerox, brought the concept of windows, icons, a mouse, and pull-down menus (the WIMP interface) to a mass market, with its Macintosh in 1984. Xerox invented a networking scheme called Ethernet and brought it to market in 1980 (in a joint effort with Digital and Intel), but it remained for smaller companies like 3-Com to commercialize Ethernet broadly. Hewlett-Packard commercialized the laser printer, another Xerox-PARC innovation. And so on.[51]

This critique of Xerox is valid but does not diminish the magnitude of what it accomplished in the 1970s. One may compare Xerox to its more nimble Silicon Valley competitors, but out of fairness one should also compare Xerox to IBM, Digital, and the other established computer companies. Most of them were in a position to dominate computing: DEC with its minicomputers and interactive operating systems, Data General with its elegant Nova architecture, Honeywell with its Multics time-sharing system, Control Data with its Plato interactive system, and IBM for the technical innovations that its research labs generated. Although they did not reap the rewards they had hoped for, each of these companies built the foundation for computing after 1980.

Within Xerox-PARC, researchers designed and built a computer, the Alto, in 1973 (figure 8.4). An architectural feature borrowed from the MIT-Lincoln Labs TX-2 gave the Alto the power to drive a sophisticated screen and I/O facilities without seriously degrading the processor's performance. Eventually over a thousand were built, and nearly all were used within the company. Networking was optional, but once available, few Alto users did without an Ethernet connection. An Alto cost about $18,000 to build. By virtue of its features, many claimed that the Alto was the first true personal computer. It was not marketed to the public, however—it would have cost too much for personal use.[52] Besides using

Figure 8.4
Xerox Alto, ca. 1973. (*Source:* Smithsonian Institution.)

a mouse and windows, the Alto also had a "bit-mapped" screen, where each picture element on the screen could be manipulated by setting bits in the Alto's memory. That allowed users to scale letters and mix text and graphics on the screen. It also meant that a text-editing system would have the feature "what you see is what you get" (WYSIWYG)—a phrase made popular by the comedian Flip Wilson on the television program "Laugh-In."[53]

In 1981 Xerox introduced a commercial version, called the 8010 Star Information System, announced with great fanfare at the National Computer Conference in Chicago that summer. Advertisements described an office environment that would be commonplace ten years later, even more capable than what office workers in 1991 had. But the product fizzled. Around the same time Xerox introduced an "ordinary" personal computer using CP/M, but that, too, failed to sell.[54]

The Star, derived from the Alto, was technically superior to almost any other office machine then in existence, including the Wang WPS. Personal computers would have some of the Star's features by 1984, but integrated networks of personal computers would not become common for another ten years. In the late 1970s, Wang had a better sense than Xerox of what an office environment was like and what its needs were. Advertisements for the Star depicted an executive calling up, composing, and sending documents at his desk; somehow Xerox forgot that business executives do not even place their own telephone calls but get a secretary to do that. By contrast, Wang aimed its products at the office workers who actually did the typing and filing. The Alto was more advanced, which explains why its features became common in office computing in the 1990s. The Wang was more practical but less on the cutting edge, which explains both Wang's stunning financial success in the late 1970s, and its slide into bankruptcy afterward.

Along with its invention of a windows-based interface, Xerox's invention of Ethernet would have other far-reaching consequences. Ethernet provided an effective way of linking computers to one another in a local environment. Although the first decade of personal computing emphasized the use of computers as autonomous, separate devices, by the mid-1980s it became common to link them in offices by some form of Ethernet-based scheme. Such a network was, finally, a way of circumventing Grosch's Law, which implied that a large and expensive computer would outperform a cluster of small machines purchased for the same amount of money. That law had held up throughout the turmoil of the minicomputer and the PC; but the effectiveness of Ethernet finally brought it, and the mainframe culture it supported, down.[55] How that happened will be discussed in the next chapter.

Personal Computers: the Second Wave, 1977–1985

Once again, these top-down innovations from large, established firms were matched by an equally brisk pace of innovation from the bottom up—from personal computer makers.

In the summer of 1977 Radio Shack began offering its TRS-80 in its stores, at prices starting at $400. The Model 1 used the Z-80 chip; it was more advanced than the Intel 8080 (although it did not copy the Altair architecture). The Model 1 included a keyboard and a monitor, and cassettes to be used for storage. A start-up routine and BASIC (not Microsoft's) were in a read-only memory. The marketing clout of Radio

Shack, with its stores all over the country, helped make it an instant hit for the company.[56] Because Radio Shack's customers included people who were not electronics hobbyists or hackers, the Model 1 allowed the personal computer to find a mass audience. Years later one could find TRS-80 computers doing the accounting and inventory of small businesses, for example, using simple BASIC programs loaded from cassettes or a floppy disk. The TRS-80 signaled the end of the experimental phase of personal computing and the beginning of its mature phase.

Two other computers introduced that year completed this transition. The Commodore PET also came complete with monitor, keyboard, and cassette player built into a single box. It used a microprocessor with a different architecture from the Intel 8080—the 6502 (sold by MOS Technologies). The PET's chief drawback was its calculator-style keyboard, and for that reason it was not as successful in the U.S. as the other comparable computers introduced that year. But it sold very well in Europe, and on the Continent it became a standard for many years.

The third machine introduced in 1977 was the Apple II (figure 8.5). The legend of its birth in a Silicon Valley garage, assisted by two idealistic young men, Steve Jobs and Steve Wozniak, is part of the folklore of Silicon Valley. According to the legend, Steve Wozniak chose the 6502 chip for the Apple simply because it cost less than an 8080. Before designing the computer he had tried out his ideas in discussions at the Homebrew Computer Club, which met regularly at a hall on the Stanford campus. The Apple II was a tour de force of circuit design. It used fewer chips than the comparable Altair machines, yet it outperformed most of them. It had excellent color graphics capabilities, better than most mainframes or minicomputers. That made it suitable for fast-action interactive games, one of the few things that all agreed personal computers were good for. It was attractively housed in a plastic case. It had a nonthreatening, nontechnical name. Even though users had to open the case to hook up a printer, it was less intimidating than the Altair line of computers. Jobs and Wozniak, and other members of the Homebrew Computer Club, did not invent the personal computer, as the legend often goes. But the Apple II came closest to Stewart Brand's prediction that computers would not only come to the people, they would be embraced by the people as a friendly, nonthreatening piece of technology that could enrich their personal lives. The engineering and design of the Apple II reflected those aims.

Wozniak wrote his own BASIC for the Apple, but the Apple II was later marketed with a better version, written by Microsoft for the 6502 and

Figure 8.5
Personal computers: Apple II, ca. 1977, with a monitor and an Apple disk drive.
(*Source:* Smithsonian Institution.)

supplied in a ROM. A payment of $10,500 from Apple to Microsoft in August 1977, for part of the license fee, is said to have rescued Microsoft from insolvency at a critical moment of its history.[57] Although it was more expensive than either the TRS-80 or the PET, the Apple II sold better. It did not take long for people to write imaginative software for it. Like the Altair, the Apple II had a bus architecture with slots for expansion—a feature Wozniak argued strenuously for, probably because he had seen its advantages on a Data General Nova.[58] The bus architecture allowed Apple and other companies to expand the Apple's capabilities and keep it viable throughout the volatile late 1970s and into the 1980s. Among the cards offered in 1980 was the SoftCard, from Microsoft, which allowed an Apple II to run CP/M. For Microsoft, a company later famous for software, this piece of hardware was ironically one of its best selling products at the time.

By the end of 1977 the personal computer had matured. Machines like the TRS-80 were true appliances that almost anyone could buy and get running. They were useful for playing games and for learning the rudiments of computing, but they were not good enough for serious applications. Systems based on the Altair bus were more sophisticated and more difficult to set up and get running, but when properly configured could compete with minicomputers for a variety of applications. The Apple II bridged those two worlds, with the flexibility of the one and the ease of use and friendliness of the other. At the base was a growing commercial software industry.

None of this was much of a threat to the computer establishment of IBM, Digital, Data General, or the BUNCH. Within a few years, though, the potent combination of cheap commodity hardware and commercial software would redefine the computer industry and the society that would come to depend on it. The trajectories of DEC, IBM, Wang, and Xerox did not intersect those of MITS, IMSAI, Apple, Radio Shack, or the other personal computer suppliers into the late 1970s. Innovations in personal computing did not seem as significant as those at places like Xerox or even IBM. But in time they would affect all of computing just as much. One of those innovations came from Apple.

APPLE II's Disk Drive and VisiCalc

By 1977 many personal computer companies, including MITS and IMSAI, were offering 8-inch floppy disk drives. These were much better than cassette tape but also expensive. The Apple II used cassette tape, but by the end of 1977 Steve Wozniak was designing a disk controller for it. Apple purchased the drives themselves (in a new 5 1/4-inch size) from Shugart Associates, but Wozniak felt that the controlling circuits then in use were too complex, requiring as many as fifty chips. He designed a circuit that used five chips. It was, and remains, a marvel of elegance and economy, one that professors have used as an example in engineering courses. He later recounted how he was driven by aesthetic considerations as much as engineering concerns to make it simple, fast, and elegant.[59]

Apple's 5 1/4-inch floppy drive could hold 113 Kbytes of data and sold for $495, which included operating system software and a controller that plugged into one of the Apple II's internal slots.[60] It was a good match for the needs of the personal computer—the drive allowed people to market and distribute useful commercial software, and not just the simple games and checkbook-balancing programs that were the limit of cassette tape capacity. Floppy disk storage, combined with operating system software that insulated software producers from the peculiarities of specific machines, brought software to the fore. Ensuing decades would continue to see advances in hardware. But no longer would computer generations, defined by specific machines and their technology, best describe the evolution of computing. With a few exceptions, new computers would cease to be pivotal—or even interesting—to the history of computing.

In October 1979 a program called VisiCalc was offered for the Apple II. Its creators were Daniel Bricklin and Robert Frankston, who had met while working on Project MAC at MIT. Bricklin had worked for Digital Equipment Corporation and in the late 1970s attended the Harvard Business School. There he came across the calculations that generations of B-school students had to master: performing arithmetic on rows and columns of data, typically of a company's performance for a set of months, quarters, or years. Such calculations were common throughout the financial world, and had been semi-automated for decades using IBM punched-card equipment. He recalled one of his professors posting, changing, and analyzing such tables on the blackboard, using figures that his assistant had calculated by hand the night before. Bricklin conceived of a program to automate these "spreadsheets" (a term already in limited use among accountants). Dan Flystra, a second-year student who had his own small software marketing company, agreed to help him market the program. Bricklin then went to Frankston, who agreed to help write it.

In January 1979 Bricklin and Frankston formed Software Arts, based in Frankston's attic in Arlington, Massachusetts (the Boston area has fewer garages than in Silicon Valley). That spring the program took shape, as Frankston and Bricklin rented time on the MIT Multics system. In June, VisiCalc was shown at the National Computer Conference. The name stood for visible calculator, although inspiration for it may have come from eating breakfast one morning at Vic's Egg on One coffee shop on Massachusetts Avenue. (Nathan Pritikin would not have approved, but such eateries are another common feature of the Boston scene not found in Silicon Valley.)[61]

Bricklin wanted to develop this program for DEC equipment, "and maybe sell it door-to-door on Route 128." Flystra had an Apple II and a TRS-80; he let Bricklin use the Apple, so VisiCalc was developed on an Apple. The price was around $200. Apple itself was not interested in marketing the program. But the product received good reviews. A financial analyst said it might be the "software tail that wags the hardware dog."[62] He was right: in many computer stores people would come in and ask for VisiCalc and then the computer (Apple II) they needed to run it. Sales passed the hundred thousand mark by mid-1981 (the year the IBM personal computer was announced, an event that led to Software Arts's demise).

An owner of an Apple II could now do two things that even those with access to mainframes could not do. The first was play games; admittedly not a serious application, but one that nevertheless had a healthy market. The second was use VisiCalc; which was as important as any application running on a mainframe. Word processing, previously available only to corporate customers who could afford systems from Wang or Lanier, soon followed.

IBM PC (1981)

Although after the Apple II and its floppy drive were available, one could say that hardware advances no longer drove the history of computing, there were a few exceptions, and among them was the IBM Personal Computer. Its announcement in August 1981 did matter, even though it represented an incremental advance over existing technology. Its processor, an Intel 8088, was descended from the 8080, handling data internally in 16-bit words (external communication was still 8 bits).[63] It used the ASCII code. Its 62-pin bus architecture was similar to the Altair's bus, and it came with five empty expansion slots. Microsoft BASIC was supplied in a read-only memory chip. It had a built-in cassette port, which, combined with BASIC, meant there was no need for a disk operating system. Most customers wanted disk storage, and they had a choice of three operating systems: CP/M-86, a Pascal-based system designed at the University of California at San Diego, and PC-DOS from Microsoft. CP/M-86 was not ready until 1982, and few customers bought the Pascal system, so PC-DOS prevailed. The floppy disk drives, keyboard, and video monitor were also variants of components used before. IBM incorporated the monitor driver into the PC's basic circuit board, so that users did not tie up a communication port. The monochrome monitor could display a full screen of 25 lines of 80 characters— an improvement over the Apple II and essential for serious office applications. A version with a color monitor was also available (figure 8.6).

With the PC, IBM also announced the availability of word processing, accounting, games software, and a version of VisiCalc. A spreadsheet introduced in October 1982, 1-2-3 from Lotus Development, took

Figure 8.6
Personal computers: IBM PC, 1981. Note the two internal floppy disk drives.
(*Source:* Smithsonian Institution.)

advantage of the PC's architecture and ran much faster than its
competitor, VisiCalc. This combination of the IBM Personal Computer
and Lotus 1-2-3 soon overtook Apple in sales and dispelled whatever
doubts remained about these machines as serious rivals to mainframe
and minicomputers. In December 1982 *Time* magazine named the
computer "Machine of the Year" for 1983.[64]

MS-DOS

Microsoft was a small company when an IBM division in Boca Raton,
Florida, embarked on this project, code named "Chess." Microsoft was
best known for its version of BASIC. IBM had developed a version of
BASIC for a product called the System/23 Datamaster, but the need to
reconcile this version of BASIC with other IBM products caused delays.
The Chess team saw what was happening in the personal computer field,

and they recognized that any delays would be fatal. As a result they would go outside the IBM organization for nearly every part of this product, including the software.[65]

Representatives of IBM approached Bill Gates in the summer of 1980 to supply a version of BASIC that would run on the Intel 8088 that IBM had chosen.[66] IBM thought it would be able to use a version of CP/M for the operating system; CP/M was already established as the standard for 8080-based systems, and Digital Research was working on a 16-bit extension. But negotiations with Gary Kildall of Digital Research stalled. When IBM visited Digital Research to strike the deal, Kildall was not there, and his wife, who handled the company's administrative work, refused to sign IBM's nondisclosure agreement. (Given the charges that had been leveled against IBM over the years, she was not being unreasonable.[67]) In any event, Digital Research's 16-bit version of CP/M was not far enough along in development, although the company had been promising it for some time. (It was eventually offered for the IBM PC, after PC-DOS had become dominant.)

In the end, Microsoft offered IBM a 16-bit operating system of its own. IBM called it PC-DOS, and Microsoft was free to market it elsewhere as MS-DOS. PC-DOS was based on 86-DOS, an operating system that Tim Paterson of Seattle Computer Products had written for the 8086 chip. Microsoft initially paid about $15,000 for the rights to use Seattle Computer Products's work. (Microsoft later paid a larger sum of money for the complete rights.) Seattle Computer Products referred to it internally by the code name QDOS for "Quick and Dirty Operating System"; it ended up as MS-DOS, one of the longest-lived and most-influential pieces of software ever written.[68]

MS-DOS was in the spirit of CP/M. Contrary to folklore, it was not simply an extension of CP/M written for the advanced 8086 chip. Paterson was familiar with a dialect of CP/M used by the Cromemco personal computer, as well as operating systems offered by Northstar and a few other descendants of the Altair. A CP/M users manual was another influence, although Paterson did not have access to CP/M source code. Another influence was an advanced version of Microsoft BASIC that also supported disk storage, which it was probably led to the use of a file allocation table by MS-DOS to keep track of data on a disk. The 86-DOS did use the same internal function calls as CP/M; actually, it used 8086 addresses and conventions that Intel had published in documenting the chip, to make it easy to run programs written for the 8080 on the new microprocessor. It used the CP/M commands "Type," "Rename," and

"Erase." MS-DOS also retained CP/M's notion of the BIOS, which allowed it to run on computers from different manufacturers with relatively minor changes.[69]

It is worth mentioning the differences between CP/M and MS-DOS, since these help explain the latter's success. A few changes were relatively minor: the cryptic all-purpose PIP command was changed to more prosaic terms like "Copy"; this made MS-DOS more accessible to a new generation of computer users but severed the historical link with the Digital Equipment Corporation, whose software was the *real* ancestor of personal computer systems. CP/M's syntax specified the first argument as the destination and the second as the source; this was reversed to something that seems to be more natural to most people. (The CP/M syntax was also used by Intel's assembler code and by the assembler for the IBM System/360).[70] More fundamental improvements included MS-DOS's ability to address more memory—a consequence of the Intel chip it was written for. MS-DOS used a file allocation table; CP/M used a less-sophisticated method. CP/M's annoying need to reboot the system if the wrong disk was inserted into a drive was eliminated. Doing that in MS-DOS produced a message, "Abort, Retry, Fail?" This message would later be cited as an example of MS-DOS's unfriendly user interface, but those who said that probably never experienced CP/M's "Warm Boot" message, which was much worse and sometimes gave the feeling of being kicked by a real boot. Several features may have been inspired by UNIX, for example, version 2, which allowed users to store files on a disk in a hierarchical tree of directories and subdirectories.[71] Tim Paterson later stated that he had intended to incorporate multitasking into DOS, but "they [Microsoft] needed to get something really quick."[72]

System software, whether for mainframes or for personal computers, seems always to require "mythical man-months" to create, to come in over budget, and to be saddled with long passages of inefficient code. Tim Paterson's initial work on 86-DOS took about two months, and the code occupied about 6 K.[73] MS-DOS was, and is, a piece of skillful programming. It was the culmination of ideas about interactive computing that began with the TX-0 at MIT. It has its faults, some perhaps serious, but those who claim that MS-DOS's success was solely due to Bill Gates's cunning, or to Gary Kildall's flying his airplane when IBM's representatives came looking for him, are wrong.

The PC and IBM

The Personal Computer was IBM's second foray into this market, after the 5100—it even had the designation 5150 in some product literature. Neither IBM nor anyone else foresaw how successful it would be, or that others would copy its architecture to make it the standard for the next decade and beyond. In keeping with a long tradition in the computer industry, IBM grossly underestimated sales: it estimated a total of 250,000 units; "[a]s it turned out, there were some *months* when we built and sold nearly that many systems.[74] MS-DOS transformed Microsoft from a company that mainly sold BASIC to one that dominated the small systems industry in operating systems. IBM found itself with an enormously successful product made up of parts designed by others, using ASCII instead of EBCDIC, and with an operating system it did not have complete rights to. It was said that if IBM's Personal Computer division were a separate company, it would have been ranked #3 in the industry in 1984, after the rest of IBM and Digital Equipment Corporation. Within ten years there were over fifty million computers installed that were variants of the original PC architecture and ran advanced versions of MS-DOS.[75]

"The Better is the Enemy of the Good"

The evolution of technological artifacts is often compared to the evolution by natural selection of living things. There are many parallels, including the way selective forces of the marketplace affect the survival of a technology.[76] There are differences, too: living things inherit their characteristics from their parents—at most two—but an inventor can borrow things from any number of existing devices. Nor does nature have the privilege that Seymour Cray had, namely, to start with a clean sheet of paper when embarking on a new computer design.

The history of personal computing shows that these differences are perhaps less than imagined. The IBM PC's microprocessor descended from a chip designed for a terminal, although Datapoint never used it for that. Its operating system descended from a "quick and dirty" operating system that began as a temporary expedient. The PC had a limit of 640 K of directly addressable memory. That, too, was unplanned and had nothing to do with the inherent limits of the Intel microprocessor. 640 K was thought to be far more than adequate; within a few years that limit became a millstone around the necks of programmers and users alike. The IBM PC and its clones allowed commercial software to come to the fore, as long as it could run on that computer or machines that were 100 percent compatible with it. Those visionaries who had predicted and longed for this moment now had mixed feelings. This was what they wanted, but they had not anticipated the price to be paid, namely, being trapped in the architecture of the IBM PC and its operating system.

Macintosh (1984)

Among those who looked at the IBM PC and asked why not something better were a group of people at Apple. They scoffed at its conservative design, forgetting that IBM had made a deliberate decision to produce an evolutionary machine. They saw the limitations of MS-DOS, but not its value as a standard. (Of course, neither did IBM at the time.) But what would personal computing be like if it incorporated some of the research done in the previous decade at Xerox's Palo Alto Research Center? The Xerox Star had been announced within months of the PC, but it failed to catch on. Some people at Apple thought they could be more successful.

For all the creative activity that went on at Xerox-PARC in the 1970s, it must be emphasized that the roots of personal computing—the microprocessor, the Altair, the bus architecture, the Apple II, BASIC, CP/M, VisiCalc, the IBM PC, the floppy disk, Lotus 1-2-3, and MS-DOS—owed *nothing* to Xerox-PARC research.

In 1979 that began to change. That fall Apple began work on a computer called the Macintosh. It was the brainchild of Jef Raskin, who before joining Apple had been a professor of computer science at UC San Diego. He had also been the head of a small computer center, where he taught students to program Data General Novas.[77] Raskin had also been a visiting scholar at Stanford's Artificial Intelligence Laboratory, and while there he became familiar with what was going on at Xerox-PARC. According to Raskin, he persuaded the Apple team then developing another text-based computer to incorporate the graphics features he had seen at PARC. Apple introduced that computer, the Lisa, in 1983. Like the Xerox Star, it was expensive (around $10,000), and sales were disappointing. Raskin's Macintosh would preserve the Lisa's best features but sell at a price that Apple II customers could afford.[78] As with so much in the history of computing, there is a dispute over who was responsible for the Macintosh.[79] Many histories describe a visit by

Figure 8.7
Personal computers: Apple Macintosh, 1984. Most Macintosh users soon found that the machine required a second, external disk drive. (*Source:* Smithsonian Institution.)

Apple cofounder Steve Jobs to PARC in 1979 as the pivotal moment in transferring PARC technology to a mass market. Work on the Macintosh was already underway at Apple by the time of that visit. The visit did result in Jobs' hiring several key people away from Xerox, however, and moving people is the best way to transfer technology. According to Raskin, the visit also resulted in Jobs' insisting that the Macintosh have features not present in the original design. Among those was the mouse (figure 8.7).[80]

In January 1984 Apple introduced the Macintosh in a legendary commercial during the Super Bowl, in which Apple promised that the Macintosh would prevent the year 1984 from being the technological dystopia forecast by Orwell's novel *1984*. The computer sold for $2,495—more than the $1,000 Raskin was aiming for, but cheaper than the Lisa. It was more expensive than an IBM PC, but no PC at that time, no matter what software or boards users added, could offer the graphical interface of the Macintosh.

The Macintosh used a Motorola 68000 microprocessor, whose architecture resembled that of the PDP-11. The computer came with a single disk drive, using the new 3 1/2-inch form, a high-resolution black-on-white monitor, a mouse, and 128K of memory. Most users found they soon had to upgrade to a 512K "Fat Mac"; they also found it necessary to purchase a second disk drive. A few programs were announced at the same time: a "paint" (drawing) program, based on work done at Xerox-PARC on a Data General Nova, and a word processor that came close to WYSIWYG.

A year later the Macintosh came with a rudimentary networking ability, called AppleTalk. This allowed the simple sharing of files and printers. Like so much about the system, it was simple, easy to use, and not challenged by the PC and its clones for years. But there was no hard disk option, so users could not effectively set up a Mac as a server to the others. A person using a Macintosh at home would not be connected to a network, and the Mac was unable to challenge the lead of IBM and its clones in an office environment, except in those offices where the graphics abilities were especially needed. Unlike the Apple II and the IBM PC, the Macintosh was "closed": users could not add boards and were discouraged from even opening up the case.[81] This was a bold—some argued foolish—departure from the prevailing wisdom, but it helped make the Macintosh cheaper, smaller, and faster than the Lisa or the Star. A version introduced in 1987 offered color and opened up the system, although Apple still tightly controlled the Macintosh's configuration.[82]

The Mac's elegant system software was its greatest accomplishment. It displayed a combination of aesthetic beauty and practical engineering that is extremely rare. One can point to specific details. When a file was opened or closed, its symbol expanded or contracted on the screen in little steps—somehow it just felt right. Ultimately this feeling is subjective, but it was one that few would disagree with. The Macintosh software was something rarely found among engineering artifacts. The system evolved as the Mac grew, and it was paid the highest compliment from Microsoft, who tried to copy it with its Windows program. One can hope that some future system will have that combination as well, but the odds are not in favor of it.

The Macintosh had more capability than the Alto, it ran faster than the Lisa, yet its software occupied a fraction of the memory of either of those predecessors. It was not just a copy of what Xerox had done at PARC. But there was a price for being so innovative: the Macintosh was difficult for programmers to develop applications software for, especially compared to MS-DOS. And though faster than the Lisa, its complex graphics meant that it could not be as fast as a DOS program, like Lotus 1-2-3, that used more primitive commands that were closer to machine code. Among sophisticated customers that created a split: one group favored the elegance and sophistication of the Mac, while others preferred the raw horsepower and access to individual bits that MS-DOS allowed. For those who were not members of the computer priesthood, the Macintosh was a godsend; whatever time was lost by its relative slowness was more than compensated for by the time the user did not have to spend reading an indecipherable users manual.

Microsoft had supplied some of the applications software for the Macintosh, but Apple developed and controlled its operating system in-house. Even before the Macintosh's announcement, other companies were trying to provide a similar interface for the IBM PC. In 1982 the creators of VisiCalc announced a product called VisiOn for the IBM PC that was similar to the Macintosh's interface but never lived up to its promise. IBM developed a program called Top View, and Digital Research developed GEM (Graphics Environment Manager) along the same lines. Microsoft came up with a product called Interface Manager, but early versions introduced in the mid-1980s sold poorly. Later versions of Interface Manager, renamed "Windows," would succeed dramatically. Version 3 of Windows, the breakthrough version, was not introduced until around 1990, so for the next seven years, IBM PCs and their clones would be known by the primitive MS-DOS interface inherited from the minicomputer world.

Like the IBM PC, the Macintosh's design created a barrier to expanding memory, only it was a more generous 4 megabytes instead of the PC's miserly 640 Kbytes. A laser printer offered in 1985 completed the transfer of Xerox-PARC innovations and allowed the Macintosh to keep a strong foothold in at least some offices. The Macintosh's equivalent of VisiCalc was a program called PageMaker from Aldus, introduced in 1985. When combined with the laser printer it allowed users to do sophisticated printing on an Apple, at a fraction of the cost of traditional methods.

The Clones

The personal computer revolution seems to have little to do with the age of mainframes that preceded it, but with the passage of time, we can find common themes. IBM's success with its System/360, and its need to give out a lot of technical information about it, led to the plug compatible industry, which in turn led to IBM's having to adjust its own product line. Something similar happened with the PC, only this time with a different outcome. Most of the IBM PCs, including the 8088 microprocessor, consisted of parts made by other manufacturers, who were free to sell those parts elsewhere. Microsoft, for instance, retained the right to sell its operating system to others. The core of what made a personal computer an "IBM PC" was the basic input-output system (BIOS), which was stored on a read-only memory chip. The idea went back to Gary Kildall's CP/M: let the BIOS be the only place where there could be code that tailored the operating system to the specifics of a particular machine. IBM owned the code in the personal computer's BIOS and prosecuted any company that used it without permission.

Around the time of the PC's announcement, three Texas Instruments employees were thinking of leaving their jobs and starting a company of their own, which they called Compaq. Legend has it that Rod Canion, Jim Harris, and Bill Murto sketched out an IBM-compatible PC on a napkin in a Houston restaurant. They conceived of the idea of reverse-engineering the IBM PC and producing a machine that would be 100 percent compatible. To get around IBM's ownership of the BIOS code, they hired people who had no knowledge of that code, put them in a "clean room," where they would not be corrupted by anyone sneaking the forbidden code to them, and had them come up with a BIOS of their own that replicated the functions of IBM's. This was expensive, but it was legal. The Compaq computer, delivered in 1983, was portable, although heavy. That was really a marketing ploy: At twenty-five pounds they "gave new meaning to the phrase pumping iron." What made it a success was its complete compatibility with the IBM PC at a competitive price. Compaq's sales propelled the company into the top 100 rankings of computer companies by 1985, one of the fastest trajectories of any start-up.[83]

Compaq's heroic efforts to break through IBM's control of its PC architecture did not have to be repeated too often. A small company named Phoenix Technologies also reverse-engineered the BIOS chip, and instead of building a computer around it, they simply offered a

BIOS chip for sale. Now building an IBM-compatible PC was easy. The trade press instituted a test for compatibility: would the machine run Lotus 1-2-3, which was written to take advantage of the PC's inner workings to gain maximum speed? Better still, would it run Flight Simulator, a program written by Bruce Artwick that exercised every nook and cranny of the IBM architecture?[84] If the answer was Yes and Yes, the machine was a true clone. The floodgates opened. Unlike its successful footwork during the times of System/360 and the plug compatibles, this time IBM lost control over its own architecture.

The introduction of IBM Compatibles and the Macintosh signaled the end of the pioneering phase of personal computing. Minicomputer and mainframe manufacturers could no longer ignore this phenomenon. In the late 1980s, companies like Novell would introduce more capable networking abilities for personal computers, which allowed networks of PCs to seriously challenge many large systems. After some hesitant

Figure 8.8
An early "transportable" computer. Osborne, ca. 1981. Just as revolutionary as its small size was the fact that the computer came with the CP/M operating system and applications software, all for less than $2,000.

Figure 8.9
An early "laptop" computer. Tandy Radio Shack TRS-80, Model 100, ca. 1983.
Like the Osborne, it used an 8-bit microprocessor. System software and the
BASIC programming language were supplied by Microsoft and included with the
machine. The machine shown here was much modified and extended and
served as the author's home computer for many years. (*Source:* Smithsonian
Institution.)

beginnings based on 8-bit designs, manufacturers developed portable
computers that were compatible with those on the desktop (figs. 8.8,
8.9). Commercial software, driven relentlessly by the marketplace
created by Microsoft, led to applications that likewise challenged the
mini and mainframe world. By 1991 the IBM-compatible computers,
based on advanced versions of the Intel 8086 chip and running Windows
3.1, brought the Macintosh's features to the business and commercial
world. For reasons having to do more with IBM's poor management
than anything else, companies like Compaq and Dell would earn more
profits selling IBM-compatible computers than IBM would. IBM
remained a major vendor, but the biggest winner was Microsoft, whose
operating system was sold with both IBM computers and their clones.

The personal computer revolutionized the office environment, but it had not become a revolutionary machine in the political or cultural sense, the sense that Stewart Brand and others had predicted and hoped for. Computers came "to the people," but for a price: corporate control.

9

Workstations, UNIX, and the Net, 1981–1995

The VAX brought the power of a scientific mainframe into the engineering division of a company. Beginning in the 1980s a new class of computers brought that power to the individual desktop. These "workstations" did that by using an inexpensive microprocessor, typically the Motorola 68000. The lower cost was relative, less than a VAX but much more than a PC. Their architecture and physical design also had much in common with personal computers. The difference was their use of the UNIX operating system, and their extensive networking abilities that allowed sharing data and expensive peripherals like plotters.

First out of the gate was Apollo, of Chelmsford, Massachusetts. Its founder, Bill Poduska, had previously cofounded Prime, the company that pioneered the 32-bit mini. In 1981 Apollo delivered a product that used the Motorola microprocessor and its own operating and networking systems, called Domain.[1] The price for a single workstation (the name apparently originated at this time) began at $40,000.[2] As Wang and Xerox had already discovered, having a computer at each worker's desk, networked to other machines, was more efficient than having a centralized time-shared computer accessed through "dumb" terminals. The workstations sold well to firms like GE-Calma and Mentor Graphics, who used them for computer-aided design and engineering of products like circuit boards, aircraft components, and automobiles. By mid-1980 Apollo had shipped 1,000 systems. It soon encountered competition, and in 1989 it was acquired by Hewlett-Packard, which had entered the market with a workstation (the 9000) of its own design in 1985.[3]

Competition soon came from a new company located just down the road from Apple in Silicon Valley. SUN Microsystems, founded in early 1982 by Vinod Khosla, continued the long tradition of effecting a transfer of technology from a publicly funded university research project to a profit-making company by moving key people. In this case the project was the Stanford University Networked workstation (hence the company's name); the person was Andy Bechtolsheim of Stanford. A parallel transfer brought its software; Bill Joy, who, with ARPA funding, had enhanced the UNIX operating system while at Berkeley (figure 9.1). Joy moved across the Bay to join SUN in June 1982.[4] SUN had already introduced a workstation in May, with the more capable SUN-2 following shortly. Prices were in the $20,000 range. Among the things that Bill Joy brought with him to Mountain View was Berkeley UNIX.

UNIX: From New Jersey to California

Bill Joy's move to SUN signified the last stop on a transcontinental journey for UNIX: it began in New Jersey, stopped in Champaign-Urbana for a while, and was extensively rewritten while at Berkeley. In Silicon Valley it would move from its academic niche into commercial use. Berkeley UNIX was a key to SUN's success and helped push the Internet out of its ARPA roots in the 1990s.

Bell Laboratories, where UNIX was created, was a part of AT&T, a regulated monopoly. Before it breakup in 1981, AT&T had agreed not to engage in commercial computing activities; in return it enjoyed steady

Figure 9.1
Bill Joy. (*Source*: SUN Microsystems.)

and regular profits from its business of providing telephone service throughout the United States. Ken Thompson and Dennis Ritchie said that they initially thought of UNIX as something to be used only within Bell Labs, but several factors conspired to all but ensure that it would "escape,"[5] most importantly, that AT&T would not offer to sell it for a profit. Thus universities could obtain a UNIX license for a nominal cost—a few hundred dollars at most (commercial customers had to pay more). Also important was that UNIX was not a complete operating system, as it was then understood, but rather a set of basic tools that allowed users to manipulate files in a simple and straightforward manner.

The result was that UNIX was a godsend for university computer science departments. For a nominal fee, AT&T's Western Electric subsidiary supplied UNIX's source code. The code was written in the C programming language, not machine language. That meant that although developed on DEC computers, UNIX could run on any machine that had a C compiler. By contrast, most computer vendors guarded source code as their family jewels, seldom gave it out, and did all they could to lock a customer into their products. And no one minded if a university modified the UNIX to enhance its capabilities. That was what graduate students—and many bright undergraduates as well—were in school for. Thus all the things needed to turn AT&T's UNIX into a practical system—for example, tailoring it for specific monitors, printers, and storage systems—got done cheaply. That work was excellent training for students as well. When these students graduated, they took these skills with them, along with an appreciation for the software that had allowed them such freedom. Some of them found after graduation and entry into the "real world" that corporate computer centers had little room for that kind of easy access to the lower levels of a machine. Those programmers turned themselves into evangelists, spreading the UNIX way of programming into the corporate world.

Bill Joy was one of many students who had tinkered with AT&T's version of UNIX hoping to make it better. The University of California at Berkeley obtained a UNIX tape in 1974, following a visit by Ken Thompson. The system was soon running on several PDP-11s on the campus. Bill Joy also arrived on the campus that year.[6]

Thompson and Ritchie's immediate goal in creating UNIX was to have a way of sharing files easily. They also were in need of programming tools, after the Labs canceled work on Multics in 1969. After the initial effort on the PDP-7, they moved UNIX to a PDP-11 and rewrote it in C. For those reasons UNIX was frugal in the extreme: two- or three-letter abbreviations were the norm for most commands. Sometimes these abbreviations corresponded to what the command did (e.g. "cp" for copy); other times the relationship was tenuous (e.g. "grep": globally search for the regular expression and print).[7] One of UNIX's tenets was that the output of any UNIX process be usable as input for another. That gave UNIX enormous power and flexibility. It also meant an absence of features like page-breaks or information that revealed the status of the file being worked on, since these would clutter up the file if "piped" to another process. It also made it easy to write programs that acted like a "virus," programs that could replicate themselves by producing executable code as output. UNIX was powerful, but not useful for the hoi polloi.[8]

Bill Joy and his fellow students at Berkeley set out to make UNIX more accessible. The initial impetus came when the primitive Model 33 Teletypes were replaced by "dumb" CRT-based terminals (Lear-Siegler ADM-3s). By 1978 Joy was offering tapes of the first Berkeley Software Distribution (BSD) at a nominal cost to his friends and colleagues around the country.[9] The enhancements to Bell Labs's UNIX strained the capabilities of the PDP-11, and work shifted over to a VAX, which DEC had just introduced. In 1980 ARPA threw its support behind Berkeley UNIX as a common system the agency could recommend for all its clients. That UNIX was, in theory, portable to computers from manufacturers other than DEC was a main reason. Among the many enhancements added to Berkeley UNIX (in version 4.2 BSD) was support for networking by a protocol known as TCP/IP, which ARPA promoted as a way to interconnect networks. This protocol, and its bundling with Berkeley UNIX, forever linked UNIX and the Internet.[10]

The Ironies of UNIX

Although UNIX was written by Thompson and Ritchie for themselves and for researchers like them, it found its way into general use. UNIX's strength came from the fact that Thompson and Ritchie had a firm sense of what they wanted and what they did not want; it did, however, sprout a number of incompatible, baroque, and feature-laden versions. Berkeley UNIX was developed on a VAX; Digital Equipment only grudgingly tolerated UNIX on the VAX and steered its customers to VMS. VAX computers, running Berkeley UNIX with TCP/IP, helped transform the restricted ARPANET to the wide-open Internet; when the Internet broke through to public use in the 1990s, Digital Equipment Corporation hardly profited. UNIX, born in a collegial environment, was best known for the way it made the sharing of files easy; that also meant that UNIX systems would be vulnerable to viruses and unauthorized intrusions by hackers. UNIX spread because AT&T gave it away; Berkeley UNIX generated enormous profits for SUN and other workstation vendors but not for AT&T. When AT&T was allowed to market computer products after divestiture, it failed to set a standard for, or profit from, its creation. AT&T touted UNIX as an "open" system; other companies introduced incompatible versions for little more reason than to be different from AT&T, a competitor. The name UNIX was a pun on Multics, chosen to imply a simpler system; it was not intended to imply unity, and after 1985 it was anything but universal. Finally, for all its qualities, UNIX never challenged the dominance of Microsoft's MS-DOS/Windows, which became the *real* desktop standard after 1981. Such were the ironies of UNIX.

SUN Microsystems took full advantage of the strategy of open systems. It not only used UNIX and the Motorola microprocessor but also a standardized version of Ethernet, and an internal bus that others were free to adopt. Apollo and Hewlett-Packard retained their proprietary systems, although each eventually offered UNIX as well. The SUN model, its profits, and the availability of venture capital spawned the JAWS phenomenon (just another work station). Many of the competitors tried to find a toehold in a specific niche: for example, Symbolics produced a workstation that was optimized to run the LISP programming language for artificial intelligence applications. But most failed in the face of SUN's open, general-purpose machines offering good performance at a low price. Besides HP and Apollo, the only serious exception was Silicon Graphics. Like SUN, Silicon Graphics commercialized a university-sponsored research project; a "geometry engine" chip that performed the calculations needed to represent three-dimensional images. Like SUN, technology transfer was effected by moving a key person—Jim Clark, who had studied under David Evans and Ivan Sutherland at Utah and had developed a prototype engine at Stanford. (Clark later left Silicon Graphics to found Netscape, a company aimed at commercializing software for the Internet.)

VAX Strategy

Just as the personal computer field was divided into the DOS and Macintosh camps, there was also a battle going on in the scientific and engineering field.

Workstation companies could not compete with mini and mainframes on the basis of the power of a single machine; they competed by selling networks of machines, whose collective power they alleged was greater than the sum of the parts. SUN stated this succinctly in its advertising slogan, "The network is the computer." Throughout the 1980s Digital Equipment Corporation had a powerful strategy of its own that combined SUN's emphasis on networking with IBM's concept of a unified family of computers. DEC's plan was to offer the customer a single architecture, the VAX, with a single operating system, VMS, in solitary or networked configurations that ranged from desktop to mainframe capability. The only part of the VAX Strategy that was not Digital's own was the networking—Ethernet, which DEC obtained in an agreement with Intel and Xerox. The VAX 11/780 was followed by smaller machines like the 11/750 in 1980 and MicroVax II in 1984; and larger machines like the 8600 (Venus) in 1984 and the 9000 (which DEC called a mainframe) in 1990.[11]

The VAX Strategy had risks of its own, risks that resembled IBM's "betting the company" with the System/360. DEC had to convince the customer that it could supply everything, from office automation software to printer ribbons, yet not convey a sense that it was charging excessively high prices. It had to design and build products with good performance across the entire line. DEC had to stop marketing its own competing architectures, including the 36-bit computers descended from the PDP-10. The PDP-10 series was based on an old design and was incompatible with the VAX; a simple conversion of PDP-10 applications to the VAX seemed straightforward.

DEC misjudged how beloved the PDP-10 was among its customers—either forgetting, or perhaps never realizing, how much modern computing was a child of that venerable machine. There was even an outcry when DEC announced it was phasing out DECtape. DEC's announcement that no PDP-10 machines would be developed after 1983 was met by strong customer resistance, but Ken Olsen stood firm. A rational decision? Yes, but how does one measure its intangibles? The PDP-10 was the system that first created the illusion of personal computing. Its TOPS-10 operating system inspired personal computer system software. It was the computer that William Gates learned to program on, and to write Microsoft BASIC on. The early ARPANET linked up more PDP-10s than any other computer. The PDP-10 hardware was long obsolete. The mental model of computing that it created is not only still alive, it greets us every time we turn on a networked personal computer or workstation.

But there *was* a deviation from the VAX strategy—at the personal computer level. There DEC introduced not one but three incompatible machines in 1982. Not only were these incompatible with the VAX, they were not fully compatible with the IBM PC either. One of them, the Rainbow, sold modestly but could not slow down the IBM-compatible juggernaut.[12] The lack of full compatibility with the IBM PC standard was a fatal error.

The VAX Strategy worked well through the 1980s. By that time the IBM 360-370 architecture was becoming top-heavy with enhancements, modifications, and extensions. Meanwhile IBM was enjoying brisk sales of its mid-range System/38 and of course its PC, neither compatible with the System/370. IBM's customers were confused. IBM salesmen, who grew up selling "big iron" (large mainframes) and who regarded anything smaller as toys, often added to the confusion by trying to steer every customer to the System/370. For a brief and glorious moment—just before the stock market crash in October 1987—it looked as if DEC was not only in a solid number two position but poised to do the unthinkable, surpass IBM and dominate the industry. But it was not to be. DEC's stock was among the heaviest losers that month, and by 1990 the drawbacks to its VAX strategy, combined with its inability to bring new VAX products to market, began a series of quarters in which DEC lost most of the money it had earned through the entire minicomputer era. DEC probably could have weathered an assault from UNIX workstations or from the IBM PC if either occurred alone, but the combination was too much.

RISC

If those two blunders were not enough, DEC made a third. This blunder involved what had been the company's strongest suit; computer architecture. Simply put, DEC failed to develop new architectures to reflect the changes in chip and software technology that had taken place in the 1970s.

Although its name implied an extension of the PDP-11 minicomputer, the VAX architecture had a lot in common with the IBM System/360 and its descendants. Like the 360, its instruction set was contained in a microprogram, stored in a read-only memory. Like the 360, the VAX presented its programmers with a rich set of instructions that operated on data in almost every conceivable way. The 370/168 had over 200 instructions, the VAX 11/780 over 250. There were sets of instructions for integers, floating-point numbers, packed decimal numbers, and character strings, operating in a variety of modes.[13] This philosophy had evolved in an environment dominated by magnetic core memory, to which access was slow relative to processor operations. Thus it made sense to specify in great detail what one wanted to do with a piece of data before going off to memory to get it. The instruction sets also reflected the state of compiler technology. If the processor could perform a lot of arithmetic on data with only one instruction, then the compiler would have that much less work to do. A rich instruction set would reduce the "semantic gap" between the English-like commands of a high-level programming language and the primitive and tedious commands of machine code. Cheap read-only memory chips meant that designers could create these rich instruction sets at a low cost if the computer was microprogrammed.[14]

Those assumptions had been long accepted. But computer science was not stagnant. In the mid-1970s John Cocke of IBM looked at the rapid advances in compilers and concluded that a smaller set of instructions, using more frequent commands to load and store data to and from memory, could operate faster than the System/370. Thomas Watson Jr. once wrote a memo describing IBM's need to have "wild ducks" among its ranks—people who were not content to accept conventional wisdom about the way things were done. Cultivating such people in the conservative culture of IBM was not easy, but Watson knew, perhaps better than any other computer executive, that IBM could not survive without them. John Cocke, with his then-radical ideas about computer design, fit that description.[15]

Cocke's ideas led to an experimental machine called the IBM 801, completed under the direction of George Radin in 1979.[16] For many reasons, including the success and profits of the 370 line and its successors, IBM held back introducing a commercial version of the design. (The IBM-RT, introduced in 1986, was a commercial failure and did not exploit the idea very well.) Still, word of the 801 project got out, along with a rumor that it could execute System/370 programs at much faster speeds although it was a smaller computer. By the late 1970s magnetic core had been replaced by semiconductor memory, whose access times matched the speeds of processors. Frequent load and store instructions no longer exacted a speed penalty. Finally, some researchers looked at the VAX and concluded that they could not extend its design any further; they began looking for alternatives.

In 1980 a group at Berkeley led by David Patterson, after hearing "rumors of the 801," started a similar project called RISC—"Reduced

Instruction Set Computer." Another project, called MIPS (Millions of Instructions Per Second), began in 1981 at Stanford under the leadership of John Hennessy.[17] As they publicized their work they were met with skepticism: RISC looked good in a university laboratory but did not address the real needs of actual customers (figure 9.2). One trade journal even worried that RISC, from the start associated with UNIX, was not well-suited for data-processing jobs written in COBOL.[18] Meanwhile, sales of Intel-based PCs, the VAX, and the System/370 family—all complex instruction-set processors—were booming. With a massive buildup of the Defense Department under President Ronald Reagan, Wall Street was enjoying another round of go-go years. Those watching the trajectory of their stocks in DEC, Data General, IBM, and Wang were not worried about RISC.

SUN Microsystems' products initially used the Motorola 68000 microprocessor, whose design was very much in the spirit of the PDP-11 and VAX. Beginning in 1987 and probably owing to Bill Joy's influence, SUN introduced a workstation with a RISC chip based on Patterson's research

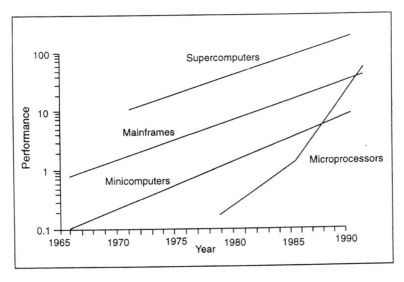

Figure 9.2
The Benefits of RISC: Computer Performance of Microprocessors. Hennessy and Patterson used graphs such as this one to show how inexpensive microprocessors would soon overtake all other classes of computers in performance. (*Source*: John L. Hennessy and Norman P. Jouppi, "Computer Technology and Architecture: an Evolving Interaction," *IEEE Computer* [September 1991]: 19.) © IEEE.

at Berkeley. Called SPARC (Scalable Processor Architecture), this design did more than anything else to overcome skepticism about RISC. Hennessy and Patterson became evangelists for RISC, buttressed by some impressive quantitative measurements that showed how a RISC design could squeeze much more processing power out of a piece of silicon than conventional wisdom had thought possible. More telling, their data showed that RISC offered a way of improving microprocessor speeds much more rapidly than mini and mainframe speeds were improving—or could improve. The unmistakable implication was that the puny, cheap microprocessor, born of a pocket calculator, would soon surpass minicomputers, mainframes, and even supercomputers, in performance. If true, their conclusions meant that the computer industry as it had been known for decades, and over which the U.S. Justice Department fought IBM throughout the 1970s, was being driven to extinction by its own offspring.[19]

SUN went a step further to promote RISC: they licensed the SPARC design so that other companies might adopt it and make SPARC a standard.[20] The combination of a license to copy the SPARC processor, plus Berkeley UNIX, made it almost as easy to enter the workstation market as it was to make an IBM compatible PC. SUN gambled that it, too, would benefit by continuing to introduce products with high performance and a low price. They succeeded, although such a philosophy meant it had to accept slim profit margins, since SUN could not own the architecture.

The Stanford MIPS project also spawned a commercial venture, MIPS Computer Systems, which also helped establish a commercial market for RISC microprocessors. Digital Equipment Corporation bought a chip from MIPS for one of their workstations in 1989—even DEC now admitted that RISC was not going away. (An internal RISC project at DEC, called Prism, had been canceled in 1988.) Silicon Graphics also based its newer workstations on MIPS microprocessors.[21] Hewlett-Packard converted its line of workstations to a RISC design called precision architecture. After failing with the RT, IBM introduced a successful RISC workstation in 1990, the R/6000. In the early 1990s Apple and IBM joined forces with Motorola to produce a RISC microprocessor called Power PC, which they hoped would topple the Intel 8086 family. IBM's role in the design of the Power PC was a fitting vindication of the ideas of John Cocke, the "wild duck" who started the whole phenomenon.

Networking I: Ethernet

A RISC architecture, UNIX, and scientific or engineering applications differentiated workstations from personal computers. Another distinction was that workstations were designed from the start to be networked, especially at the local level, for example, within a building or a division of an engineering company. That was done using Ethernet, one of the most significant of all the inventions that came from the Xerox Palo Alto Research Center. If the Internet of the 1990s became the "Information Superhighway," then Ethernet became the equally important network of local roads to feed it. As a descendent of ARPA research, the global networks we now call the Internet came into existence before the local Ethernet was invented at Xerox. But Ethernet transformed the nature of office and personal computing before the Internet had a significant effect. How Ethernet did that will therefore be examined first.

In his autobiography, Herb Grosch notes with pride that Grosch's Law (see chapter 6), conceived in 1950 before there were even commercial installations, held through waves of technical innovation.[22] In the late 1970s it was especially frustrating to realize that even though one could get all the functions of an IBM 7090 on a fifty-dollar chip, buying an ensemble of cheap systems did not give as much computing power as spending the same money on one large system. Ethernet changed that equation, by enabling small clusters of workstations and, later, PCs to work together effectively.

Ethernet was invented at Xerox-PARC in 1973 by Robert Metcalfe and David Boggs. Metcalfe was an alumnus of MIT's Project MAC, and in 1969 he helped link MIT to ARPANET, connecting a PDP-10 to it. He moved to Xerox-PARC in 1972; one of his first tasks there was to hook up PARC's PDP-10 clone, the MAXC, to ARPANET. "As of June 1972, I was the networking guy at PARC."[23] Metcalfe connected Xerox's MAXC to ARPANET, but the focus at Xerox was on *local* networking: to connect a single-user computer (later to become the Alto) to others like it, and to a shared, high-quality printer, all within the same building. The ARPANET model, with its expensive, dedicated Interface Message Processors was not appropriate.

When Metcalfe arrived at PARC there was already a local network established, using Data General minicomputers linked in a star-shaped topology.[24] Metcalfe and his colleagues felt that even the Data General network was too expensive and not flexible enough to work in an office setting, where one may want to connect or disconnect machines frequently. He also felt it was not robust enough—the network's operation depended on a few critical pieces not failing. He recalled a network he saw in Hawaii that used radio signals to link computers among the Hawaiian islands, called ALOHAnet.[25] With this system, files were broken up into "packets," no longer than 1000 bits long, with an address of the intended recipient attached to the head of each. Other computers on the net were tuned to the UHF frequency and listened for the packets, accepting the ones that were addressed to it and ignoring all the others.

What made this system attractive for Metcalfe was that the medium—in this case radio—was passive. It simply carried the signals, with the computers at each node doing the processing, queuing, and routing work. The offices at Xerox PARC were not separated by water, but the concept was perfectly suited for a suite of offices in a single building. Metcalfe proposed substituting a cheap coaxial cable for the "ether" that carried ALOHAnet's signals.[26] A new computer could be added to the "Ethernet" simply by tapping into the cable. To send data, a computer first listened to make sure there were no packets already on the line; if not, it sent out its own. If two computers happened to transmit at the same time, each would back off for a random interval and try again. If such collisions started to occur frequently, the computers themselves would back off and not transmit so often.[27] By careful mathematical analysis Metcalfe showed that such a system could handle a lot of traffic without becoming overloaded. He wrote a description of it in May 1973 and recruited David Boggs to help build it. They had a fairly large network running by the following year. Metcalfe recalled that its speed, around three million bits per second, was unheard of at the time, when "the 50-kilobit-per-second (Kbps) telephone circuits of the ARPANET were considered fast."[28]

Those speeds fundamentally altered the relationship between small and large computers. Clusters of small computers now, finally, provided an alternative to the classic model of a large central system that was time-shared and accessed through dumb terminals.

Ethernet would have its biggest impact on the workstation, and later PC, market, but its first success came in 1979, when Digital Equipment Corporation, Intel, and Xerox joined to establish it as a standard, with DEC using it for the VAX. Gordon Bell believes that it was fortunate in becoming rooted firmly enough to withstand the introduction of a competing scheme, Token Ring, by IBM.[29] UNIX-based workstations nearly all adopted Ethernet, although Token Ring and a few alternate schemes are also used.

DOS-based personal computers were late in getting networking. Neither the Intel processors they used, nor DOS, was well-suited for it. There was a social factor at work, too: it was, after all, a *personal* computer—why should one want to connect it with someone else's, or even worse, have to share resources like mass storage or a printer? IBM's entry into the market made personal computers respectable. But many users had not forgotten the idealistic notions of empowerment that had spawned the PC in the first place. Personal computers eventually became networked, too, though by a circuitous route.

Workstations and "VAXen" found a market among engineers and scientists, but with only a few exceptions the commercial office environment continued to use office automation systems from Wang, IBM, and others.[30] Good word-processing programs, and especially the spreadsheet program 1-2-3, introduced for the IBM Personal Computer by Lotus Development Corporation in 1982, helped bring the IBM PC and its clones into that market. Lotus 1-2-3 was like VisiCalc, with some additional features. (The "2" and "3" implied the features of graphing and database capabilities.) Because it was written specifically for the IBM PC, using assembly language to gain faster performance, it ran much faster than other spreadsheets. In corporate offices the little stream of personal computers that began with the Apple II became a flood of IBM PCs and PC-compatibles, running Lotus 1-2-3, word processing software like Word Perfect, and database programs like dBase III.

The people running corporate information services departments saw this flood as a Biblical plague. Purchasers of PCs and PC software were driven by personal, not corporate, needs. These personal needs were often driven by advertising hyperbole in the trade journals, which promised digital Utopia for anyone smart enough to buy the latest piece of software, and smart enough to bypass the bureaucracy of corporate purchasing and evaluation. Information services people, wedded to Wang word processors or IBM mainframes, were losing control of what they were hired to manage.

Both sides were right. The PC and DOS standards led to commercial software that was not only inexpensive but also better than what came with centralized systems. The PC also led to poor quality software that wasted more company time than it saved. The "personal" in the PC meant also that a worker's choices of software, based on personal satisfaction, did not always mesh with the goals of his or her employer.

By the mid-1980s it was clear that no amount of corporate policy directives could keep the PC out of the office, especially among those employees who already had a PC at home. The solution was a technical fix: network the PCs to one another, in a local-area network (LAN). By 1984 there were over twenty products announced or on the market that claimed to connect IBM PCs in a local-area network.[31] Many of these performed poorly, given the limits of the Intel processor that powered early IBM Personal Computers and clones. After more advanced versions of the Intel chip became common (especially the 80386, introduced in 1985), there was a shakeout, and networking for PCs became practical. The company that emerged with over half the business by 1989 was Novell, located in the Salt Lake City area. Novell's Netware was a complex—and expensive—operating system that overlaid DOS, seizing the machine and directing control to a "file server"—typically a PC with generous mass storage and I/O capability (the term "server" originated in Metcalfe and Boggs's 1976 paper on Ethernet). By locating data and office automation software on this server rather than on individual machines, some measure of central control could be reestablished.

Networking of PCs lagged behind the networking that UNIX workstations enjoyed from the start, but the personal computer's lower cost and better office software drove this market. Some predicted a convergence of PC and UNIX workstations, but that did not occur. Office workers used computers for word processing, basic accounting using a spreadsheet, and those who filed data used simple database programs. There was an abundant selection of good software for these applications, at reasonable prices, for the PC. Those who needed graphics, say for desktop publishing, could buy a Macintosh. There was no incentive to spend more money for a UNIX workstation, which offered less workaday software. Recall that UNIX was more a set of tools than a complete operating system. And it was even harder to use and understand than DOS. Nor was UNIX as standardized. Several windowing systems, including X-Windows, came to the UNIX world in the late 1980s, but no standard emerged that compared to Microsoft Windows for DOS machines.[32]

For all the criticisms of the IBM PC architecture and of MS-DOS, it is a measure of their quality that those standards could evolve so far and remain not just viable but dominant for so long. By 1995 the standard still included a descendant of the Intel 8086 processor, an advanced version of DOS enhanced by Microsoft Windows, version 3.1 or later, and a networking scheme from Novell or a handful of other companies. The Macintosh also evolved and remained competitive, although its share of the market declined in the face of competition from advanced versions of Windows for the IBM-compatible machines.

Local networking took the "personal" out of personal computing, at least in the office environment. (One could still do whatever one wanted at home.) PC users in the workplace accepted this Faustian bargain. The more computer-savvy among them resisted, but the majority of office workers hardly even noticed how much this represented a shift away from the forces that drove the invention of the personal computer in the first place. The ease with which this transition took place shows that those who believed in truly autonomous, personal computing were perhaps naive. Still, the networked office computers of the 1990s gave their users a lot more autonomy and independence than the time-shared mainframes accessed through "dumb terminals" or "glass Tele-types" in the 1970s. It was just not how the people at *Byte* magazine or the Homebrew Computer Club had imagined things would evolve.

Networking II: Internet

Most benefits of connecting office workers to a LAN went to adminis-trators and managers. For their part, users no longer had to worry about backing up files—something few PC owners ever learned to do faithfully anyway—and they could now exchange files and messages with one another using electronic mail. But there was one unanticipated, very important thing that users connected to a LAN got in return—access to the Internet.

The present-day Internet, though well known, is hard to define. It is descended from the ARPANET described in chapter 6 (figure 9.3). Like ARPANET and the other networks described earlier, the Internet uses "packet switching." Sending a message does not require a dedicated connection from one computer to another, as, say, one has when calling someone on the telephone.[33] There are however several major differ-ences. The Internet is not a single network but rather the connection of many different networks across the globe; hence the name. Some of those networks are open to the public, not just to a restricted or privileged community. (Note there are still many networks that are restricted, e.g., one used by a bank for its internal operations.) Finally, the Internet allows communication across these different networks by its use of a common protocol, TCP/IP (transmission control protocol/ internet protocol). This interconnection of networks to one another, using the glue of TCP/IP, constitutes the present-day Internet.[34]

The Internet made its way into general use by a combination of social and technical factors. Among the former was the shift of financial and administrative support from ARPA, to the National Science Foundation

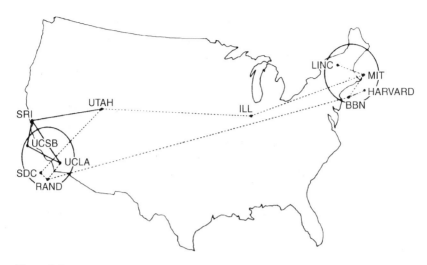

Figure 9.3
ARPANET, as of December 1970. The map shows 11 nodes, mainly concentrated in the Los Angeles and Boston regions. (*Source*: DARPA.)

in the 1980s, and then in the 1990s to entities that allowed Internet access to anyone, including those who would use it for commercial purposes. As recently as 1992, Internet users were about evenly distributed among governmental, educational, military, net-related, commercial, and nonprofit organizations (identified in their addresses, respectively, by the sequences ".gov," ".edu," ".mil," ".net," ".com," and ".org"). By 1995 commercial users overwhelmed the rest, and the phrase "'X' dot com," where "X" is the name of a corporation, has entered our vocabulary. Documenting this sequence of events is a work still in progress and that aspect of the Internet's growth will not be discussed further here.[35]

The technical factors behind the emergence of the Internet are better known. One has already been mentioned: ARPA's support for the development of, and its decision in 1980 to adopt, the TCP/IP protocol. ARPA's support, and the protocol's inclusion in Berkeley UNIX, meant that it would become widely available in universities and would not be held as a proprietary product by IBM, DEC, or any other computer company.[36]

Another factor was the rise of local area networks. The initial goal for ARPANET was the sharing of expensive computer resources; individuals would gain access to these resources through terminals that were connected to mainframes by time-sharing. (As the ARPANET took shape one could also, in a few places, connect a terminal directly to the network through a terminal interface processor (TIP)—a variation of the IMP concept.) With the invention of Ethernet in 1973, and the personal computer the following year, the economics of computing changed. Computing power was no longer scarce. Time-sharing matured and became available on many mainframes, but it was supplanted by client-server computing that descended from the work at Xerox-PARC. And throughout this era Moore's Law ruled: computing power, as measured by the density of the silicon chips that went into these machines, was doubling about every eighteen months.

Local area networks made it possible for large numbers of people to gain access to the Internet. Ethernet's speeds were fast enough to match the high speeds of the dedicated lines that formed the Internet's backbone. High-speed networking had always been among the features workstation companies wanted to supply—recall SUN's marketing slogan: "The Network is the Computer." What had not been anticipated was how advances in personal computers, driven by ever more powerful processors from Intel, brought that capability to offices and other places outside the academic and research worlds. By the late 1980s those with UNIX workstations, and by 1995 those with personal computers on a LAN, all had access to the Internet, without each machine requiring a direct connection to the Internet's high-speed lines.

Ethernet's high data rates thus provided a way of getting around the fact that communication speeds and data capacity had not kept up with the advances in computer processing speeds and storage. Gordon Moore's colleague at Intel, Andrew Grove, came up with his own "law" to describe this disparity: while chip density doubles every eighteen months (Moore's Law), telecommunications bandwidth doubles every 100 years (Grove's Law).[37] Bandwidth growth has picked up since 1990, but it still lags well behind the growth of chip density. Grove believes the disparity is a result of overregulation of the telecommunications industry. Whatever the cause, it is true that the telephone lines coming into homes and offices cannot handle data at more than about 50 thousand bits per second—usually less, and in any case well below the speeds achieved by Ethernet and *required* by many Internet activities.

Since the mid-1990s modem manufacturers have made heroic efforts to improve data rates for ordinary telephone connections. Plans have also emerged to use other wires that come into the home for Internet traffic: the line that carries cable television or even the power lines. Various satellite or microwave wireless technologies are also being developed. One of these approaches will probably break the bottleneck. Meanwhile, people do connect their home computers to the Internet by dialing a local telephone number, but the access they get is a fraction of what they can find at the office, laboratory, or university. For now, Grove's Law seems to hold.

Networking III: the World Wide Web

As the Internet emerged from its roots in ARPA, it began to change. The initial activities on the Internet were ARPANET derived: users could log on to a remote computer, transfer large files from one machine to another, and send mail. The first two activities later known as ("Telnet" and "FTP") were explicit goals of the original ARPANET; mail was not, but it emerged soon after the first few nodes were working. Early e-mail facilities were grafted onto the file-transfer operation, but before long dedicated e-mail software was developed, with most of the features found in modern e-mail systems already in place (e.g. the ability to reply to someone, save a message, or send a message to a list).[38]

The first serious extension to that triad gave a hint of what the popular press calls a "virtual community" based on the Internet. Whether that phrase has any meaning, and if so, what it is, will be discussed later, but what evoked it was the development of news or discussion groups on the early Internet. Although these groups are associated with the Internet, for years only those with access to UNIX systems had access to them. For the general public they were anticipated in the personal computer arena by so-called bulletin-board systems (BBSs), which as the name implies, acted like bulletin boards, on which anyone could post a note for all to read. BBSs typically ran on limited facilities, using public-domain software running on an IBM XT or the equivalent. Key technical developments were the introduction in 1981 of an inexpensive modem by Hayes Microcomputer Products, and of the XT itself in 1983, with its 10-megabyte hard disk. Users dialed into these bulletin boards with a local telephone call, at a rate of a few hundred bits per second.[39] But they worked and were well liked, and some remain in use in the mid-1990s.

UNIX-based new groups first appeared after 1979, somewhat independently of the mainstream ARPANET-Internet activities and under the general name of Usenet.[40] These were arranged into a set of major categories, for example, "comp" for computers, or "rec" for hobbies. A category called "alt" ("alternative") included subcategories that tended to the bizarre; of course these were the ones that newspaper reporters focused on in the first stories about the Internet to appear in print. Reporters knew that a story about alternative sexual preferences would attract more readers than a story about UNIX file structures, even if the latter were far more important to the operation of the Net. And perhaps editors were somewhat fearful that, if unchecked, news groups would put them out of business (they probably will). The resulting stories were hardly accurate but did kindle a general interest in the Internet, although at a time when few people outside universities could gain access to these groups.

Another bulletin board system emerged at university and industrial computer centers that used IBM mainframes. Beginning around 1981, IBM system administrators at the City University of New York and Yale developed BITNET, which linked large IBM systems by a clever bit of programming that treated messages as though they were data punched on decks of 80-column cards. With BITNET came LISTSERV, a system for managing discussion groups. As might be expected among IBM mainframe users, LISTSERV discussions were a little more formal and serious than those on Usenet, although they also had a wide range. After about a decade and a half of parallel operations, all three of these streams blended into a community of news, discussions, real-time chat, and other venues for exchanging information, now found on the World Wide Web.

Gopher, WAIS

File transfer and remote log-in were fine if users already knew where a desired file was located and what it was called. Perhaps that information might be posted on a news group in response to a query, but as the Internet grew, the amount of information available on it overwhelmed these facilities. The Internet began to feel like a large library that had no card catalog, or a used bookstore that had an incredible number of great books at bargain prices but with the books piled at random on the shelves. In 1990 or early 1991, programmers at the University of Minnesota responded by creating Gopher, in honor of the university mascot. Gopher at first allowed students and faculty, including those with little experience using computers, to query campus computers for information such as class schedules, administrative policy statements, and sporting events, Gopher would then "go fer" the data and deliver it to the person seated at the terminal. It soon spread elsewhere, where system administrators installed the program on machines connected to the Internet. If Gopher was installed at a local site, users could call the program and use it directly; if not, they could telnet to a distant site that had Gopher and allowed outsiders to use it (several places soon emerged). The program displayed information as menus of menus, in a hierarchical tree. Eventually users could get to the desired data (e.g., the local weather forecast, starting at the top of the tree with, say, "National Weather Service"). Using Gopher was tedious, but it did work and was quickly accepted.

Gopher's lists of menus sometimes led to a dead end, from which it was hard to backtrack. That left more than one person "lost in cyberspace." The file name that Gopher listed might not be the name a user had in mind (e.g., "local weather," listed on Gopher as "regional radar image"). At the Cambridge supercomputer company Thinking Machines, Brewster Kahle and his colleagues developed a system called WAIS (Wide Area Information Service) that allowed users to search the contents of files directly. WAIS searched documents for which an index of every word was constructed. This may have at first seemed absurdly daunting, but by 1990 there were already several word-processing programs available for PCs that constructed such indexes, and the advanced computers sold by Thinking Machines were intended to do just that kind of heavy-duty processing. WAIS was not much harder to learn or to use than Gopher, but it never become as widespread. When it worked and retrieved exactly what a user wanted to find, the fortunate user got an immediate sense that *this* was what the Internet was all about.

Like the early news groups, Gopher and WAIS were rendered obsolete by the World Wide Web and its system of information retrieval. They could still be found in 1995, but by then, as with Telnet, FTP, and LISTSERV, these indexing programs were no longer central to using the Internet. Most people who used the Net after 1995 had probably not even heard of them. One could easily write a history of the Internet without mentioning Gopher at all, so brief was its time on the stage. But it is mentioned, if only briefly, in this chapter, for the same reason that earlier chapters dealt (at greater length) with punched cards. Perhaps someday the World Wide Web will be forgotten after it, too, has been replaced by something new.

World Wide Web, Mosaic

It is difficult to discuss the World Wide Web without confronting a general feeling that this is the culmination of all the developments in computing since von Neumann's EDVAC Report of 1945. Or is it the culmination of advances in communication since the invention of printing with movable type? Or the invention of writing? Take your pick. Giving a history of the World Wide Web, the initial code for which was written in 1990, will be much more difficult than writing about the early days of computing. Historians avoid writing about the recent past for good reasons. What would readers think of a history that ended with a statement that the culmination of all computing history was Gopher?

At the same time, the Web attracts the historian because its roots are so deep. It is attractive also for the way it illustrates a central theme of this narrative, of how computing progresses at times by almost random events. We have seen how computing was moved along at times by the vision, energy, and determination of inventors like Eckert and Mauchly, who almost willed the future into being. At other times it has proceeded as if by accident, for example, the introduction of the Altair by a model-rocket hobby shop in Albuquerque.

The development of the World Wide Web has elements of both randomness and planning. It was invented at an unforeseen and totally unexpected place: the high-energy physics laboratory CERN, on the Swiss-French border. It did not come from the research laboratories of IBM, Xerox, or even Microsoft, nor did it come out of the famed Media Lab at MIT, which stated with great fanfare that its primary goal was to transform the world through the integration of computers and communications. Yet the Web's fundamental concept, of structuring information as "hypertext," goes back to a seminal essay by Vannevar Bush in 1945, about the coming glut of information and how technology might be applied to handle it.[41] Bush's essay influenced subsequent generations of computer researchers. Two of those people have been mentioned in previous chapters: Doug Engelbart, who demonstrated a prototype information retrieval system at the 1968 Fall Joint Computer Conference, and Ted Nelson, the author of *Computer Lib/Dream Machines* and developer of a similar system called Xanadu.[42]

In his self-published manifesto, Nelson defined "hypertext" as "forms of writing which branch or perform on request; they are best presented on computer display screens."[43] Nelson praised Engelbart's On-Line System (NLS) but noted that Engelbart believed in tightly structuring information in outline formats.[44] Nelson wanted something closer to Vannevar Bush's earlier concept, which Bush hoped would replicate the mind's ability to make associations across subject boundaries. Nelson worked tirelessly through the 1970s and 1980s to bring Xanadu to life.

He remained close to, but always outside of the academic and research community, and his ideas inspired work at Brown University, led by Andries van Dam.[45] Independently of these researchers, Apple introduced a program called HyperCard for the Macintosh in 1987. HyperCard implemented only a fraction of the concepts of hypertext as van Dam or Nelson understood the concept, but it was simple, easy to use, and even easy for a novice to program. For all its limits, HyperCard brought the notion of nonlinear text and graphics out of the laboratory setting.

In the midst of all that sprouted the Internet, with a sudden and unexpected need for a way to navigate through its rich and ever-increasing resources.[46] It is still too early to write the history of what happened next. Tim Berners-Lee, who wrote the original Web prototype in late 1990, has written a brief memoir of that time, but the full story has yet to be told.[47] Berners-Lee developed the Web while at CERN, the European particle physics laboratory. He stated that "[t]he Web's major goal was to be a shared information space through which people and machines could communicate. This space was to be inclusive, rather than exclusive."[48] He was especially concerned with allowing communication across computers and software of different types. He also wanted to avoid the structure of most databases, which forced people to put information into categories before they knew if such classifications were appropriate or not. To these ends he devised a Universal Resource Identifier (later called the Uniform Resource Locator or URL) that could "point to any document (or any other type of resource) in the universe of information."[49] In place of the File Transfer Protocol then in use, he created a more sophisticated Hypertext Transfer Protocol (HTTP), which was faster and had more features. Finally, he defined a Hypertext Markup Language (HTML) for the movement of hypertext across the network. Within a few years, these abbreviations, along with WWW for the World Wide Web itself, would be as common as RAM, K, or any other jargon in the computer field.

The World Wide Web got off to a slow start. Its distinctive feature, the ability to jump to different resources through hyperlinks, was of little use until there were at least a few other places besides CERN that supported it. Until editing software was written, users had to construct the links in a document by hand, a very tedious process. To view Web materials one used a program called a "browser" (the term may have originated with Apple's Hypercard). Early Web browsers (including two called Lynx and

Viola) presented screens that were similar to Gopher's, with a lists of menu selections.

Around the fall of 1992 Marc Andreesen and Eric Bina began discussing ways of making it easier to navigate the Web. While still a student at the University of Illinois, Andreesen took a job programming for the National Center for Supercomputing Applications, a center set up with NSF money on the campus to make supercomputing more accessible (cf. the impetus for the original ARPANET). By January 1993 Andreesen and Bina had written an early version of a browser they would later call Mosaic, and they released a version of it over the Internet.[50] Mosaic married the ease of use of Hypercard with the full hypertext capabilities of the World Wide Web. To select items one used a mouse (thus circling back to Doug Engelbart, who invented it for that purpose). One knew an item had a hyperlink by its different color. A second feature of Mosaic, the one that most impressed the people who first used it, was its seamless integration of text and images.

With the help of others at NCSA, Mosaic was rewritten to run on Windows-based machines and Macintoshes as well as workstations. As a product of a government-funded laboratory, Mosaic was made available free or for a nominal charge. As with the UNIX, history was repeating itself. But not entirely: unlike the developers of UNIX, Andreesen managed to commercialize his invention quickly. In early 1994 he was approached by Jim Clark, the founder of Silicon Graphics, who suggested that they commercialize the invention. Andreesen agreed, but apparently the University of Illinois objected to this idea. Like the University of Pennsylvania a half-century before it, Illinois saw the value of the work done on its campus, but it failed to see the much greater value of the people who did that work. Clark left Silicon Graphics, and with Andreesen founded Mosaic Communications that spring. The University of Illinois asserted its claim to the name Mosaic, so the company changed its name to Netscape Communications Corporation. Clark and Andreesen visited Champaign-Urbana and quickly hired many of the programmers who had worked on the software. Netscape introduced its version of the browser in September 1994. The University of Illinois continued to offer Mosaic, in a licensing agreement with another company, but Netscape's software quickly supplanted Mosaic as the most popular version of the program.[51]

On August 8, 1995, Netscape offered shares to the public. Investors bid the stock from its initial offering price of $28 a share to $58 the first day; that evening the network news broadcast stories of people who had managed to get shares at the initial price. The public now learned of a crop of new "instant billionaires," adding that knowledge to their awareness of "dot.com," "HTTP," and "HTML." Within a few months Netscape shares were trading at over $150 a share, before falling back. Reading the newspaper accounts and watching the television news, one had the feeling that the day Netscape went public marked the real beginning of the history of computing, and that everything else had been a prologue. For this narrative, that event will mark the end.

Conclusion

Since 1945 computing has never remained stagnant, and the 1990s were no exception. The emergence of the Internet was the biggest story of these years, although it was also a time of consolidation of the desktop computer in the office. Desktop computing reached a plateau based on the Intel, DOS, Macintosh, and UNIX standards that had been invented earlier. Most offices used personal computers for word processing, spreadsheets, and databases; the only new addition was communications made possible by local-area networking. A new class of computer emerged, called the laptop (later, as it lost more weight, the notebook), but these were functionally similar to PCs. Indeed, they were advertised as being software-compatible with what was on the desk. The basic architectural decisions made in the late 1970s, including the choice of a microprocessor and the structure of a disk operating system, remained (with RISC a small but significant exception). Networking promised for some a potential conceptual shift in computing, but as of 1995 it had not replaced the concept of an autonomous, general-purpose computer on individual desks. As the World Wide Web matures, some argue that all the consumer will need is a simple Internet appliance—a reincarnation of the dumb terminal—not a general-purpose PC. But the numerous examples cited in this study—the IBM 650, the 1401, the PDP-8, the Apple II—all support the argument that the market will choose a good, cheap, general-purpose computer every time.

The biggest story of the 1990s was how the Altair, a $400 kit of parts advertised on the cover of *Popular Electronics*, managed to bring down the mighty houses of IBM, Wang, UNIVAC, Digital, and Control Data Corporation. IBM almost made the transition with its personal computer, but its inability to follow through on the beachhead it established led to multi-*billion*-dollar losses between 1991 and 1993.[52] Personal computer profits went increasingly to new companies like Dell, Compaq, and above all Microsoft. IBM recovered, but only after abandoning its no-layoff policy (which it had held to even through the 1930s), and when it emerged from that crisis it found Microsoft occupying center stage. Even the American Federation of Information Processing Societies (AFIPS), the umbrella trade organization of computer societies, perished on December 31, 1990.[53]

Of course it was not simply the $400 Altair that changed computing. DEC and Data General had a lot to do with that as well, but neither DEC nor Data General were able to build on the foundations they had laid. One could understand IBM's failings, with its tradition of punched-card batch processing, and its constant courtroom battles against plaintiffs charging that it was too big. It is not as easy to understand how the Route 128 minicomputer companies failed to make the transition. These were the companies that pioneered in processor and bus architectures, compact packaging, interactive operation, and low unit costs. Led by General Doriot of the Harvard Business School, they also were the first to do what later became a defining characteristic of Silicon Valley: to start up a technology-based company with venture capital. Netscape generated so much public interest because it showed that this tradition was still alive. There was even a possibility that this company, founded to exploit a model of computing centered on the Internet, might be able to do to Microsoft what Microsoft had just done to DEC, IBM, and the others who were founded on earlier, now-outdated models of computing.

As of 1995 Digital and Data General were still in business, although both were struggling and much-reduced in size. Data General's decline began in the early 1980s, just when Tracy Kidder's *The Soul of a New Machine* became one of the first books about the computer industry to get on the best-seller list. That book chronicled Data General's attempt to chase after the VAX and regain its leadership in minicomputers. It captured the youth, energy, and drive of the computer business, and it remains an accurate description of the computer business today. Lacking the 20-20 hindsight that we now all have, Kidder did not, however, mention how Data General's Nova, the "clean machine," had inspired the designers of personal computers, including Ed Roberts and Steve Wozniak. Someone at Data General may have recommended an alternate course: that it ignore the VAX and concentrate instead on the small systems it had helped bring into being. If so, Kidder's book does not record it.

In 1992, Ken Olsen resigned as head of Digital, as the company he founded was heading toward bankruptcy. A typical news story contrasted Olsen and a tired DEC with the young Bill Gates and his vibrant Microsoft. Few saw the irony of that comparison. Gates learned how to program on a PDP-10, and we have seen DEC's influence on Microsoft's software. More than that: Digital Equipment Corporation set in motion the forces that made companies like Microsoft possible. One person was quoted stating that were it not for Olsen we would still be programming with punched cards. That sounded like a generous overstatement made out of sympathy; in fact, one could credit him with doing that and much more. Modern computing is a story of how a vision of "man-machine symbiosis," in J. C. R. Licklider's term, came to fruition. That happened through the efforts of people like Licklider himself, as well as Doug Engelbart, Ted Hoff, Ed Roberts, Steve Jobs, Steve Wozniak, Bill Gates, Gary Kildall, Tim Berners-Lee, and many others. To that list, perhaps near the top, should be added the name Ken Olsen. The "creative destruction" of the capitalist system had worked wonders, but the process was neither rational nor fair.

Conclusion: The Digitization of the World Picture

Using Your VCR: 3 Two-hour Sessions. Do you have trouble connecting and using your VCR? Can you set a timed recording or record one show while watching another? The solution to these questions and many others will be covered in this "how to" class.

—*Montgomery County (Maryland) College, Adult Education Course Catalog, Summer 1990.*

Between 1945 and 1995 the computer transformed itself over and over again, each time redefining its essence. "The computer" started out as a fast scientific calculator; Eckert and Mauchly transformed it into UNIVAC, a machine for general data processing. Ken Olsen made it into a real-time information processor that worked symbiotically with its users. Ed Roberts transformed it into a device that anyone could own and use. Steve Jobs and Steve Wozniak turned it into an appliance that was both useful and fun. Gary Kildall and William Gates transformed it into a standardized platform that could run a cornucopia of commercial software sold in retail stores. Bob Metcalfe, Tim Berners-Lee, and others turned it into a window to a global network.

Each transformation was accompanied by assertions that further transformations were unlikely, yet each time someone managed to break through. The latest transformation, to the World Wide Web, was also preceded by statements that the computer industry was stagnating, that there was, to paraphrase a software salesman, "no more low-hanging fruit." He was wrong, and those who predict that the World Wide Web is the ultimate resting place for computing will no doubt be wrong as well.

By 1995 personal computers had become a commodity, allowing commercial software to come to the fore as the central place where innovation was conveyed to users. The layering of software, a process that began with the first "automatic coding" schemes developed for the UNIVAC, continued. That was the only way to broaden the market to include users who had no inclination or talent to write programs. Again, with each software advance, one heard that the "end of programming" had come, that "anyone" could now get a computer to do what he or she wished. As new markets opened up, the end proved elusive. The difficulty many people have in programming a VCR is a minor but real example of the problem: getting a computer to do what users want it to do is as difficult as ever and requires talent, hard work, and a commitment by developers to the user's needs.

The ease of use that the Macintosh interface brought to personal computing, which Microsoft copied with Windows, has led to a new set of frustrations. Users now find interfaces laid over these desktop interfaces (such as the so-called intelligent agents) that are supposed to make computing even easier. In fact, they have made things more difficult. This process will doubtless continue. Will we be able to build a computer as complex as HAL, the infamous star of Stanley Kubrick's *2001: A Space Odyssey*, anytime soon? Probably not. Many people came away from the movie thinking that the problem with HAL was that it was somehow out of control; but a closer viewing shows that HAL's real problem was that it worked perfectly. It broke down because it was trying to obey two conflicting instructions that were part of its programming: to obey the humans on board but to conceal from them the true nature of their mission.[1] When a real version of a HAL-like intelligent interface appears, it will probably not be as robust and reliable as the fictional one.

The Digitization of the World Picture

In 1948 a book appeared with the bold title *The Mechanization of the World Picture*. The author, a Dutch physicist named E. J. Dijksterhuis, argued that much of history was best understood as an unfolding of the "mechanistic" way of looking at the world that actually began with the Greeks and culminated in the work of Isaac Newton.[2] Dijksterhuis's work found a willing audience of readers who had experienced the power and the horrors of a mechanized world view after six years of world war.

It took a millennium and a half for a mechanistic view to take hold, but it has taken less time—about fifty years—for a view equally as revolutionary to take hold. The "digitization of the world picture" began in the mid-1930s, with the work of a few mathematicians and engineers. By 1985 this world view had triumphed. It began in an obscure corner of mathematics. Alan Turing's "machine," introduced in a paper in 1936, was a theoretical construction.[3] The invention of the stored-program electronic computer breathed life into his idea and made it more real than he probably thought possible. The ensuing decades saw one field after another taken over, absorbed, or transformed by the computer as if it were a universal solvent.[4] A special issue of the trade journal *Electronics*, in October 1973 described as "The Great Takeover," the way traditional analog electronic circuits were replaced by miniature digital computers programmed to emulate them; most ordinary radios, for example, had lost their tuning dial by 1973 and were "tuned" by digital keypads. Ten years later, *Time* proclaimed the computer "Machine of the Year" for 1983, with the opening headline "The Computer Moves In."[5]

The latest manifestation of this takeover is the Internet, embraced across the political and cultural spectrum, by Newt Gingrich, Al Gore, Stewart Brand, the late Timothy Leary, "Generation X," and numerous people in between. Most accounts describe it as a marriage of communications and computing.[6] The evidence presented here suggests otherwise; that the Internet simply represents yet another takeover, by digital computing of an activity (telecommunications) that had a long history based on analog techniques.

Those who so glowingly describe the World Wide Web as the culmination of fifty years of prologue either do not know or have forgotten history. The very same statements were made when the first UNIVACs were installed, when minicomputers and time-sharing appeared, and when the personal computer was introduced (figure C.1). This will not be the last time these words are spoken. But promises of a technological Utopia have been common in American history, and at least a few champions of the Internet are aware of how naive these earlier visions were.[7] Silicon Valley has some of the most congested real highways in the country, as people commute to work with a technology that Henry Ford invented to *reduce* urban congestion. Most people have some sense of the fact that the automobile did not fulfill many of Ford's promises simply because it was too successful. The word "smog" crept into the English language around the time of Ford's death in the late 1940s; "gridlock," "strip malls," and "suburban sprawl" came later. What equivalent will describe the dark side of networked digital computing? And will those "side effects" become evident only fifty years from now, as was the case with automobiles? Can we anticipate them before it is too late or too difficult to manage them?

Figure C.1
Digital Utopia, as depicted on the cover of *Byte* magazine (January 1977). *Byte's* cover illustrations stood out among all the computer publications. (*Source*: Robert Tinney.)

Each transformation of digital computing was propelled by individuals with an idealistic notion that computing, in its new form, would be a liberating force that could redress many of the imbalances brought on by the smokestack of the "second wave," in Alvin Toffler's phrase. UNIVAC installations were accompanied by glowing predictions that the "automation" they produced would lead to a reduced workweek. In the mid-1960s enthusiasts and hackers saw the PDP-10 and PDP-8 as machines that would liberate computing from the tentacles of the IBM octopus. The Apple II reflected the Utopian visions of the San Francisco Bay area in the early 1970s. And so it will be with universal access to the Internet.

In each case the future has turned out to be more complex, and less revolutionary, than its proponents imagined. The UNIVAC did not solve the problem of unemployment. Personal computers did not put ordinary individuals on an equal footing with those in positions of power. It did find a market that exceeded all expectations—but in the office and not the home, as a tool that assisted the functions of the corporate workplace.[8] Looking out over the polluted and decayed landscape of the 1970s-era industrial Rustbelt, young people programmed their personal computers to model a middle landscape; one that gave its inhabitants all the benefits of industrialization with none of the drawbacks. But the social problems of the outside world remained. Utopia stayed inside the computer screen and stubbornly refused to come out. Computer modeling evolved into "virtual reality"—a new variant of the mind-altering drugs in vogue in the 1960s. Timothy Leary argued that virtual reality was more effective than LSD as a way to bring humans back to the Garden of Eden. So far that is not happening, and perhaps this is a good thing, given the level of thought that characterizes most visions of what Digital Utopia ought to look like.

We have seen that political and social forces have always shaped the direction of digital computing. Now, with computing among the defining technologies of American society, those forces are increasingly out in the open and part of public discussion. Politicians and judges as much as engineers decide where highways and bridges get built, who may serve a region with telephone service, and how much competition an electric utility may have. These legislators and jurists rely upon industry lobbyists or specialists on their staff to guide them through the technical dimension of their policies. All the while, new technologies (such as direct broadcast satellite television) disrupt their plans. But that does not stop the process or shift decision-making away from these centers.

Computing is no different. The idea of politicians directing technology is still distasteful to computer pioneers, many of whom are still alive and retain a vivid memory of how they surmounted technical, not political, challenges. But when a technology becomes integrated into the affairs of ordinary daily life, it must acknowledge politics. Some groups, such as the Electronic Frontier Foundation (founded by Mitch Kapor), are doing this by stepping back to try to identify the digital equivalents of "smog" and "gridlock." But historically the United States has promoted as rapid a deployment of technology as possible, and has left it to future generations to deal with the consequences. It is not surprising, therefore, that attempts to regulate or control the content of the Internet have so far been clumsy and have failed. How that plays out remains to be seen.

A century and a half ago, Henry David Thoreau observed with suspicion the technophilic aspect of American character. Railroads were the high technology of his day, but he did not share the public's enthusiasm for the Fitchburg line, whose tracks ran behind Walden Pond. "We do not ride on the railroad; it rides on us," he said. What the nation needs is "a stern and more than Spartan simplicity of life." A few miles west of Thoreau's cabin, the Fitchburg railroad built a branch to serve the Assabet Mills, which by the time of the Civil War was one of the country's largest producers of woolen goods. A century later these same mills were blanketing the Earth with PDP-8s. One wonders what Thoreau would have made of this connection.[9] Would he have seized the opportunity to set up his own Walden Pond home page, to let others know what he was up to? Or would he have continued to rely on the pencils he made for himself?

We created the computer to serve us. The notion that it might become our master has been the stuff of science fiction for decades, but it was always hard to take those stories seriously when it took heroic efforts just to get a computer to do basic chores. As we start to accept the World Wide Web as a natural part of our daily existence, perhaps it is time to revisit the question of control. My hope is that, with an understanding of history and a dash of Thoreauvian skepticism, we can learn to use the computer rather than allowing it to use us.

Notes

Preface

1. William Aspray, ed., *Computing Before Computers* (Ames, Iowa: Iowa State University Press, 1990).

Introduction

1. Dictionaries, even editions published in the 1970s, define computer as "a calculator especially designed for the solution of complex mathematical problems." This is the first definition given in *Webster's Third International Dictionary, Unabridged*; this definition is then qualified as "specifically: a programmable electronic device that can store, retrieve, and process data."

2. Some of the early automatic machines were called "calculators," as in the Harvard Mark I, or "Automatic Sequence Controlled Calculator." But the letter "C" in ENIAC, designed at the Moore School in the early 1940s and dedicated in 1946, stood for "Computer."

3. Amy Friedlander, *Natural Monopoly and Universal Service: Telephones and Telegraphs in the U.S. Communications Infrastructure, 1837–1940* (Reston, VA: CNRI, 1995).

4. If anything, it might go the other way: historians of technology are turning their attention to the mundane; and studies of computing are so common they surprise no one. See, for example, Henry Petroski, *The Pencil: a History of Design and Circumstance* (New York: Knopf, 1990), and Robert Friedel, *Zipper: an Exploration in Novelty* (New York: W.W. Norton, 1994).

5. See, for example I. Bernard Cohen, *Revolution in Science* (Cambridge: Harvard University Press, 1985).

6. I had not heard of the World Wide Web when I began working on this study, although I was aware of the existence of the Internet. Although touted as revolutionary by *Time* on its cover in 1983, the personal computer is now disparaged as crippled by its crude user interface and lack of connectivity to the Web.

7. My unscientific basis for this observation is the vigorous activity in the history of technology being undertaken by scholars on the World Wide Web. I have also noted that historians are among the first to adopt the latest word processing and scholars' database tools.

8. See, for example, Alvin Toffler, *The Third Wave* (New York: Morrow, 1980).

9. Even the best sociological studies of computing ignore its historical evolution, as if the technology sociologists observe is a given and not something that is rapidly evolving; for example, Shoshanna Zuboff, *In the Age of the Smart Machine* (New York: Basic Books, 1988), and Sherry Turkle, *The Second Self, Computers and the Human Spirit* (New York: Simon & Schuster, 1984).

10. See, for example, James W. Cortada, *Before the Computer* (Princeton: Princeton University Press, 1993); Arthur Norberg, "High Technology Calculation in the Early Twentieth Century: Punched Card Machinery in Business and Government," *Technology and Culture* 31 (October 1990): 753–779; and JoAnne Yates, *Control Through Communication: the Rise of System in American Management* (Baltimore: John Hopkins University Press, 1989).

11. See, for example, James R. Beniger, *The Control Revolution: Technological and Economic Origins of the Information Society* (Cambridge: Harvard University Press, 1986).

12. For example, the *Annals of the History of Computing*, the journal of record for the study of this subject, seldom publishes papers that connect computing with, say, radar, ballistic missiles, or nuclear weapons history, other than on the role of the computer as an aide to those technologies. On the other side, one finds histories of modern 20th century technology that make no mention of the computer at all, as in Thomas Parke Hughes, *American Genesis: A Century of Invention and Technological Enthusiasm, 1870–1970* (New York: Viking, 1989).

13. Thomas Parke Hughes, *Networks of Power: Electrification in Western Society, 1880–1930* (Baltimore: Johns Hopkins University Press, 1983).

14. The most accessible of the many works written on this topic is Wiebe E. Bijker, Thomas P. Hughes, and Trevor Pinch, eds., *The Social Construction of Technological Systems* (Cambridge: MIT Press, 1987).

15. The most important is Donald MacKenzie. See, for example, *Inventing Accuracy: A Historical Sociology of Nuclear Missile Guidance* (Cambridge: MIT Press, 1990), and "Negotiating Arithmetic, Constructing Proof: the Sociology of Mathematics and Information Technology," *Social Studies of Science* 23 (1993): 37–65. Another practitioner is Bryan Pfaffenberger; see his "The Social Meaning of the Personal Computer, or Why the Personal Computer Revolution was no Revolution," *Anthropological Quarterly* 61: 1 (1988): 39–47.

16. See, for example Steven Levy, *Hackers: Heroes of the Computer Revolution* (Garden City, NY: Doubleday, 1984); and Paul Freiberger, *Fire in the Valley: the Making of the Personal Computer* (Berkeley: Osborne/McGraw-Hill, 1984).

17. William Aspray, *John von Neumann and the Origins of Modern Computing* (Cambridge: MIT Press, 1990).

18. Tomas J. Misa, "Military Needs, Commercial Realities, and the Development of the Transistor, 1948–1958," in Merritt Roe Smith, ed., *Military Enterprise and Technological Change* (Cambridge: MIT Press, 1985): 253–287.

19. Michael A. Dennis, "A Change of State: the Political Cultures of Technological Practice at the MIT Instrumentation Lab and the Johns Hopkins University Applied Physics Laboratory, 1930–1945" (Ph.D. diss., Johns Hopkins University, 1990).

20. Manuel DeLanda, *War in the Age of Intelligent Machines* (Cambridge: MIT Press, 1991); also Chris Hables Gray, "Computers as Weapons and Metaphors: The U.S. Military 1940–1990 and Postmodern War," (Ph.D. diss., University of California, Santa Cruz, 1991).

21. Charles Bashe, Lyle R. Johnson, John H. Palmer, and Emerson Pugh, *IBM's Early Computers* (Cambridge: MIT Press, 1986); Emerson Pugh et al., *IBM's 360 and Early 370 Systems* (Cambridge: MIT Press, 1991); and Emerson Pugh, *Building IBM: Shaping an Industry and Its Technology* (Cambridge: MIT Press, 1995).

22. The term seems to have come into use around 1959.

23. George H. Mealy, "Operating Systems," in Saul Rosen, ed., *Programming Systems and Languages* (New York: McGraw-Hill, 1967): 517–518.

24. JoAnne Yates, *Control Through Communication: the Rise of System in American Management* (Baltimore: Johns Hopkins University Press, 1989); David F. Noble, *Forces of Production* (New York: Knopf, 1984); James R. Beniger, *The Control Revolution: Technological and Economic Origins of the Information Society* (Cambridge: Harvard University Press, 1986).

25. Brian Randell, ed., *The Origins of Digital Computers: Selected Papers*, 2nd ed. (Berlin: Springer-Verlag, 1975): 327–328; Peter J. Bird, *LEO: the First Business Computer* (Berkshire, UK: Hasler Publishing Ltd., 1994).

26. Kenneth Flamm, *Creating the Computer: Government, Industry, and High Technology* (Washington, DC: Brookings Institution, 1988): 134; see also Martin Campbell-Kelly, *ICL: a Business and Technical History* (Oxford: Oxford University Press, 1989).

27. Edward Feigenbaum and Pamela McCorduck, *The Fifth Generation: Artificial Intelligence and Japan's Computer Challenge to the World* (Reading, MA: Addison-Wesley, 1983); Michael Cusumano, "Factory Concepts and Practices in Software Development," *Annals of the History of Computing* 13: 1 (1991): 3–32.

28. Seymour E. Goodman, "Soviet Computing and Technology Transfer: an Overview," *World Politics* 31: 4 (July 1979): 539–570.

29. Using the computer for centralized planning has been touted by American "futurists" such as Herman Kahn and R. Buckminster Fuller. After completing Project Whirlwind, J. Forrester turned to an application he called "System Dynamics." In 1996, large-scale computer modeling of the U.S. government was vigorously promoted by presidential candidate H. Ross Perot, the founder of Electronic Data Systems.

Chapter 1

1. Testimony by Cannon, Hagley Museum, *Honeywell v. Sperry Rand* papers, Series III, Box 140, p. 17,680; see also Harold Bergstein, "An Interview with Eckert and Mauchly," *Datamation* (April 1962): 25–30. A more detailed analysis of Aiken's observation is discussed in a forthcoming book by I. Bernard Cohen on the life and work of Howard Aiken. I am grateful to Professor Cohen for making drafts of the relevant chapters of this book available to me before its publication.

2. Note that in 1994 the U.S. government suspended support for the Super-conducting Super Collider (SSC). So it appears there that the total world "market" has peaked at about a dozen cyclotrons, a scientific instrument invented around the same time as the electronic computer with about the same cost and complexity.

3. For example, in an address by von Neumann to the Mathematical Computing Advisory Panel of the U.S. Navy in May 1946, he compares the electronic computers then under development to "... what is at present still the major practical mode of computing, namely, human procedure with an electromechanical desk multiplier." Published in the *Annals of the History of Computing* 10 (1989): 248.

4. For an account of the early development of this activity, see JoAnne Yates, *Control Through Communication: the Rise of System in American Management* (Baltimore: Johns Hopkins University Press, 1989); also James Beniger, *The Control Revolution* (Cambridge: Harvard University Press, 1986).

5. Arthur Norberg, "High Technology Calculation in the Early Twentieth Century: Punched Card Machinery in Business and Government," *Technology and Culture* 31 (1990): 753–779; also Martin Campbell-Kelly, *ICL: a Business and Technical History* (Oxford: Oxford University Press, 1989).

6. The following discussion on punched-card computation is derived from Martin Campbell-Kelly, "Punched-Card Machinery," in William Aspray, ed., *Computing Before Computers* (Ames: Iowa State University Press, 1990), chapter 4; also Campbell-Kelly, *ICL*; and Edmund C. Berkeley, *Giant Brains, or Machines that Think* (New York: Wiley, 1949), chapter 4.

7. Campbell-Kelly, in Aspray, ed., *Computing Before Computers*; also the review of Campbell-Kelly's *ICL* by Kenneth Flamm in *Annals of the History of Computing* 13: 1 (1991).

8. Wallace J. Eckert, *Punched Card Methods in Scientific Computation* (New York: The Thomas J. Watson Astronomical Computing Bureau, Columbia University, 1940): 22.

9. Ibid., 1.

10. Ibid., 108–110. Of the twelve steps, only the first six were performed automatically; the rest required some human intervention.

11. J. Lynch and C. E. Johnson, "Programming Principles for the IBM Relay Calculators," Ballistic Research Laboratories, Report No. 705, October 1949, IBM Archives, Valhalla, New York.

12. Ibid., 8.

13. Brian Randell, ed., *The Origins of Digital Computers: Selected Papers*, 2d ed. (Berlin: Springer-Verlag, 1975): 188.

14. Lynch and Johnson, "Programming Principles," 4; also Wallace Eckert, "The IBM Pluggable Sequence Relay Calculator," *Mathematical Tables and Other Aids to Computation* 3 (1948): 149–161; also Ballistic Research Laboratories, "Computing Laboratory," undated 15-page brochure, probably 1952, National Air and Space Museum, NBS Collection.

15. Charles J. Bashe, Lyle R. Johnson, John H. Palmer, and Emerson Pugh, *IBM's Early Computers* (Cambridge: MIT Press, 1986): 44–46, 59–68.

16. Ibid., 67.

17. William Woodbury, "The 603-405 Computer," in *Proceedings of a Second Symposium on Calculating Machinery; Sept. 1949* (Cambridge: Harvard University Press, 1951): 316–320; also Michael R. Williams, *A History of Computing Technology* (Englewood Cliffs, NJ: Prentice-Hall, 1985): 256.

18. G. J. Toben, quoted in Bashe et al., *IBM's Early Computers*, 69.

19. Bashe et al., *IBM's Early Computers*, 68–72; also John W. Sheldon and Liston Tatum, "The IBM Card-Programmed Electronic Calculator," *Review of Electronic Digital Computers*, Joint IRE-AIEE Conference, February 1952, 30–36.

20. Paul Ceruzzi, *Beyond the Limits: Flight Enters the Computer Age* (Cambridge: MIT Press, 1989), chapter 2; see also Smithsonian Videohistory Program, RAND Corporation interviews, June 12–13, 1990; interview with Clifford Shaw, 12 June 1990, 13.

21. In France, Compagnie des Machines Bull introduced, in 1952, a machine having a similar architecture. Called the "Gamma 3," it was very successful and was one of the first products produced in France to achieve a critical mass of

sales. See Bruno LeClerc, "From Gamma 2 to Gamma E.T.: The Birth of Electronic Computing at Bull," *Annals of the History of Computing* 12: 1 (1990): 5–22.

22. See, for example, Computer Research Corporation, "Comparison of the Card-Programmed Computer [sic] with the General-Purpose Model CRC 102-A," 16 page pamphlet (1953) National Air & Space Museum, NBS archive.

23. David Alan Grier, "The ENIAC, the Verb 'to program' and the Emergence of Digital Computers," *Annals of the History of Computers* 18: 1 (1996): 51–55.

24. "Historical Comments," in L. R. Johnson, *System Structure in Data, Programs, and Computers* (Englewood Cliffs, NJ: Prentice-Hall, 1970): 185. A copy of the original memorandum is in the University of Pennsylvania archives.

25. For a discussion of the fate of the EDVAC see Michael Williams, "The Origins, Uses, and Fate of the EDVAC," *Annals of the History of Computing* 15 (1993): 22–38.

26. A copy of the First Draft is in the National Air and Space Museum Archives, NBS Collection.

27. It also comes from the fact that the IAS machine was copied in a number of locations.

28. Herman Goldstine, *The Computer from Pascal to von Neumann* (Princeton: Princeton University Press, 1972): 182.

29. Because the term "programming" had not yet come into use, I use "set up" in this section.

30. John Mauchly, "Preparation of Problems for EDVAC-Type Machines," Harvard University, *Proceedings of a Symposium on Large-Scale Digital Calculating Machinery* (Cambridge: Harvard University Press, 1948): 203–207.

31. Eckert, "Disclosure . . . ," written January 29, 1944; reprinted in Herman Lukoff, *From Dits to Bits: A Personal History of the Electronic Computer* (Portland, OR: Robotics Press): 207–209.

32. J. Presper Eckert, "Disclosure of a Magnetic Calculating Machine," University of Pennsylvania, Moore School of Electrical Engineering, memorandum of January 29, 1944; in Lukoff, *From Dits to Bits*, 207–209; also J. Presper Eckert and John Mauchly, "Automatic High Speed Computing: A Progress Report on the EDVAC," portions reprinted in Johnson, *System Structure in Data, Programs, and Computers*, 184–187. There are many accounts of the relationship of von Neumann with Eckert and Mauchly, and of the relative contributions each made to the EDVAC report. See Goldstine, *Computer From Pascal to von Neumann*; Mauchly's own account is told in "Amending the ENIAC Story," *Datamation* (October 1979): 217–220. The details of these events were covered in the trial *Honeywell v. Sperry Rand, Inc.*, concluded in 1974.

33. The above discussion makes no mention of the possible contribution of Alan Turing, who stated something very much like this principle in theoretical terms long before. Turing may have inspired those working on the EDVAC design, but that is unknown at this time.

34. For more on who invented the Stored-Program Principle, see *Annals of the History of Computing* 4 (October 1982): 358–361.

35. Moore School of Electrical Engineering, "Theory and Techniques for Design of Electronic Digital Computers; Lectures Given at the Moore School of Electrical Engineering, July 8–August 31, 1946" (Cambridge: MIT Press, 1985 [reprint]).

36. The acronym EDSAC stands for "Electronic Delay Storage Automatic Calculator"; BINAC stands for "Binary Automatic Computer."

37. The term "word" as applied to a chunk of digits handled in a computer probably also came from von Neumann.

38. A typical modern description of the architecture is described in "Digital Computer Architecture," in Jack Belzer, Albert Holzman, and Allen Kent, eds., *Encyclopedia of Computer Science and Technology*, vol. 7 (New York: Dekker, 1970): 289–326.

39. For an example of how this subject is treated in introductory college-level textbooks, see Helene G. Kershner, *Introduction to Computer Literacy* (Lexington, MA: D. C. Heath, 1990), Jack B. Rochester and Jon Rochester, *Computers for People* (Homewood, IL: Richard D. Irwin, 1991). See also W. Danniel Hillis, "The Connection Machine," *Scientific American* (June 1987): 108–115, for an explicit statement of the "non-von-Neumann" nature of parallel designs. Ironically, Hillis incorrectly identifies the ENIAC as a sequential machine; in fact, the ENIAC was a parallel processing machine.

40. Alan Perlis, "Epigrams on Programming," *ACM Sigplan Notices* (October 1981): 7–13.

41. In the following discussion I rely heavily on the arguments and data in Nancy Stern, *From ENIAC to UNIVAC: an Appraisal of the Eckert-Mauchly Computers* (Bedford, MA: Digital Press, 1981), especially chapter 5.

42. Mauchly, memorandum of 3/31/1948; Hagley Museum, Sperry Univac Company Records, Series I, Box 3, folder "Eckert-Mauchly Computer Corporation; Mauchly, John."

43. Von Neumann to Herman Goldstine, letter of May 8, 1945; Hagley Museum, Sperry Univac Corporate Records; Series II, Box 74, folder "May 1945–October 1945."

44. Howard H. Aiken, "The Future of Automatic Computing Machinery," Elektronische Rechenmaschinen und Informationsverarbeitung; Beihefte der NTZ, Band 4, 1956, pp. 31–35. Incredibly, he made this statement in 1956, by

which time there was ample evidence that such machines not only existed but were becoming rather common. Wallace Eckert commented on computers vs. punched card machines at one of several conferences on high speed computing machinery held at Harvard in the late 1940s.

45. Stern, *From ENIAC to UNIVAC*, 91. Stern believes that "the two men . . . were, in fact, fired."

46. Mauchly to J. P. Eckert Jr. et al, 1/12/1948; Hagley Museum, Sperry UNIVAC Company Records, Series I, Box 3, folder "Eckert-Mauchly Computer Corporation; Mauchly, John."

47. *Computers and Their Future* (Lladudno, Wales, 1970): 7–8

48. Stern, *From ENIAC to UNIVAC*, 148–151. In Britain, the LEO computer ran test programs on the premises of the J. Lyons Catering Company of London by February 1951, a month before the UNIVAC delivery. However, it was not until late 1953 before LEO was considered finished. See S. H. Lavington, *Early British Computers* (Bedford, MA: Digital Press, 1980): 72–73; also Peter Bird, "LEO, the Pride of Lyons," *Annals of the History of Computing* 14: 2 (1992): 55–56.

49. Oral History Session, UNISYS Corporation, May 17–18, 1990, Smithsonian Institution, Washington, DC.

50. L. R. Johnson, "Installation of a Large Electronic Computer," in *Proceedings of the ACM Meeting* (Toronto, September 8–10, 1952): 77–81.

51. J. Presper Eckert, "Thoughts on the History of Computing," *IEEE Computer* (December 1976): 64.

52. Luther A. Harr, "The Univac System, a 1954 Progress Report" (Remington Rand Corporation, 1954): 6.

53. James C. McPherson, "Census Experience Operating a UNIVAC System," *Symposium on Managerial Aspects of Digital Computer Installations* (30 March, 1953, Washington, DC): 33.

54. Lukoff, *From Dits to Bits*, chapter 9.

55. Roddy F. Osborn, "GE and UNIVAC: Harnessing the High-Speed Computer," *Harvard Business Review* (July–August 1954): 102.

56. McPherson, "Census Experience Operating a UNIVAC System," 30–36.

57. Remington Rand Corporation, "Univac Fac-Tronic System," Undated brochure, ca. 1951, Unisys Archives; also McPherson, "Census Experience Operating a UNIVAC System."

58. Paul E. Ceruzzi,"The First Generation of Computers and the Aerospace Industry," National Air and Space Museum *Research Report* (1985): 75–89; also Robert Dorfman, "The Discovery of Linear Programming," *Annals of the History of Computing* 6: 3 (1984): 283–295, and Mina Rees, "The Computing Program at

the Office of Naval Research," *Annals of the History of Computing* 4: 2 (1982): 102–120.

59. L. R. Johnson, "Installation of a Large Electronic Computer," In *Proceedings of ACM Meeting* (Toronto, 1952): 77–81.

60. Luther Harr, "The UNIVAC System, a 1954 Progress Report," (Remington Rand Corporation, 1954) UNISYS Archives.

61. Ibid., 7.

62. Lawrence Livermore Laboratory, "Computing at Lawrence Livermore Laboratory," UCID Report 20079, 1984.

63. Roddy F. Osborn, "GE and UNIVAC: Harnessing the High-Speed Computer," *Harvard Business Review* (July–August 1954): 99–107.

64. John Diebold, *Automation* (New York: Van Nostrand, 1952).

65. John Diebold, "Factories Without Men: New Industrial Revolution," *The Nation* (1953): 227–228, 250–251, 271–272. See also David F. Noble, *Forces of Production: A Social History of Industrial Automation* (New York: Knopf, 1986), chapter 4.

66. Roddy Osborn, "GE and UNIVAC," 99.

67. Ibid., 103; also UNIVAC Oral History, Smithsonian Institution Archives.

68. Ibid., 104.

69. Ibid., 107.

70. Harr, "The UNIVAC System: A 1954 Progress Report," 1. UNISYS Archives.

71. Smithsonian/UNISYS UNIVAC Oral History Project, May 17–18 (1990).

72. Emerson W. Pugh, *Memories that Shaped an Industry* (Cambridge: MIT Press, 1984): 30.

73. Saul Rosen, "Electronic Computers: a Historical Survey," *Computing Surveys* 1 (March 1969): 7–36.

74. Bashe et al., *IBM's Early Computers*, 161–162.

75. Cuthbert C. Hurd, "Edited Testimony," *Annals of the History of Computing* 3 (April 1981): 168.

76. Ibid., 169.

77. Ibid., 169.

78. Ibid., 169.

79. Bashe et al., *IBM's Early Computers* 129.

80. Ibid., 173–178.

81. Erwin Tomash and Arnold A. Cohen. "The Birth of an ERA: Engineering Research Associates, Inc., 1946–1955," *Annals of the History of Computing* 1: 2 (1979): 83–97; also Charles J. Murray, *The Supermen: The Story of Seymour Cray and the Technical Wizards behind the Supercomputer* (New York: Wiley, 1997).

82. Seymour Cray, "What's All This About Gallium Arsenide?" Videotape in the author's possession of a talk given at the "Supercomputing '88" conference in Orlando, Florida, November 1988; also Murray, *The Supermen*, 44–45.

83. Tomash and Cohen, "The Birth of an ERA," 90.

84. Seymour R. Cray, "Computer-Programmed Preventative Maintenance for Internal Memory Sections of the ERA 1103 Computer System," in *Proceedings of WESCON* (1954): 62–66.

85. Samuel S. Snyder, "Influence of U.S. Cryptologic Organizations on the Digital Computer Industry," *Journal of Systems and Software* 1 (1979): 87–102.

86. Ben Ferber, "The Use of the Charactron with the ERA 1103," *Proceedings of WJCC* (February 7–9, 1956): 34–35.

87. Alice R. and Arthur W. Burks, *The First Electronic Computer: The Atanasoff Story* (Ann Arbor: University of Michigan Press, 1988), chapter 1; also J. Presper Eckert, "A Survey of Digital Computer Memory Systems," in *Proceedings of IRE* (October 1953): 1393–1406.

88. Perry O. Crawford Jr., "Automatic Control by Arithmetic Operations" (Master's thesis, MIT, 1942). The bulk of Crawford's thesis discussed photographic rather than magnetic storage techniques, but it appears that this thesis was partially the inspiration for work that Eckert later did at the Moore School.

89. Tomash and Cohen, "The Birth of an ERA," 83–97.

90. Engineering Research Associates, Inc., *High-Speed Computing Devices* (New York: McGraw Hill, 1950; reprinted 1983, Los Angeles: Tomash Publishers): 322–339.

91. Richard E. Sprague, "The CADAC," U.S. Navy, *Symposium on Commercially Available General-Purpose Electronic Digital Computers of Moderate Price* (Washington, DC: 1952): 13–17.

92. Richard E. Sprague, "A Western View of Computer History," *Comm ACM* 15 (July 1972): 686–692.

93. E. D. Lucas, "Efficient Linkage of Graphical Data with a Digital Computer," in *Proceedings of WESCON* (1954): 32–37.

94. Willis E. Dobbins, "Designing a Low-Cost General Purpose Computer," in *Proceedings of ACM Meeting* (Toronto, September, 1952): 28–29.

95. Sprague, "A Western View of Computer History," *CACM*: 686–692.

96. C. Gordon Bell and Allen Newell, *Computer Structures: Readings and Examples* (New York: McGraw-Hill, 1971): 217.

97. B. E. Carpenter and R. W. Doran, "The Other Turing Machine," *Computer Journal* 20 (August 1977): 269–279.

98. Martin Campbell-Kelly, "Programming the Pilot ACE: Early Programming at the National Physical Laboratory," *Annals of the History of Computing* 3 (1981): 133–162.

99. Bashe et al., *IBM's Early Computers*, 168.

100. Tomash and Cohen, "The Birth of an ERA," 83–97.

101. Bashe et al., *IBM's Early Computers*, chapters 3 and 5.

102. Ibid., 170–172.

103. Thomas J. Watson Jr., *Father, Son & Co: My Life at IBM and Beyond* (New York: Bantam Books, 1990): 224.

104. Lancelot W. Armstrong, UNIVAC Conference, Smithsonian Institution, May 17–18, 1990 (Washington, DC: Smithsonian Institution Archives): 105.

105. F. C. Williams and T. Kilburn, "A Storage System for Use with Binary-Digital Computing Machines," *Institution of Electrical Engineers* in Proceedings III: 96 (March 1949): 81–100.

106. J. C. Chu and R. J. Klein, "Williams Tube Selection Program," in *Proceedings ACM Meeting* (Toronto, September, 1952): 110–114.

Chapter 2

1. Joanne Yates, *Control Through Communication: The Rise of System in American Management* (Baltimore: Johns Hopkins University Press, 1989).

2. Francis J. Murray, *Mathematical Machines*, vol. 1: Digital Computers (New York: Columbia University Press, 1961): 43–44; also Edwin Darby, *It All Adds Up: The Growth of the Victor Comptometer Corporation* (Chicago: Victor Comptometer Corporation, 1968).

3. The property is called "hysteresis," after the Greek word to denote "deficiency" or "lagging," because the rate at which the material becomes magnetized lags behind the rate at which a magnetizing current can affect it.

4. Emerson Pugh, *Memories that Shaped an Industry: Decisions Leading to IBM System/360* (Cambridge: MIT Press, 1984), chapter 2.

5. Kent C. Redmond and Thomas M. Smith, *Project Whirlwind: The History of a Pioneer Computer* (Bedford, MA: Digital Press, 1980): 206; "Engineering Report on the Static Magnetic Memory for the ENIAC," Burroughs Corporation, Report prepared for the Ballistics Research Laboratory, Aberdeen Proving Ground, Philadelphia, 31 July, 1953; also *Electronics* (May 1953): 198 and 200. The Smithsonian's National Museum of American History has one of the surviving remnants of a Mark IV shift register that used Wang's invention.

6. Pugh, *Memories*, 59.

7. "SAGE (Semi-Automatic Ground Environment)," Special Issue, *Annals of the History of Computing* 5: 4 (1983).

8. Morton M. Astrahan and John F. Jacobs, "History of the Design of the SAGE Computer—the AN/FSQ-7," *Annals of the History of Computing* 5: 4 (1983): 343–344. One witness to those events remembered that as the team was visiting an IBM facility, a technician who was adjusting a drum accidentally let the screwdriver hit the drum's surface, causing a sound not unlike fingernails run across a blackboard. To everyone's surprise, no data stored on the drum were lost. In any event, the orderly production facilities that IBM had set up for its commercial computers was judged to be ahead of what the other companies could show.

9. Gordon Bell, "The Computer Museum Member's First Field Trip: The Northbay [sic] AN/FSQ SAGE Site," *CACM* 26: 2 (1983): 118–119.

10. Edmund Van Deusen, "Electronics Goes Modern," *Fortune* (June 1955): 132–136, 148.

11. Pugh, *Memories*, 102–117.

12. Ibid., 126.

13. Katherine Fishman, *The Computer Establishment* (New York: Harper & Row, 1981): 44.

14. Saul Rosen, "Electronic Computers: a Historical Survey," *Computing Surveys* 1 (March 1969): 7–36.

15. Ibid.; also Fishman, *The Computer Establishment*, 161–162. For details on the RAYDAC, see Engineering Research Associates, *High Speed Computing Devices* (New York, McGraw-Hill, 1950): 206–207.

16. For an insider's view of Honeywell's decision, see interview with Richard Bloch, Smithsonian Computer History Project; also W. David Gardner, "Chip off the Old Bloch," *Datamation* (June 1982): 241–242.

17. *Fortune* 52 (July 1955), supplement, 2–5.

18. Robert W. House, "Reliability Experience on the OARAC," in *Proceedings Eastern Joint Computer Conference* (1953): 43–44; also Robert Johnson, Interview, *Annals of the History of Computing* 12: 2 (1990): 130–137; and George Snively,

"General Electric Enters the Computer Business, *Annals of the History of Computing* 10: (1988): 74–78.

19. Homer R. Oldfield, *King of the Seven Dwarfs: General Electric's Ambiguous Challenge to the Computer Industry* (Los Alamitos, CA: IEEE Computer Society Press, 1996).

20. Fishman, *Computer Establishment*, 164–165.

21. W. K. Halstead et al., "Purpose and Application of the RCA BIZMAC System," in *Proceedings Western Joint Computer Conference* (1956): 119–123; also Rosen, "Electronic Computers," 16–17.

22. Rosen, "Electronic Computers," 16–17; W. K. Halstead, et al., "Purpose and Application of the RCA BIZMAC System," 119–123.

23. R. P. Daly, "Integrated Data Processing with the UNIVAC File Computer," *Proceedings Western Joint Computer Conference* (1956): 95–98.

24. Franklin Fisher, James McKie, and Richard Mancke, *IBM and the U.S. Data Processing Industry* (New York, Praeger, 1983): 53.

25. See, for example, T. A. Heppenheimer, "How von Neumann Showed the Way," *American Heritage of Invention & Technology* (Fall 1990): 8–16.

26. The following description is taken from several texts, including John L. Hennessy and David A. Patterson, *Computer Architecture: a Quantitative Approach* (San Mateo, CA: Morgan Kaufmann, 1990); Simon Lavington, *Early British Computers* (Bedford, MA: Digital Press, 1980): 106–119; C. Gordon Bell and Allen Newell, *Computer Structures: Readings and Examples* (New York: McGraw-Hill, 1971).

27. Bell and Newell, *Computer Structures*, 224; also Adams Associates, "Computer Characteristics Chart," *Datamation* (November/December, 1960): 14–17.

28. Simon Lavington, *History of Manchester Computers* (Manchester, UK: NCC Publications, 1975): 12.

29. C. C. Hurd, "Early IBM Computers: Edited Testimony," *Annals of the History of Computing* 3: 2 (April 1981): 185–176.

30. Lavington, *History of Manchester Computers*, 78–82; also Hennessy and Patterson, *Computer Architecture*, 91–92.

31. Werner Buchholz, ed., *Planning a Computer System: Project Stretch* (New York: McGraw-Hill, 1962), chapter 9.

32. The stack concept in computer architecture probably originated with Professor F. L. Bauer of the Technical University of Munich; see for example, F. L. Bauer, "The Formula-Controlled Logical Computer STANISLAUS," *MTAC* 14 (1960): 64–67.

33. Richard E. Smith, "A Historical Overview of Computer Architecture," *Annals of the History of Computing* 10: 4 (1989): 286.

34. T. H. Myer and I. E. Sutherland, "On the Design of Display Processors," *Communications of the ACM* 11: 6 (June 1968): 410–414; also C. Gordon Bell, J. Craig Mudge, and John McNamara, *Computer Engineering: a DEC View of Hardware Systems Design* (Bedford, MA: Digital Press, 1978): 202.

35. Bell, Mudge, and McNamara, *Computer Engineering*, 256–257.

36. Gerald Brock, *The Telecommunications Industry: the Dynamics of Market Structure* (Cambridge, MA: Ballinger, 1981): 187–194. Little of the litigation had to do with the transistor; much, however, concerned Bell's Western Electric subsidiary's role as manager of the Sandia Corporation, a military installation in New Mexico that manufactured atomic bombs.

37. Thomas J. Misa, "Military Needs, Commercial Realities, and the Development of the Transistor, 1948–1958," in Merritt Roe Smith, ed., *Military Enterprise and Technological Change* (Cambridge: MIT Press, 1985): 253–287; also J. H. Felker, "Performance of TRADIC System," in *Proceedings Eastern Joint Computer Conference* (1954): 46–49. These machines used point-contact transistors, which were not only unreliable but difficult to produce in large quantities.

38. John Allen, "The TX-0: its Past and Present," *Computer Museum Report* #8 (Spring 1984): 2–11.

39. Samuel Snyder, "Influence of U.S. Cryptologic Organizations on the Digital Computer Industry," *Journal of Systems and Software* 1 (1979): 87–102.

40. Ibid.; also telephone conversation with Ray Potts, March 31, 1995.

41. Herman Lukoff, *From Dits to Bits: a Personal History of the Electronic Computer* (Portland, OR: Robotics Press, 1979); J. L. Maddox et al., "The TRANSAC S-1000 Computer," in *Proceedings Eastern Joint Computer Conference* (December 1956): 13–16; also Saul Rosen, "Electronic Computers: a Historical Survey."

42. Ibid., 89.

43. Fisher, *IBM and the U.S. Data Processing Industry*, 87.

44. L. P. Robinson, "Model 30-201 Electronic Digital Computer," *Symposium on Commercially-available General-purpose Digital Computers of Moderate Price* (Washington, DC, 14 May 1952): 31–36.

45. Fisher, *IBM and the U.S. Data Processing Industry*, 83.

46. For many years this computer was on display at the Smithsonian Institution's National Museum of American History. The claim is taken from a brochure Burroughs made available for that exhibit. Oral histories of several Burroughs employees are on file with the Smithsonian Computer History Project, including an interview with Robert Campbell, 11 April 1972.

47. William Rodgers, *Think: a Biography of the Watsons and IBM* (New York, Stein & Day, 1969): 58–60.

48. Ibid., 213.

49. The most colorful and vehement attack on IBM is found in Ted Nelson, *Computer Lib* (South Bend, IN: Ted Nelson, 1974): 52–56.

50. Eric Weiss, "Obituary: Robert B. Forest," *Annals of the History of Computing* 19: 2 (1997): 70–73.

51. Bashe et al., *IBM's Early Computers*, 280.

52. Ibid., 286.

53. Ibid., also T. Noyes and W. E. Dickenson, "Engineering Design of a Magnetic Disk Random Access Memory," in *Proceedings, Western Joint Computer Conference* (February 7, 1956): 42–44.

54. Mitchell E. Morris, "Professor RAMAC's Tenure," *Datamation* (April 1981): 195–198.

55. This story has been independently told to the author by several people, including at least one person who worked on the BMEWS system at the time. Charles Bashe, et al, in IBM's official history, mentions a "very taut schedule" but does not repeat this story. See Bashe, et al., *IBM's Early Computers* (Cambridge: MIT Press, 1986): 446–449.

56. For a period of two years at its Federal Systems Division, IBM did not allow interactive terminals at the sites where its programmers were working. All programs had to be submitted on the standard forms. A more typical situation was to have the programmers use the forms, letting keypunchers prepare the decks, but providing a single 029 punch so that he or she could punch one or two cards from time to time. See Robert N. Britcher, "Cards, Couriers, and the Race to Correctness," *Journal of Systems and Software* 17 (1992): 281–284.

57. Bashe et al., *IBM's Early Computers*, 468–469.

58. Bell and Newell, *Computer Structures*, chapter 18; also Fisher et al., *IBM and the U.S. Data Processing Industry*, 53.

Chapter 3

1. I wish to thank Professor J. A. N. Lee, of Virginia Tech, for bringing this citation to my attention.

2. The story is told by Merrill Flood, a research mathematician at the RAND Corporation, in a letter to the editor, *Datamation* (December 1, 1984): 15–16. Flood says that he heard Ike's comment secondhand.

3. National Academy of Engineering, Washington, DC. Press release dated 6 October 1993.

4. Susan Lammers, ed., *Programmers at Work* (Redmond, WA: Microsoft Press, 1989), 278.

5. Margaret Hamilton, interview with the author, 6 April 1992; J. David Bolter, *Turing's Man: Western Culture in the Computer Age* (Chapel Hill: University of North Carolina Press, 1984).

6. Alan Perlis has said of this, "Beware of the Turing tar-pit in which everything is possible but nothing is easy." In Perlis, "Epigrams on Programming," *ACM SIGPLAN Notices* (October 1981): 10.

7. Barry Boehm, "Software Engineering," *IEEE Transactions on Computers* C25 (December 1976), 1226–1241; for a refutation of Boehm's thesis, see Werner Frank, "The History of Myth #1," *Datamation* (May 1983): 252–263.

8. Brian Randell, "Epilogue," in Charles and Ray Eames, *A Computer Perspective: Background to the Computer Age*, new edition, foreword by I. Bernard Cohen (Cambridge: Harvard University Press, 1990): 163.

9. Harvard University Computation Laboratory, *A Manual of Operation for the Automatic Sequence Controlled Calculator*, Charles Babbage Institute Reprint series, vol. 8 (Cambridge: MIT Press, 1985; originally published 1946).

10. Hopper, quoted in "Computers and Their Future: Speeches Given at the World Computer Pioneer Conference," Lladudno, Wales, July 1970 (Lladudno: Richard Williams and Partners): 7/3–7/4.

11. Grace Hopper, Log Book for the ASCC, April 7–Aug. 31, 1944; Smithsonian Institution Archives; a description of Baker's problem is found in Herbert Grosch, *Computer: Bit Slices from a Life* (Novato, CA: Third Millenium Books, 1991): 51–52.

12. Some of these concepts predate the digital era and can be traced to, for example, the MIT Differential Analyzer, an analog machine that was programmed by sequences of tapes. See interview with Perry Crawford, Smithsonian Computer History Project, 29 October 1970.

13. Konrad Zuse, "Planfertigungsgeräte," 1944; Bonn, Gesellschaft für Mathematik und Datenverarbeitung, Zuse Archive, folder 101/024.

14. Konrad Zuse, "Der Programmator," *Zeitschrift für Angewandte Mathematik und Mechanik* 32 (1952): 246; Heinz Rutishauser, "Rechenplanfertigung bei programmgesteuerten Rechenmaschinen," *Mitteilungen* aus dem Institut für angewandte Mathematik der ETH, no. 3, 1952; also F. L. Bauer, "Between Zuse and Rutishauser: the Early Development of Digital Computing in Central Europe," in N. Metropolis, J. Howlett, and Gian-Carlo Rota, eds., *A History of Computing in the Twentieth Century* (New York: Academic Press, 1980): 505–524.

15. Rutishauser, "Automatische Rechenplanfertigung bei programmgesteuerten Rechenmaschinen," *ZAMP* 3 (1952): 312–313 (author's translation).

16. Martin Campbell-Kelly, "Programming the EDSAC: Early Programming Activity at the University of Cambridge," *Annals of the History of Computing* 2 (1980): 7–36; Also Maurice Wilkes, D. J. Wheeler, and Stanley Gill, *The Preparation of Programs for an Electronic Digital Computer* (Cambridge: Addison-Wesley, 1951).

17. Maurice Wilkes, *Memoirs of a Computer Pioneer* (Cambridge: MIT Press, 1985): 142–144.

18. Smithsonian Institution, NMAH Archives, Grace Hopper Papers, Box 5, folder #1, program of ACM Meeting, Oak Ridge, April 1949.

19. J. W. Mauchly, memorandum to EMCC executive committee, 16 August 1948; Hagley Museum, Accession 1825, Series I, Box 3: Mauchly, John; also Richard Pearson, "Admiral Hopper Dies; Pioneer in Computers," *Washington Post* (January 2, 1992).

20. Grace Hopper, "Compiling Routines," internal memorandum, Eckert-Mauchly Computer Corporation, Philadelphia, 31 December 1953; Box 6, folder 9, Hopper Papers, Smithsonian Archives.

21. Ibid.

22. Hopper papers, NMAH, Box 5, folder 7. Here she uses the term "generator" in a sense that one might use "compiler" today.

23. For a concise definition of the terms "compiler" and "interpreter" as they were initially used, see Grace M. Hopper, "Compiling Routines," *Computers and Automation* 2 (May 1953): 1–5. In Hopper's words: "The interpretive method of using subroutines consists of fixing the location of the subroutine in the computer memory, and causing the main program to interpret what may be called a "pseudo-code,' and thus refer to the subroutine and perform it. . . . The compiling method of using subroutines consists of copying the subroutine in to the main routine, adjusting memory locations as necessary to position the subroutine properly in the program and to supply arguments and results" (p. 2).

24. Wilkes, Wheeler, and Gill discuss a program they call an "assembly subroutine," which does the same thing. They also describe something similar that they call "interpretive" routines: See Wilkes, Wheeler, and Gill, *The Preparation of Programs for an Electronic Digital Computer,* 26–37. These terms survive to the present day but with different meanings. Suffice it to say that in the early 1950s it had become clear that the computer could be instructed to take over many of the chores of producing programs, but just how much, and in what form, was less certain. See Also Martin Campbell-Kelly, "Programming the EDSAC: Early Programming Activity at the University of Cambridge," *Annals of the History of Computing,* 2 (1980): 7–36.

25. Michael Mahoney, "Software and the Assembly Line," paper presented at Workshop on Technohistory of Electrical Information Technology, Deutsches Museum, Munich, December 15–19, 1990.

26. J. H. Laning and N. Zierler, "A Program for Translation of Mathematical Equations for Whirlwind I," (Cambridge, MA: MIT Engineering Memorandum no. E-364, January 1954). National Air and Space Museum, NBS Collection, Box 39, Folder 8.

27. See John Backus, "Programming in America in the 1950s—Some Personal Recollections," in Metropolis, Howlett, and Rota, eds. *History of Computing in the Twentieth Century*, 125–135, especially pp. 133–134.

28. Laning and Zierler, "A Program for Translation of Mathematical Equations For Whirlwind I," (1954) frontispiece. In the copy that I examined, from the library of the National Bureau of Standards, someone—possibly Samuel Alexander—crossed out the word "interpretive" and wrote in pencil above it "translated."

29. Backus, "Programming in America."

30. Charles W. Adams and J. H. Laning Jr., "The MIT Systems of Automatic Coding: Comprehensive, Summer Session, and Algebraic," in *Symposium on Automatic Programming of Digital Computers* (Washington, DC: U.S. Navy, 1954): 64.

31. Donald Knuth, "The Early Development of Programming Languages," in Metropolis, Howlett, and Rota, eds., *History of Computing in the Twentieth Century*, 239.

32. Paul Armer, "SHARE—a Eulogy to Cooperative Effort," *Annals of the History of Computing*, 2: 2 (April 1980): 122–129. Armer says that "SHARE" did not stand for anything; others say it stood for "Society to Help Avoid Redundant Effort."

33. Donald Knuth, *Sorting and Searching* (Reading, MA: Addison-Wesley, 1973): 3.

34. See, for example, the work of C. A. R. Hoare. See also Knuth, *Sorting and Searching*.

35. Von Neumann to Herman Goldstine, 8 May 1945; Hagley Museum, Accession 1825, Box 74, series II, Chron File, May 1945–October 1945.

36. Donald Knuth, "Von Neumann's First Computer Program," *Computing Surveys* 2 (December 1970): 247–260.

37. Between 1947 to June of 1950 Holberton went by the surname Snyder. See interview with Holberton, UNIVAC Oral History Conference, Smithsonian Institution, May 17–18, 1990, 52; Smithsonian Institution Archives. Soon after Remington Rand acquired EMCC, she left for the U.S. Navy's David Taylor Model Basin just outside Washington, D.C., where she continued work as a programmer on UNIVAC #6.

38. Knuth, *Sorting and Searching*, 386. Holberton's routine is described in U.S. Naval Mathematics Advisory Panel, *Symposium on Automatic Programming for Digital Computers* (Washington, DC, 1954): 34–39.

39. Eric Weiss, ed., *Computer Usage—Fundamentals* (New York: McGraw-Hill, 1969).

40. Holberton, interview in UNIVAC oral history conference, Smithsonian Institution, May 17–18, 1990, p. 52; Smithsonian Institution Archives.

41. The most recent version, accepted by the International Standards Organization in 1992, is called "Fortran 90," (now spelled as a proper noun, no longer as an acronym). A version of Fortran aimed at parallel, non-von Neumann architecture computers, is also under development; it is presently called High Performance Fortran. See Ian Stockdale, "Vendors, International Standards Community Adopt Fortran 90," *NAS News* (NASA Ames Research Center, September 1992): 1–8.

42. Backus, "Programming in America." Professor F. L. Bauer of the Technical University of Munich, who was one of the developers of ALGOL, pointed out to the author that many ALGOL compilers were also fast and efficient, so this, in his view, was obviously not the sole reason for FORTRAN's success.

43. Backus, "Programming in America," 131.

44. The committee was called "CODASYL": Conference On DAta SYstems Languages. It was not a committee in the normal sense of the word; actual work on COBOL was done by two other committees that included representatives of computer manufacturers and government agencies. See Jean Sammett, *Programming Languages: History and Fundamentals* (Englewood Cliffs, NJ: Prentice Hall, 1969), section V.3.

45. "One Compiler, Coming Up!" *Datamation* (May/June 1959): 15; also Saul Rosen, "Programming Systems and Languages: a Historical Survey," *Proc. SJCC* 25 (1964): 1–15.

46. The example is taken from Sammett, *Programming Languages*, 337.

47. The movie was released in 1968. In a later scene, HAL states that he was created in Urbana, Illinois, which in 1968 was the location of an ambitious supercomputer project known as "Illiac-IV."

48. Paul Ceruzzi, "Aspects of the History of European Computing," Paper presented at the 12th European Studies Conference, Omaha, NE, 8–10 October, 1987.

49. Grace Hopper, "Keynote Address," in Richard L. Wexelblatt, ed., *History of Programming Languages* (New York: Academic Press, 1981): 7–20.

50. Ibid.; see also Sammet, *Programming Languages*.

51. Sammet, *Programming Languages*, 11.

52. G. F. Ryckman, *Proc WJCC* 1960, 341–343; also William Orchard-Hays, "The Evolution of Programming Systems," *IRE Proc.* 49 (January 1961): 283–295.

53. The slash-asterisk command was designed by Larry Josephson, who later had a second career as an underground disk jockey with the New York alternative radio station WBAI. Larry Josephson, private communication to the author.

54. Sammet, *Programming Languages*, 205–215; Bob Rosin, personal communication, 23 June 1995.

55. The PL/I programming language, described later, was also a part of this unifying effort; it was to replace COBOL for business, and FORTRAN for scientific applications.

56. Fred Brooks, *The Mythical Man-Month: Essays on Software Engineering* (Reading, MA: Addison-Wesley, 1975).

57. See, for example, University of Michigan, College of Engineering, "Applications of Logic to Advanced Digital Computer Programming," Notes for an Intensive Course for the Engineer and Scientist, 1957 Summer Session. National Air and Space Museum, NBS Collection.

58. Paul Ceruzzi, "Electronics Technology and Computer Science," *Annals of the History of Computing*, 10: 4 (1989): 257–275; also Louis Fein, "The Role of the University in Computers, Data Processing, and Related Fields," *Communications, ACM* 2 (1959): 7–14.

59. Stanford University Archives, George Forsythe papers; also Donald E. Knuth, "George Forsythe and the Development of Computer Science," *Communications, ACM* 15: 8 (1972): 721–726.

60. Newell, Perlis, and Simon, letter to the editor, *Science* 157 (22 September, 1967): 1373–1374.

61. Herbert Simon, *The Sciences of the Artificial*, 2nd edition (Cambridge: MIT Press, 1981).

62. Newell, Perlis, and Simon, letter to the editor, 1374.

63. *Communications, ACM* 11 (March 1968): 147.

64. Seymour V. Pollack, "The Development of Computer Science," in Seymour Pollack, ed., *Studies in Computer Science* (Washington, DC: Mathematical Association of America, 1982): 1–51.

65. Ibid., 35.

66. Donald E. Knuth, *The Art of Computer Programming*, vol. 1: *Fundamental Algorithms* (Reading, MA: Addison-Wesley, 1968). Knuth published volume 2 in 1969 and volume 3 in 1973; he has thus far not produced the remaining four volumes. In a recent interview he expressed a hope that he would be able to publish volume 4 (in three parts) in 2003, and volume 5 in 2008.

67. *Communications, ACM* 11 (March 1968): 147.

68. "Language Protection by Trademark Ill-advised," *Communications, ACM* 11: 3 (March 1968): 148–149.

69. According to Mooers, in an interview with the author, the TRAC language was "stolen" and issued, without restrictions, under the name "MINT"—"Mint Is Not TRAC."

70. One person who agreed with Mooers was Bill Gates, who wrote to him expressing sympathy and an interest in following Mooer's crusade. The letter is now among Mooers's papers at the Charles Babbage Institute.

71. Peter Naur and Brian Randell, "Software Engineering; Report on a Conference Sponsored by the NATO Science Committee," 7–11 October 1968 (Garmisch, Germany: NATO, 1969).

72. James Tomayko, "Twenty Year Retrospective: the NATO Software Engineering Conferences," *Annals of the History of Computing*, 11: 2 (1989): 132–143.

73. Thomas J. Bergin and James (Jay) Horning, "Dagstuhl Conference," *Annals of the History of Computing* 19: 3 (1997): 74–76.

74. Franklin Fisher, James W. McKie, Richard B. Mancke, *IBM and the U.S. Data Processing Industry* (New York: Praeger, 1983): 176–177.

75. David C. Mounce, *CICS: a Light Hearted Chronicle* (Hursely, UK: IBM, 1994): i.

76. Peter Salus, *A Quarter Century of UNIX* (Reading, MA: Addison-Wesley, 1994); UNIX will be discussed more fully later on.

77. Dennis M. Ritchie, "The Development of the C Programming Language," in Thomas J. Bergin Jr. and Richard G. Gibson Jr., eds., *History of Programming Languages—II* (New York: ACM Press, 1996): 671–698.

78. Ibid., 673.

79. Rosen, "Programming Systems and Languages: a Historical Survey," 12.

80. C. H. Lindsey, "A History of ALGOL-68," *ACM Sigplan Notices* 28: 3 (March 1993): 97–132.

81. For example, there was an acrimonious debate among committee members over how ALGOL-68 should handle the use of a comma to separate thousands and a point for the decimal, or the opposite, as is common in Germany; for example, 2,555.32 in Germany is written 2.555,32.

82. Nicholas Wirth, "Recollections about the Development of PASCAL," *ACM Sigplan Notices* 28: 3 (March 1993): 333–342.

83. Merrill Flood, a mathematician with the RAND Corporation, claimed that he coined the term in a memorandum written in the late 1940s, but apparently the term was re-created among computer people ten years later. Flood, letter to the editor, *Datamation* (December 1, 1984): 15–16. Flood says that he heard Ike's

comment secondhand. "Software" did not appear in print until around 1959, when the term gradually began appearing in the computer journals and trade press.

Chapter 4

1. "History of ADP in IRS Service Centers," videotape, IRS Austin, Texas, Service Center, December 11, 1991.

2. Ibid.

3. U.S. Internal Revenue Service, "Data Processing History and Evolution in IRS," undated typescript in author's possession.

4. Digital Equipment Corporation, "PDP-8 Typesetting System," brochure, ca. 1966, in author's possession; also Eric Weiss, ed., *Computer Usage—Applications* (New York: Computer Usage Corporation, 1969), chapter 6.

5. Gerald Brock, *The U.S. Computer Industry: A Study in Market Power* (Cambridge, MA: Ballinger, 1975).

6. IBM's research labs were the origins of floppy disk storage, FORTRAN, and relational databases, among other innovations.

7. Samuel Snyder, "Influence of U.S. Cryptologic Organizations on the Digital Computer Industry," *Journal of Systems and Software* 1 (1979): 87–102.

8. Saul Rosen, "Electronic Computers: a Historical Survey, *Computing Surveys* 1 (March 1969): 7–36.

9. Brock, *The U.S. Computer Industry.*

10. Paul Forman, "Behind Quantum Electronics: National Security as Basis of Physical Research in the United States, 1940–1960," HSPS 18: 1 (1987): 149–229.

11. George Ballester, "A Brief History of Computers at Massachussetts Blue Cross/Blue Shield," unpublished paper, ca. 1984, in author's possession; also Charles J. Bashe et al., *IBM's Early Computers* (Cambridge: MIT Press, 1986): 50, 115.

12. Ballester, "A Brief History of Computers at Massachussetts Blue Cross/Blue Shield."

13. Ibid.

14. Ibid.

15. NASA, Ames Research Center, "A Justification of the Need to Replace the IBM 7040/7094 Direct Couple System," Moffett Field, California (March 31, 1967): 8. National Air and Space Museum Archives.

16. Ibid.; also "Computation Division Staff Meeting Minutes," February 13, 1968, NASA-Ames Research Center. NASM Archives.

17. NASA-Ames Research Center, "ADPE Acquisition Plan—category A," memorandum September 25, 1967; also Bill Mead, "Computation Division Contact Report, June 20, 1969," NASM Archives.

18. NASA-Ames Research Center, "IBM 1401 System," correspondence file, NASM Archives.

19. NASA-Ames Research Center, "Findings of the Evaluation Committee on the RFP of the Central Computer Facility," memorandum of 9/15/69; NASA-Ames file, Folder SYS-AA: IBM 360/67 SYSTEM, FY-69-70, NASM Archives.

20. Emerson W. Pugh, Lyle R. Johnson, and John H. Palmer, *IBM's 360 and Early 370 Systems* (Cambridge: MIT Press, 1991): 630–631.

21. NASA-Ames Research Center, Memorandum, Wayne Hathaway to Chief, Computer Systems Branch, March 31, 1971; folder SYS-AA: IBM 360/67 (ARPA Network), NASM Archives.

22. NASA-Ames Research Center, folder #24: IBM 360/50 Computer, Category A, NASM Archives.

23. Material for the following account has been drawn from a number of materials supplied to the author from the IRS, including a film, "History of ADP in IRS Service Centers," IRS Austin, Texas, Service Center, December 11, 1991; Daniel Capozzoli, "The Early Years of Data Processing," *Computer Services Newsletter*, US Internal Revenue Service, Washington, DC, July 1987; and "ADP-History," unpaginated, undated typescript (ca. 1970), IRS, Washington, DC. The author wishes to acknowledge the support of Shelly Davis, IRS historian, for assistance.

24. Capozzoli, "The Early Years of Data Processing," 1.

25. Ibid.

26. For this comparison I am assuming a tape capacity of about 5 megabytes per 2,400-foot tape, a suitcase containing ten tapes, and a four-hour flight. See Bashe et al., *IBM's Early Computers* (Cambridge: MIT Press, 1986) 215; Pugh et al., *IBM's 360 and Early 370 Systems*, 530.

27. Internal Revenue Service, "History of Automatic Data Processing."

28. U.S. General Accounting Office, "Safeguarding Taxpayer Information—An Evaluation of the Proposed Computerized Tax Administration System," United States Comptroller General, Report to the Congress, LCD-76-115 [January 17, 1977].

29. Ibid., 4.

30. Ibid. See also *Computerworld* 11: 8 (February 21, 1977): 1, 6.

31. Ibid.

32. U.S. Senate, 95th Congress, Hearings on Treasury, Postal Service, and General Government Appropriations, FY 1978, Part 1; U.S. House of Representatives, Hearings Before a Subcommittee of Appropriations, Subcommittee on the Treasury, Postal Service, and General Government Appropriations, Part 1, Department of the Treasury, U.S. National Archives. The quotations are from the hearing held in the House, and were posed by Rep. Clarence Miller of Ohio.

33. IRS History Office, private communication; also U.S. Senate, Committee on Appropriation, Subcommittee on the Department of the U.S. Treasury, U.S. Postal Service, and General Government Appropriations, Hearings before the 95th Congress, on H.R. 12930, March 7, 1978, part 2, 269–297.

34. U.S. House of Representatives, Hearings before a Subcommittee of the Committee on Appropriations, 95th Congress, March 12, 1978, 438.

35. The whole story is well told in David Burnham, *The Rise of the Computer State* (New York: Random House, 1983).

36. In this discussion I emphasize alternatives to sequential batch operation on standard commercial mainframe computers. That does not mean that real-time (NASA-Houston), on-line (proposed IRS), and time-shared (NASA-Ames) uses are all the same. They are quite different, but they all depart from the true model of computing of that era, which was a descendent of the precomputer tabulator era.

37. James Tomayko, *Computers in Spaceflight: the NASA Experience* (Washington, DC: NASA, 1988): 244; S. E. James, "Evolution of Real-time Computer Systems for Manned Spaceflight," *IBM J. Res. Dev.* 25 (181): 418.

38. Paul Ceruzzi, *Beyond the Limits: Flight Enters the Computer Age* (Cambridge: MIT Press, 1989), chapter 9.

39. S. E. James, "Evolution of Real Time Computer Systems for Manned Spaceflight," *IBM J Res Dev* 25 (1981): 245; also Saul I. Gass, "Project Mercury Real-Time Computational and Data-flow System," *Proc. Eastern JCC* (Washington, DC): 33–46.

40. Gass, "Project Mercury," 37.

41. Marilyn Scott, and Robert Hoffman, "The Mercury Programming System," *Proc. Eastern JCC* 20 (December 12–14, 1961): 47–53. A variation of this scheme was independently developed by minicomputer designers to eliminate the special trap processor. It has since become known as a "priority-driven interrupt" and is a cornerstone of modern computing systems of all sizes.

42. Tomayko, *Computers in Spaceflight*, 246.

43. Bashe, *IBM's early Computers*. IBM distinguished between "on-line" and "real-time" operation, in that the former did not have to process data at the same or

faster rate than data were being fed into the system. An airline reservation system is "on-line": a user wants the system to respond quickly but is willing to tolerate brief delays, up to a few seconds. Such delays, however brief, would be fatal to a real-time system, such as the one controlling a missile launch.

44. Mark Mergen (IBM Corporation), private communication, July 8, 1994. I also want to thank David Larkin of IBM for assisting me with tracking down this thread of history. 'SPOOL' stood for "simultaneous peripheral operations on line," as in the ability to print while the processor was doing another job.

45. The situation was analogous to the U.S. phone system, which before the breakup of AT&T in the early 1980s was a regulated monopoly. Residential customers typically rented, not owned, their telephones, and they were forbidden to make any modifications to the instruments or to the wiring in their homes.

46. David E. Lundstrom, *A Few Good Men from UNIVAC* (Cambridge: MIT Press, 1987): 136; also David K. Allison, "U.S. Navy Research and Development Since World War II," in Merritt Roe Smith, ed., *Military Enterprise and* Technological Change (Cambridge: MIT Press, 1985): 289–328. The NTDS had a 30-bit word length.

47. Lundstrom, *A Few Good Men*, 67.

48. Seymour Cray, interview with David K. Allison (May 1995), Smithsonian Institution Computer History Project. The term "minicomputer" is misleading because it suggests only physical size; however, it has become so accepted that I will continue to use it in this chapter.

49. Montgomery Phister Jr., "Quotron II: An Early Multiprogrammed Multiprocessor for the Communication of Stock Market Data," *Annals of the History of Computing* 11 (1989): 109–126; also New York Stock Exchange, undated brochure in the author's collection.

50. Jamie Parker Pearson, *Digital at Work* (Bedford, MA: Digital Press, 1992): 6–9.

51. C. Gordon Bell, J. Craig Mudge, and John McNamara, *Computer Engineering: a DEC View of Hardware Systems Design* (Bedford, MA: Digital Press, 1978): 125–127.

52. "The TX-0: Its Past and Present," Computer Museum Report 8 (Spring 1984): 2–11.

53. Bell et al., *Computer Engineering*, 126–127.

54. Ibid., 129.

55. Fisher, *IBM*, 34, 308.

56. Bell et al., *Computer Engineering*, 136–139; Pearson, *Digital at Work*, 16–21.

57. Steven Levy, *Hackers: Heroes of the Computer Revolution* (New York: Anchor/Doubleday, 1984).

58. Pearson, *Digital at Work*, 143.

59. Glenn Rifkin, and George Harrar, *The Ultimate Entrepreneur: The Story of Ken Olsen and Digital Equipment Corporation* (Chicago: Contemporary Books, 1988).

60. Bell et al., *Computer Engineering*, 181.

61. Ibid., 141.

62. A. Clark Wesley, "The LINC was Early and Small," in Adele Goldberg, ed., *A History of Personal Workstations* (New York: ACM Press, 1988): 347–391.

63. P. R. Morris, *A History of the World Semiconductor Industry* (London: Peter Peregrinus, 1990): 42.

64. C. Gordon Bell and Allen Newell, *Computer Structures: Readings and Examples* (New York: McGraw-Hill, 1971); Bashe et al., *IBM's Early Computers*, 448–449.

65. Digital Equipment Corporation, "PDP-8 Typesetting System," advertising brochure, 1966, Computer Museum Archives, Boston, MA, "Milestones of a Revolution" file.

66. Bell et al., *Computer Engineering*, p. 215; Wes Clark, "The LINC Was Early and Small," in Adele Goldberg, ed., *A History of Personal Workstations*: 345–391.

67. Bell et al., 64, 180, 198–199.

68. Ibid., 198; also Rifkin and Harrar, *The Ultimate Entrepreneur.*

69. Teletype Corporation, "Teletype, Model 33 Equipment for Fast, Economical 8-level Data Communications," Product Literature, ca. 1966, AT&T Archives.

70. C. E. MacKenzie, *Coded Character Sets: History & Development* (Reading, MA: Addison Wesley, 1980); also Friden Corporation, "Technical Manual for 2201 Flexowriter Automatic Writing Machine," ca. 1968, copy in author's possession.

71. John Leng, telephone interview with the author, April 13, 1993.

72. Another term that has come into common use, which describes a similar relationship, is "Value-Added Reseller" (VAR).

73. Digital Equipment Corporation, "PDP-8 Typesetting System," brochure, ca. 1966; Computer Museum, Boston, "Milestones of a Revolution" Archive.

74. "A Chorus Line: Computerized Lighting Control Comes to Broadway," *Theatre Crafts* (Nov./Dec. 1975): cover, 6–11, 26–29. Some theater lighting professionals have privately told this author that the computer created a new set of headaches on many a night, when the system acted up.

75. Computer Museum, Boston, "Timeline of Computer History" exhibition, 1984–1989.

76. Bob Cumberford, "An Interview with Alec Issigonis," *Road & Track* (Road Test Annual, 1962): 18–19.

77. DEC's management style attained the status of myth, especially after the publication of *The Change Masters* by Rosabeth Moss Kanter (1983). DEC's troubles in the 1990s obviously lead one to question this myth. The author's view on the causes of DEC's downfall will be examined in a later chapter.

78. Pearson, *Digital at Work*.

79. Robert Sobel, *IBM: Colossus in Transition* (New York, 1981): 228.

80. Truett Allison to Rachel Hellenga, letter September 14, 1990. Computer Museum, Boston, "Milestones of a Revolution" Archive.

81. Ben Ross Schneider, *Travels in Computerland* (Reading, MA: Addison-Wesley, 1974): 73.

82. See Ted Nelson, *Computer Lib* (Ted Nelson, 1974); ironically, in his preface to the second edition, Nelson noted that as DEC grew larger, "it was getting just like IBM."

83. Sobel, *IBM*.

84. Max Palevsky, Oral History Interview, February 15, 1973, 20; Smithsonian Computer History Project; Smithsonian Archives.

85. Pearson, *Digital at Work*, chapter 1. Chapter 2 is an edited transcript of an interview with Forrester and Robert Everett, of the MITRE Corporation. Neither Forrester nor Everett became DEC employees, although both were, or still are, on the DEC Board of Directors.

86. The entrepreneurial companies of Silicon Valley were always, and remain, heavily staffed by MIT graduates, it should be noted. The failure of Digital Equipment Corporation to remain competitive will be discussed in a later chapter, but that alone should not have caused the fall of Route 128 as a center of innovation. Silicon Valley firms fail all the time, yet the region continues to prosper. Perhaps it was a failure by the Boston venture capitalists to recognize and risk money on innovative ideas; or it may have been a failure of MIT to sustain the excellence it had developed during the 1940s and 1950s.

87. John Kenneth Galbraith, *The New Industrial State* (Boston: Houghton Mifflin, 1967); William H. Whyte, *The Organization Man* (New York: Simon & Schuster, 1956).

Chapter 5

1. Robert Sobel, *IBM: Colossus in Transition* (New York: Bantam, 1981): 159, 188, 274.

2. Franklin Fisher, James W. McKie, and Richard B. Mancke, *IBM and the U.S. Data Processing Industry* (New York: Praeger, 1983): 65.

3. Ibid.; also Sobel, *IBM*, and Katherine Fishman, *The Computer Establishment* (New York: McGraw-Hill, 1981): 470.

4. Watts S. Humphrey, "MOBIDIC and Fieldata," *Annals of the History of Computing* 9: 2 (1987): 137–182.

5. Bob Evans, quoted in T. A. Wise, "IBM's $5,000,000,000 Gamble," *Fortune* (September 1966): 118–123, 224, 226, 228.

6. Ibid., 44.

7. For comparison, in 1975, Xerox dropped out of the large computer business, writing off an investment in the former Scientific Data Systems of close to $1 billion—about what IBM had risked on System/360. If was one of the biggest write-offs ever faced by an American corporation. Yet Xerox survived. Xerox's foray into computing will be discussed shortly.

8. Emerson Pugh, Lyle R. Johnson, and John H. Palmer, *IBM's 360 and Early 370 Systems* (Cambridge: MIT Press, 1991): 169.

9. Ibid., 169–174. See also Bob O. Evans, "Introduction to SPREAD Report," *Annals of the History of Computing* 5 (1983): 4–5; also Oscar H. Rothenbuecher, "The Top 50 Companies in the Data Processing Industry," *Datamation* (June 1976): 48–59.

10. Bob O. Evans, "SPREAD Report: the Origin of the IBM System/360 Project," *Annals of the History of Computing* 5 (1983): 4–44.

11. Bob O. Evans, "IBM System/360," Computer Museum Report (Summer 1984): 11.

12. Ibid., 11.

13. Minghui Hu, "Maintaining NASTRAN: The Politics and Techniques of Aerospace Computing" (Masters Thesis, Blacksburg, VA: Virginia Polytechnic and State University, February 1995).

14. The SPREAD Committee report was signed by 13 men: John W. Haanstra, Bob O. Evans, Joel D. Aron, Frederick P. Brooks Jr., John W. Fairclough, William P. Heisling, Herbert Hellerman, Walter H. Johnson, Michael J. Kelly, Douglas V. Newton, Bruce G. Oldfield, Deymour A. Rosen, and Jerrold Svigals.

15. Evans, "Spread Report," 31.

16. A. Padegs, "System/360 and Beyond," *IBM Journal of Research and Development* 25 (1981): 377–390.

17. Maurice Wilkes, "The Best Way to Design an Automatic Calculating Machine," *Manchester Inaugural Conference*, Manchester, England (July 1951): 16–18. Reprinted in *Annals of the History of Computing* 8 (April 1986): 118–121.

18. This does add a layer of complexity, and one cannot get something for nothing. The SPREAD Committee specified that a model of the 360 could be "hard-wired" instead of microprogrammed if its designers could demonstrate that by doing so one gained at least a 1.25 cost-performance advantage over the microprogrammed design. The highest performance models of the 360, like most supercomputers, were not microprogrammed.

19. Wilkes, "The Best Way."

20. Evans, "SPREAD Report," 33.

21. Maurice Wilkes, *Automatic Digital Computers* (London: Methuen, 1956): 139–140.

22. Gordon Bell, the designer of the PDP-8 minicomputer, called the 360's ability to emulate other computers through changing its microprogram "probably the most significant real innovation": C. Gordon Bell and Allen Newell, *Computer Structures* (New York: McGraw-Hill, 1971): 562.

23. Pugh et al., *IBM's 360*, 163, 214–217.

24. The phrase is attributed to Alan Perlis.

25. Pugh et al., *IBM's 360*, 162–163.

26. Buchholz, Werner, ed; *Planning a Computer System: Project Stretch* (New York, 1962); see also Fisher et al., *IBM and the U.S. Data Processing Industry*, 47–50.

27. Buchholz, *Project Stretch*, 40; also Buchholz, "Origin of the Word Byte," *Annals of the History of Computing* 3 (1981): 72.

28. C. E.MacKenzie, *Coded Character Sets: History & Development* (Reading, MA: Addison Wesley, 1980): 126 ff. Mackenzie was an employee of IBM who was a member of the committee developing ASCII.

29. Padegs, *IBM Journal of Research and Development* (September 1981): 379.

30. Compare the two standard personal computer operating systems, MS-DOS and the Macintosh System, after 1984; with VHS and Beta, who battled for the standards in home videotapes, until the former prevailed. See Paul David, "Understanding the Economics of QWERTY: the Necessity of History," in *Economic History and the Modern Economist* (Oxford, 1986): 30–49.

31. Although not directly related, there is another difference. Some computers fetch bytes from memory with the most significant first; others do the reverse.

One group is called "big-endian," the other "little-endian." The terms are from *Gulliver's Travels*, which describes a conflict over which side of a boiled egg to open. See Danny Cohen, "On Holy Wars and a Plea for Peace," *IEEE Computer* 14 (October 1981): 48–54.

32. This phenomenon stems only in part from a short-sightedness on the designers' part. The 360 proved not only to be more popular than expected; its architecture proved much more long-lasting as well. The need for more address bits is thus a measure of an architecture's success. The same course of events happened with the Digital Equipment Corporation's PDP-11 minicomputer, and with the Intel Corporations's 8086 series of microprocessors. See N. S. Prasad, *IBM Mainframes: Architecture and Design* (New York: McGraw Hill, 1989).

33. Evans, "SPREAD Report," 31; also Evans, "IBM System/360."

34. One I/O device also included in the initial announcement was a video display terminal. The Model 2250 was not a successful product, but a later VDT, the Model 3270, was. Video terminals were not used as much as card punches and readers for the 360.

35. Fernando Corbato, "An Experimental Time-sharing System," *Proc. SJCC*, San Francisco, May 1–3, 1962 (AFIPS Press, 1962): 335–344.

36. Judy E. O'Neill, "The evolution of interactive computing through time-sharing and networking" (Ph.D. diss., University of Minnesota, 1992).

37. Pugh et al., *IBM's 360 and Early 370 Systems*, 360–363.

38. Ibid., 360–363.

39. In the oral histories of the SPREAD Report (Evans, "The SPREAD Report," 1983), there is a vigorous exchange between the interviewers and the former members of the SPREAD Committee over this point. My arguments follow in part from this discussion but also from other evidence, as given later.

40. Arthur Norberg and Judy O'Neill, *Transforming Computer Technology: Information Processing for the Pentagon, 1962–1986* (Baltimore: Johns Hopkins University Press, 1996): 100. Norberg and O'Neill do not elaborate on this criterion; among those interviewed for the *Annals of the History of Computing*, cited above, there were several who vigorously disagreed with MIT's assessment of the System/360's alleged deficiencies.

41. Melinda Varian, "VM and the VM Community: Past, Present, and Future," unpublished typescript (Princeton, NJ, April 1991): 23–24.

42. Sobel, *IBM: Colossus in Transition*, 284–286.

43. Pugh et al., *IBM's 360 and Early 370*, 448–450.

44. Ibid., 451.

45. The suffix "tron" is descended from the Greek word "to plow," and typically referred to an instrument or probe.

46. John Brooks, *The Go-Go Years* (New York: Dutton, 1984). He also gives an alternate definition: "a method of operating the stock market ... characterized by rapid in-and-out trading of huge blocks of stock."

47. Brooks, *Go-Go*, 231; Fishman, *Computer Establishment*, 246–249; Sobel, *IBM*, 224–225.

48. Evans, "IBM System/360," Computer Museum Report (Summer 1984): 18. The general bear market, as well as the Federal antitrust suit, were also factors.

49. Brooks, quoted in Evans, "The SPREAD Report," 31.

50. Boelie Elzen and Donald MacKenzie, "The Social Limits of Speed," *Annals of the History of Computing* 16: 1 (1994): 46–61.

51. Ibid., also Seymour Cray, "What's all this about Gallium Arsenide?" video-tape of a talk, Orlando, FL, November 1988.

52. Saul Rosen, "Electronic Computers: a Historical Survey," *Computing Surveys* 1 (March 1969): 21.

53. Fishman, *The Computer Establishment*, 182.

54. Sobel, *IBM*, 288; Fishman, *Computer Establishment*, 165.

55. Jamie Pearson, *Digital at Work* (Bedford, MA: Digital Press, 1992): 127–128.

56. Pugh et al., *IBM's 360 and Early 370 Systems*, 551.

57. Fisher et al., *IBM and the U.S. Data Processing Industry*, 379.

58. Kenneth Flamm, *Creating the Computer: Government, Industry, and High Technology* (Washington, DC: Brookings Institution, 1988): 132, 195–196.

59. For a view of this era quite different from Franklin Fisher's, see Thomas DeLamarter, *Big Blue: IBM's Use and Abuse of Power* (New York: Dodd, Mead, 1986).

60. The nomenclature "1106," "1108," and so on was a reminder of UNIVAC's legacy as the supplier of one of the first scientific computers in the 1950s: the 1103.

61. Sperry Rand Corporation, "Sperry Facts," brochure dated 1979.

62. Max Palevsky, interview with R. Mapstone, February 15, 1973, Smithsonian Computer History Project, Smithsonian Archives.

63. Ibid., 12–13. The PB-250 sold for $25,000 to $30,000. Its most distinctive architectural feature was its use of a delay line memory—possibly the only transistorized machine to use one.

64. Ibid., also Fishman, *The Computer Establishment*, 216–219.

65. Douglas Smith, *Fumbling the Future: How Xerox Invented, Then Ignored, the First Personal Computer* (New York: William Morrow, 1988): 122.

66. Fisher et al., *IBM and the U.S. Data Processing Industry,* 267.

67. Smithsonian Institution, Computer History Project, Max Palevsky Papers, Smithsonian Archives. The SDS-940 time-sharing system was based on the quarter-million-dollar SDS-930, to which the company added specialized memory, Input/Output units, and software.

68. Hoelzer, telephone interview with author, 25 June 1993; Charles Bradshaw, letter to the author 21 July 1993; also L. A. Lanzarotta, "Computing Power at Huntsville," *Datamation* (July/August 1960): 18–21.

69. Smith, *Fumbling the Future;* also Fishman, *Computer Establishment,* 222; also Max Palevsky deposition, 10 July 1974; Hagley Museum, Accession #1912, box 31, folder 0990–0993.

70. A service bureau is a company or division of a company that assists a customer with software and otherwise helps to get a computer to solve a customer's specific problems.

71. Elmer Kubie, "Reflections of the First Software Company," *Annals of the History of Computing,* 16: 2 (1994): 65–71; also Walter F. Bauer, interview with Arthur Norberg, 16 May 1983, Charles Babbage Institute Archives.

72. "Automatic Data Processing, Inc.," *Hoover's Handbook* (1991): EMM-102.

73. "ADP," *International Directory of Company Histories* (1988): 117–119.

74. Fishman, *The Computer Establishment,* 273.

75. Franklin Fisher, John J. McGowan, and Joen E. Greenwood, *Folded, Spindled, and Mutilated: Economic Analysis and U.S. vs. IBM* (Cambridge: MIT Press, 1983): 204–218.

76. Fred Brooks, *The Mythical Man-Month: Essays on Software Engineering* (Reading, MA: Addison-Wesley, 1975).

77. Fisher et al., *IBM and the U.S. Data Processing Industry,* 176–178.

78. Robert Slater, *Portraits in Silicon* (Cambridge: MIT Press, 1992); chapter 12.

79. Claude Baum, *The System Builders: the Story of SDC* (Santa Monica, CA: System Development Corporation, 1981).

80. Annual Reports of TRW, NASM Tech Files.

81. *The Washington Post,* February 9, 1994, c-1.

82. Fisher, McGowan, and Greenwood, *Folded, Spindled, and Mutilated,* chapter 1.

83. *Datamation,* Industry surveys, June 16, 1986; June 15, 1987.

84. Kenneth Flamm, *Targeting the Computer: Government Support and International Competition* (Washington, DC: Brookings Institution, 1987): 77–78.

85. Lundstrom, *A Few Good Men From UNIVAC*, 225. Portions of Commercial Credit were later sold to Dun & Bradstreet.

86. James Worthy, "Control Data Corporation: the Norris Era," *Annals of the History of Computing* 17: 1 (1995): 47–53.

87. Lundstrom, *A Few Good Men From UNIVAC*.

88. The best description of PLATO is found in Ted Nelson, *Computer Lib* (Ted Nelson, 1974): DM26–27.

Chapter 6

1. Herb Grosch, *Computer: Bit Slices from a Life* (Novato, CA: Third Millenium Books, 1991), chapter 13. The strict statement of the law is that "Computing power increases as the square of the cost."

2. For some of the flavor of the debate over whether this law was in fact true, see Herb Grosch, "Grosch's Law Revisited," *Computerworld* (April 16, 1975): 24; also Phillip Ein-Dor, "Grosch's Law Re-Revisited," *CACM* 28: 2 (February 1985): 142–151.

3. The Institute for Advanced Study Computer, built in Princeton at that time, is a good example of how difficult things are when this approach is not used. Because it was based on von Neumann's principles, it required far fewer tubes than the ENIAC. But when a tube burned out, it was often necessary for an engineer to cut through a spaghetti tangle of wires simply to get at the tube, unplug it, and put in a fresh one. Hardly an elegant implementation, however elegant the design. The IAS computer is on exhibit at the Smithsonian Institution.

4. Charles J. Bashe, Lyle R. Johnson, John H. Palmer, and Emerson W. Pugh, *IBM's early Computers* (Cambridge: MIT Press, 1986): 408–413.

5. Ernest Braun and Stuart Macdonald, *Revolution in Miniature: the History and Impact of Semiconductor Electronics*, 2nd ed. (Cambridge, UK: Cambridge University Press, 1982); also Michael F. Wolff, "The Genesis of the IC," *IEEE Spectrum* (August 1976): 45–53.

6. Braun and Macdonald, *Revolution in Miniature*, passim; also Herbert Kleiman, "The Integrated Circuit: a Case Study in Product Innovation in the Electronics Industry" (Ph.D. diss., George Washington University, 1966).

7. Bell is quoted in *IEEE Spectrum* 25 (November 1988): 87.

8. Robert Noyce, "Microelectronics," *Scientific American* (September 1977): 64.

9. Cledo Brunetti and Roger W. Curtis, "Printed Circuit Techniques," Circular #468 (Washington, DC: National Bureau of Standards, 1947).

10. Paul Eisler, *My Life with the Printed Circuit*, edited with notes by Mari Williams (Bethlehem, PA: Lehigh University Press, 1989); but see the review of Eisler by Charles Susskind in *Science* 247 (23 February 1990): 986.

11. Eisler, patent #2,441,960, May 25, 1948; applied for February 3, 1944.

12. Eisler, *My Life*; also Thomas Misa, "Military Needs, Commercial Realities, and the Development of the Transistor, 1948–1958," in Merritt Roe Smith, ed., *Military Enterprise and Technological Change* (Cambridge: MIT Press, 1985): 253–287.

13. Jack S. Kilby, "Invention of the Integrated Circuit," *IEEE Transactions on Electron Devices* 23 (July 1976): 648–654.

14. Braun and Macdonald, *Revolution in Miniature*, 95; also Kleiman, "The Integrated Circuit: a Case Study," 111.

15. *Aviation Week* (May 29, 1961): 82–83.

16. Larry Waller, "Clean-Room Inventor Whitfield Leaves a Spotless Legacy," *Electronics* (February 4, 1985): 38.

17. "The Minuteman High Reliability Component Parts Program: a History and Legacy," Rockwell International, Autonetics Strategic Systems Division, Anaheim, CA, Report C81-451/201, July 31, 1981; NASM Archives.

18. Philip J. Klass, "Reliability is Essential Minuteman Goal," *Aviation Week* (October 19, 1959): 13F.

19. James Webb, quoted by Kleiman, "The Integrated Circuit," 72; also *Newsweek* (July 26, 1971): 13; and *Missiles and Rockets* (June 24, 1963). It is not practical to narrow the per-unit cost of a single Minuteman beyond that range.

20. Braun and Macdonald quote an engineer who stated that if all military electronic components were acquired with the care lavished on Minuteman, their combined costs would exceed the U.S. GNP. (*Revolution in Miniature*, 99).

21. Donald MacKenzie, *Inventing Accuracy: A Historical Sociology of Nuclear Missile Guidance* (Cambridge, MIT Press, 1990): 155.

22. Jack S. Kilby, "Invention of the Integrated Circuit," *IEEE Transactions on Electron Devices* 23 (July 1976): 648–654; also Kilby, "Biographical Data," MS in the author's collection.

23. Kilby, "Invention of the Integrated Circuit," 650.

24. Kilby, "Biographical Data."

25. T. R. Reid, *The Chip* (New York: Simon & Schuster, 1985): 57–58.

26. RCA was the primary recipient of funds for Micro-Module. See Braun and Macdonald, *Revolution in Miniature*, 95.

27. U.S. Patent 3,138,743 for "Miniaturized Electronic Circuits."

28. Kilby, "Invention of the I.C.," 650–651; also Kilby, "Biographical Data."

29. Tom Wolfe, "The Tinkerings of Robert Noyce," *Esquire* (December 1983: 346–374; also Robert Knapp and H. B. Goodrich, *The Origins of American Scientists* (Chicago: University of Chicago Press, 1952).

30. Michael Wolff, "The Genesis of the IC," *IEEE Spectrum* (August 1976): 45–53.

31. Carolyn Caddes, *Portraits of Success: Impressions of Silicon Valley Pioneers* (Palo Alto, CA: Tioga, 1986): 44–45.

32. Wolff, "The Genesis of the IC," 51.

33. Letter from Thomas J. Watson to Sherman Fairchild, February 27, 1926; Sherman Fairchild Papers, Library of Congress, Box 29.

34. Quoted in Caddes, *Portraits of Success*, 44.

35. Eugene S. Ferguson, *Engineering and the Mind's Eye* (Cambridge: MIT Press, 1992); this book is an expanded version of an article by Ferguson, "The Mind's Eye: Nonverbal Thought in Technology," *Science* 197 (26 August 1977): 827–836.

36. Ferguson laments what he sees as the loss of this skill among engineering students, in part fostered by professors who wish to make the engineering curriculum more "scientific." Ferguson gives a number of examples of recent engineering failures to make his point, but he does not recognize the technology of integrated circuit design as a possible exception. My understanding of the design of computers like the Apple II and Data General Nova, discussed later, may contradict Ferguson's argument.

37. Kleiman, "The Integrated Circuit."

38. Marth Smith Parks, "Microelectronics in the 1970's" (Anaheim, CA, Rockwell International, 1974): 64. The Minuteman I was retargeted by physically rotating the entire missile in its silo; from that one can infer what those targets might have been, and how infrequently they were changed.

39. Jack Kilby, letter to Gwen Bell, June 26, 1984; Boston; Computer Museum Archives.

40. *Missiles and Rockets* (March 2, 1964): 35; *Aviation Week and Space Technology* (August 26, 1965).

41. "Nineteen Sixty-Four: the Year Microcircuits Grew Up," *Electronics* 37 (March 13, 1964): 10–11.

42. Quoted in Dirk Hanson, *The New Alchemists: Silicon Valley and the Microelectronics Revolution* (Boston: Little, Brown, 1982): 93.

43. Robert Noyce, "Integrated Circuits in Military Equipment," *IEEE Spectrum* (June 1964): 71.

44. Paul E. Ceruzzi, *Beyond the Limits: Flight Enters the Computer Age* (Cambridge: MIT Press, 1989), chapter 6.

45. Donald C. Fraser, and Philip Felleman, "Digital Fly-by-Wire: Computers Lead the Way," *Astronautics and Aeronautics* 12: 7/8 (1974): 24–32.

46. Eldon C. Hall, *Journey to the Moon: the History of the Apollo Guidance Computer* (Reston, VA: AIAA, 1996): 82; also A. Michal McMahon, "The Computer and the Complex: a Study of Technical Innovation in Postwar America," October 1986, NASA History Office, Washington, DC, 30.

47. Eldon Hall, "The Apollo Guidance Computer: a Designer's View," Computer Museum, Boston, *Report* (Fall 1982): 2–5.

48. A NOR-gate with three inputs. The chip contained three transistors and four resistors.

49. James Tomayko, "Computers in Spaceflight: the NASA Experience," (Washington, DC: NASA Contractor Report 182505, 1988): 28–30.

50. A. Michal McMahon, "The Computer and the Complex: a Study of Technical Innovation in Postwar America."

51. Pugh et al., *IBM's 360 and Early 370 Systems*, 76–83; also E. M. Davis et al., "Solid Logic Technology: Versatile, High Performance Microelectronics," *IBM Journal*, 8 (April 1964): 102–114.

52. The first quotation is from Bob Henle, quoted in Pugh et al., 105. The second is from John Haanstra, "Monolithics and IBM," report of September 1964, unpaginated, IBM Archives, Valhalla, NY. I am grateful to Emerson Pugh for providing me with a copy of this document.

53. Pugh et al., *IBM's 360 and Early 370*; C. Gordon Bell, "The Mini and Micro Industries," IEEE *Computer* (October 1984): 14–29; *Datamation* (November 1968): 72–73; *Datamation* (July 1974): 50–60.

54. Bell, "The Mini and Micro Industries," 14–29.

55. Don Lancaster, *TTL Cookbook* (Indianapolis: Howard Sams, 1974).

56. *IEEE Spectrum* 25: 11 (1970): 70; also Tom Monte and Ilene Pritikin, *Pritikin: the Man who Healed America's Heart* (Emmaus, PA: Rodale Press, 1988).

57. "SYMBOL: A Large Experimental System Exploring Major Hardware Replacement of Software," in Daniel Siewiorek, C. Gordon Bell, and Allen Newell, eds. *Computer Structures: Principles and Examples* (New York: McGraw-Hill, 1982): 489–507.

58. Major changes have been the advent of "Complementary Metal-Oxide Semiconductor" (CMOS) in place of TTL, and the gradual replacement of the DIP housing to "Single In-Line Memory Modules" (SIMM), and flat packaging for products like laptops.

59. W. Buchholz, "Origins of the Word Byte," *Annals of the History of Computing* 10: 4 (1989): 340.

60. Gardner Hendrie, "From the First 16-bit Mini to Fault-Tolerant Computers," Computer Museum Report (Spring 1986): 6–9.

61. Arthur Norberg, Judy O'Neill, and Kerry Freedman, "A History of the Information Processing Techniques Office of the Defense Advanced Research Projects Agency" (Minneapolis, MN: Charles Babbage Institute, 1992).

62. Siewiorek, Bell, and Newell, *Computer Structures*, chapter 24.

63. Adele Goldberg, ed., *A History of Personal Workstations* (Reading, MA: Addison-Wesley, 1988): 151; also Siewiorek, Bell, and Newell, *Computer Structures*, 396–397.

64. Goldberg, *History of Personal Workstations*, 150–151.

65. As of this writing, the Smithsonian Institution is not among the museums that has collected an IMP.

66. Glenn Rifkin and George Harrar, *The Ultimate Entrepreneur: the Story of Ken Olsen and Digital Equipment Corporation* (Chicago: Contemporary Books, 1988): 86–92.

67. Tom Wolfe, "The Tinkerings of Robert Noyce," *Esquire* (December 1983): 356.

68. Fred Brooks, "Keynote Address: Language Design as Design," in Thomas J. Bergin and Richard G. Gibson, eds., *History of Programming Languages—II* (Reading, MA: Addison-Wesley, 1996): 4–16. Steve Wozniak, designer of another "elegant" computer, the Apple II, also acknowledged the Nova as an influence. The Radio Shack TRS-80 Model 100, and the IBM 7090 are also regarded as "elegant"; but few other computers are spoken of that way.

69. Michael Hord, *The Illiac IV: the First Supercomputer* (Rockville, MD: Computer Science Press, 1982). It is worth noting that HAL, the famous computer of the film *2001: a Space Odyssey*, was "born" in Urbana, Illinois. That film was being made as the Illiac IV was being built. Arthur C. Clarke, the author of the screenplay, later claimed that he chose Urbana because one of his professors had moved there from England.

70. "A Revolution in Progress: a History of Intel to Date," brochure (Santa Clara, CA: Intel Corporation, 1984).

71. Ibid.

72. These included offerings from two companies much larger than DEC: Divisions of both Lockheed and Hewlett-Packard were offering 16-bit minicomputers, as was a start-up, Interdata, that was later purchased by the military contractor Perkin-Elmer.

73. C. Gordon Bell, interview with author, 16 June 1992, Los Altos, CA; also "Decworld," newsletter from Digital Equipment Corporation, May 1980, Special Edition on the PDP-11, copy in the author's collection.

74. Harvard University Computation Laboratory, *A Manual of Operation for the Automatic Sequence Controlled Calculator,* reprint of 1946 edition (Cambridge: MIT Press, 1985), chapter 2.

75. Digital Equipment Corporation, *PDP-11 Processor Handbook* (Maynard, MA: Digital Equipment Corporation, 1981), chapter 2; according to Braun and McDonald, Texas Instruments had taken out patents on the concept of a bus architecture; see Braun and Macdonald, *Revolution in Miniature* 109.

76. Personal computers built around the Intel 8086-series of microprocessors have data buses that are not as general as the PDP-11's. They do resemble the PDP-11 in other ways. The Motorola 68000 series of microprocessors was more closely patterned on the PDP-11.

77. Jamie Pearson, *Digital at Work* (Bedford, MA: Digital Press, 1992): 47, 59, 67.

78. Digital Equipment Corporation, *PDP-11 Processor Handbook,* (Maynard, MA: Digital Equipment Corporation, 1981): v.

79. Dick Rubenstein, telephone interview with the author, February 5, 1993.

80. C. Gordon Bell, J. Craig Mudge, and John E. McNamara, *Computer Engineering: a DEC View of Hardware Systems Design* (Bedford, MA: Digital Press, 1978): 383.

81. James W. Cortada, *Historical Dictionary of Data Processing: Technology* (Westport, CT: Greenwood Press, 1987): 142.

82. Pugh et al., *IBM's 360 and Early 370 Systems,* chapter 9.

83. Jim Geraghty, *CICS Concepts and Uses* (New York: McGraw-Hill, 1994).

84. Saul Rosen, "PUFFT—the Purdue University Fast Fortran Translator," in Saul Rosen, ed., *Programming Systems and Languages* (New York: McGraw-Hill, 1967): 253–263; also *Encyclopedia of Computer Science,* 3rd ed. (New York: McGraw-Hill, 1993): 768.

85. "25th Anniversary Issue," University of Waterloo, Department of Computing Services Newsletter (October 1982): 2.

86. Ray Argyle, "Industry Profile . . . Wes Graham of Waterloo U," *Computer Data: the Canadian Computer Magazine* (May 1976): 29–30.

87. Paul Cress, Paul Dirkson, and J. Wesley Graham, *Fortran IV with WATFOR and WATFIV* (Englewood Cliffs, NJ: Prentice Hall, 1970).

88. John G. Kemeny, *Man and the Computer* (New York: Scribner's, 1972): vii.

89. Thomas E. Kurtz, "BASIC," in Richard Wexelblat, ed., *History of Programming Languages* (New York: Academic Press, 1981): 518–519.

90. The choice of the name is obvious; however, it is also an acronym for "Beginner's All-purpose Symbolic Instruction Code." Hence it is written in all capitals.

91. William Aspray and Bernard O. Williams, "Arming American Scientists: NSF and the Provision of Scientific Computing Facilities for Universities, 1950–1973," *Annals of the History of Computing* 16: 4 (1994): 60–74.

92. Mark Bramhall, telephone interview with the author, 10 May, 1997.

93. Some of the "sacred" principles abandoned were the mandatory use of "Let" in an assignment statement; having only one statement on a line; and not allowing a statement to continue beyond a single line. Kemeny later developed "True Basic" to return the language to its pure roots, but it never caught on.

Chapter 7

Portions of this chapter first appeared in *History and Technology*, 13/1 (1996): 1–32. Used with permission.

1. Stewart Brand, "Spacewar: Fanatic Life and Symbolic Death Among the Computer Bums," *Rolling Stone* (Dec. 7, 1972): 50–58.

2. Douglas Engelbart, in Adele Goldberg, ed., *History of Personal Workstations* (Reading, MA: Addison-Wesley, 1988): 187.

3. Acknowledgements to "A Horse's Tale," in Mark Twain, *The Mysterious Stranger and Other Stories* (New York: Harper, 1922): 143–144.

4. The following description of the PDP-10 is taken mainly from C. Gordon Bell, J. Craig Mudge, and John McNamara, *Computer Engineering: a DEC View of Hardware Systems Design* (Bedford, MA: Digital Press, 1978), chapter 21.

5. See, for example, advertisements in *Datamation* from that period. A Digital spokesperson called the PDP-6 "the first of what might be called a 'personal' mainframe." Jamie Pearson, *Digital at Work* (Bedford, MA: Digital Press, 1992): 54–55.

6. Models introduced in mid-1972 used integrated circuits for logic, but memory was still implemented with magnetic core.

7. The PDP-6 was introduced in 1964, the PDP-10 in 1967, and the DEC-System 10 in 1971. 23 PDP-6s were sold. By 1976 around 450 PDP-10s and DEC-System 10s had been installed.

8. Bill Gosper, quoted in Steven Levy, *Hackers: Heroes of the Computer Revolution* (New York: Anchor Doubleday, 1984): 67.

9. Edward Fredkin interview with the author, 21 May, 1993.

10. Bell et al., *Computer Engineering*, table on p. 507.

11. Digital Equipment Corporation, "TOPS-10 Technical Summary" (Maynard, MA, n.d.). Computer Museum, Boston, PDP-10 archives.

12. Computer Museum, Boston, PDP-10 archives, box A-242.

13. The Cambridge consulting firm Bolt Beranek and Newman (BBN) also developed another operating system for the PDP-10, for the Advanced Research Projects Agency. "TENEX" was more polished than TOPS-10.

14. David Ahl, "Computer Games," in Anthony Ralston and Edwin Reilly, eds., *Encyclopedia of Computer Science*, third edition (New York: Van Nostrand, 1993): 285–287.

15. The MAXC was software-compatible with a DP-10 but used semiconductor memory instead of core. See Douglas Smith et al., *Fumbling the Future: How Xerox Invented, Then Ignored, the First Personal Computer* (New York: William Morrow, 1988): 144–145; also Tekla Perry, "Inside the PARC: the 'Information Architects'," *IEEE Spectrum* (October 1985): 62–75.

16. Deposition by Max Palevsky, 10 July 1974, *U.S. v. IBM*, Accession 1912, Box 31, transcript; Hagley Museum Archives.

17. One of them evolved into the Compuserve network. As of this writing, Compuserve subscribers are still identified by a number written in the octal notation, which uses only the digits zero through seven. That reflects its descent from PDP-10 operating systems, which also used octal rather than decimal numbers.

18. Mark Halpern, "Dreams That Get Funded," *Annals of the History of Computing* 16/3 (1994): 61–64.

19. Stephen Manes and Paul Andrews, *Gates: How Microsoft's Mogul Reinvented an Industry, and Made Himself the Richest Man in America* (New York: Doubleday, 1993): 28–36.

20. Brand, "Spacewar: The Fanatic Life"; also Freiberger, *Fire in the Valley*, 100–102.

21. Chuck House, "Hewlett-Packard and Personal Computing Systems," in Adele Goldberg, ed., *History of Personal Workstations* (Reading, MA: Addison-Wesley, 1988): 413–414; C. Gordon Bell, interview with the author, 16 June 1992.

22. Wesley A. Clark, "The LINC was Early and Small," in Goldberg, *History of Personal Workstations*, 347–391.

23. Pearson, *Digital at Work*, 52.

24. Another place was in computer games. I do not discuss these, however, mainly because I feel their customers, however important as a mass market for semiconductors, were not the critical community that calculator users were.

25. Peggy A. Kidwell and Paul Ceruzzi, *Landmarks in Digital Computing: a Smithsonian Pictorial History* (Washington, DC: Smithsonian Press, 1994); Edwin

Darby, *It All Adds Up: the Growth of Victor Comptometer Corporation* (Chicago: Victor Comptometer Corp., 1968).

26. An Wang, *Lessons* (Reading, MA: Addison-Wesley, 1986): 126–159. The key to Wang's breakthrough was his ability to perform all the functions using only a few hundred transistors. He did this by exploiting tricks that Howard Aiken and his Harvard team had developed in the 1940s with the Harvard Mark I.

27. Chuck House, "Hewlett-Packard and Personal Computing Systems," 401–427; also "The Olivetti Programma 101 Desk Calculator," in Bell and Newell, *Computer Structures*, 237–242.

28. The chief suppliers of ICs for calculators were Texas Instruments and Rockwell, whose Autonetics Division built the Minuteman Guidance systems. Calculators using their chips were sold under a variety of names, although eventually TI and Rockwell entered the market with machines under their own names. Rockwell later returned to being only a chip supplier.

29. National Museum of American History, calculator collections. A Bowmar was even mentioned in the hit Broadway play "Same Time Next Year," which was about an accountant's once-a-year affair with a woman. The accountant (played by Alan Alda in the movie version) used the Bowmar to keep track of the affair.

30. *Electronics* (April 17, 1980): 397–398; also "Microelectronics in the 1970s," booklet from Rockwell International, 1974, 39.

31. Keufell & Esser, the U.S. manufacturer of the most popular engineer's slide rules, stopped production in 1975. For about the next decade, other companies continued to make cheaper slide rules that had only the basic arithmetic and log scales.

32. Chung C. Tung, "The 'Personal Computer': a Fully Programmable Pocket Calculator," *Hewlett-Packard Journal* (1974): 2–7.

33. Gordon Moore, "Microprocessors and Integrated Electronics Technology," *Proceedings of the IEEE* 64 (June 1976): 837–841.

34. Joseph Weizenbaum, *Computer Power and Human Reason: From Judgment to Calculation* (San Franciso: W. H. Freeman, 1976), chapter 4.

35. *65-Notes* (Newsletter of the HP-65 Users' Club) 2: 1 (January 1975): 7. HP-65 customers were overwhelmingly male; the newsletter made a special note of the first female member to join the users club, a year after its founding.

36. Weizenbaum, *Computer Power*, 116.

37. Paul Freiberger, *Fire in the Valley: the Making of the Personal Computer* (Berkeley, CA: Oxborne/McGraw-Hill, 1984).

38. In addition to a regular column that appeared in "HP-65 Notes," cited above, the author has found similar comparisons in a Texas Instruments users club newsletter, as well as in "Display," a newsletter for calculator owners published in Germany in the late 1970s.

39. Ted Nelson, *Computer Lib* (South Bend, IN: Ted Nelson, 1974).

40. The "von Neumann" argument came from the fact that most calculators, unlike general-purpose computers, stored their programs in a memory deliberately kept separate from data. In fact, the program was stored on the same chips as the data, but the calculator manufacturers erected a "wall" to prevent the twain from meeting. This was done to make the machine easier to use by nonspecialists. A common memory is often regarded as a central defining feature of a true computer. Another property, which most programmable calculators *did* have, was "conditional branching": the ability to select alternate sequences of instructions based on the results of a previous calculation. That was a property lacking in the machines of the immediate precomputer era: the Harvard Mark I, the early Bell Labs relay computers, and the early Zuse computers.

41. "The Programmable Pocket Calculator Owner: Who Does He Think He Is?" *HP-65 Notes* 3: 6 (1976): 2.

42. *HP-65 Notes* 2: 1 (1975): 4–7.

43. Gordon E. Moore, "Progress in Digital Integrated Electronics," *Proceedings International Electron Devices Meeting* (December 1975): 11–13. Robert Noyce stated that Moore first noticed this trend in 1964: Noyce, "Microelectronics," *Scientific American* (September 1977): 63–69. Moore predicted that the rate would flatten out to a doubling every *two* years by 1980. That has led to confusion in the popular press over what exactly is meant by "Moore's Law." Bell, Mudge, and MacNamara (*Computer Structures*, 90) state the law as doubling every year from 1958 until 1972, then every eighteen months thereafter. Memory chip density, from the 1970s to the time of this writing, has been doubling every eighteen months.

44. Clifford Barney, "He Started MOS From Scratch," *Electronics Week* (October 8, 1984): 64.

45. Hoff recalls a book by Adi J. Khambata, *Introduction to LSI*, published in 1969, as very influential. The book gave modern version of the dilemma faced by Henry Ford and his Model T: the very same mass-production techniques that made the Model T a high-quality, low-priced car made it difficult if not impossible for Ford to change the Model T's design as the market evolved.

46. Trudy E. Bell, "Patented 20 Years Later: the Microprocessor's True Father," *The Institute* (IEEE) 14: 10 (November 1990): 1; also National Museum of American History, Division of Electricity, curatorial files, Texas Instruments collection; also Don Clark, "High-Stakes War Over Chip Patents," San Francisco *Chronicle* (September 8, 1990): b1–b3; also Michael Antonof, "Gilbert Who?" *Popular Science* (February 1991): 70–73.

47. See for example Robert Noyce and Marcian Hoff, "A History of Microprocessor Design at Intel"; *IEEE Micro* 1 (February 1981): 8–22.

48. Kenneth A. Brown, interview with Hoff, in Brown, *Inventors at Work* (Redmond, WA: Tempus Books): 283–307.

49. William Barden Jr., *How to Buy and Use Minicomputers and Microcomputers* (Indianapolis: Howard Sams, 1976): 101–103.

50. Intel Corporation, Corporate Communications Department, "A Revolution in Progress: a History of Intel to Date" (Santa Clara, CA: Intel, 1984): 12.

51. *Electronic News* (November 15, 1971).

52. Intel, "A Revolution in Progress," 21.

53. Elvia Faggin, "Faggin Contributed to First Microprocessor," letter to the Editor, *San Jose Mercury News* (October 3, 1986): 6b; reply by Marcian Hoff, "Patents Don't Tell Whole Microprocessor Tale," ibid. (October 12, 1986): 10b; also "If Hyatt Didn't Invent the Microprocessor, Who Did?" ibid. (December 2, 1990): 27.

54. Hoff, "Patents Don't Tell Whole Microprocessor Tale," 106.

55. Noyce and Hoff, "A History of Microprocessor Design"; also Lamont Wood, "The Man Who Invented the PC," *American Heritage of Invention & Technology* (Fall 1994): 64.

56. Intel, "A Revolution in Progress," 14.

57. Noyce and Hoff, "A History of Microprocessor Development."

58. Computer Museum *Report* 17 (Fall 1986): 10–11.

59. Intel Corporation, "A Revolution in Progress," 13.

60. Robert Slater, *Portraits in Silicon* (Cambridge: MIT Press, 1992): 251–261.

61. Noyce and Hoff, "A History of Microprocessor Design at Intel," 14.

62. This statement is based on conversations with several Intel employees who were involved with early microprocessor development, including Ted Hoff and John Wharton. Intel systems were used to keep scores during the 1976 Summer Olympics. That was the year Nadia Comaneci received a perfect "10" in gymnastics, a score that the system was unable to display, as it had not been programmed to display anything over "9.99." That limit, however, had nothing to do with the fact that the Intel systems had a shorter word length than the minicomputers it replaced.

63. Susan Douglas, "Oppositional Uses of Technology and Corporate Competition: the Case of Radio Broadcasting," in William Aspray, ed., *Technological Competitiveness* (New York: IEEE, 1993): 208–219.

64. The construction of the World Trade Center obliterated Radio Row, but by then integrated electronics was well underway. A single microprocessor might contain more circuits than the entire contents of *every* store on Radio Row.

65. *QST* (March 1974): 154.

66. Stan Veit, *Stan Veit's History of the Personal Computer* (Asheville, NC: World-Comm, 1993): 11; also Thomas Haddock, *A Collector's Guide to Personal Computers* (Florence, AL: Thomas Haddock, 1993): 20.

67. "Build the Mark-8, Your Personal Minicomputer," *Radio-Electronics* (July 1974): cover, 29–33.

68. Ibid. The users club became the Digital Group, an influential company in personal computing for the next several years. See Jonathan Titus, letter to the Computer Museum, June 18, 1984, Computer Museum, Boston, Personal Computer archives.

69. NMAH Collections; also Steve Ditlea, ed., *Digital Deli* (New York: Workman, 1984): 37.

70. "Build the Mark-8," 33.

71. Don Lancaster, "TV-Typewriter," *Radio-Electronics* (September 1973): cover, 43–52; Felsenstein is quoted in the Computer Museum Report 17 (Fall 1986): 16.

72. H. Edward Roberts and William Yates, "Exclusive! Altair 8800: the Most Powerful Minicomputer Project Ever Presented—Can be Built for Under $400," *Popular Electronics* (January 1975): cover, 33–38.

73. Not long after the Altair's introduction, journalists began calling these machines "microcomputers," an accurate but also ambiguous term, as it could imply two different things. A microcomputer used a microprocessor, and minicomputers did not. That was true at the time, although eventually nearly every class of machine would use microprocessors. The other definition was that a microcomputer was smaller and/or cheaper than a minicomputer. The Altair was both, but its low cost was more important than its small size.

74. Intel, "A Revolution in Progress," 14; also Veit, *Stan Veit's History of the Personal Computer*, 43; Veit stated that Roberts obtained chips that had cosmetic flaws, but Roberts and Intel both state flatly that the 8080 chips used in the Altair were not defective in any way; see "Computer Notes," MITS 1: 3 (August 1975): 2 (National Museum of American History, Mims-Altair file). The fact was that the 8080 cost Intel very little to manufacture, and it had little sense of what a fair market price to the PC market should be.

75. Spelled "buss" in the *Popular Electronics* article.

76. Veit, in *Stan Veit's History*, argues that it is to Railway Express's ineptitude that we owe the momentous decision to have a bus; others claim the decision came from Roberts's finding a supply of 100-slot connectors at an especially good price. The design change made the Altair more like the minicomputers of the day, though it made it more difficult to assemble.

77. Roberts and Yates, "Exclusive!" 34.

78. See, for example Steven Manes and Paul Andrews, 64.

79. Jim Warren, "Personal Computing: an Overview for Computer Professionals," *NCC Proceedings* 46 (1977): 493–498.

80. These included "Multichannel data acquisition system," "Machine controller," "Automatic controller for heat, air conditioning, dehumidifying," as well as "Brain for a robot," and others.

81. Veit, *Stan Veit's History*, 57–64, gives the main differences between the IMSAI and the Altair.

82. This is the reason that the acronyms TTY for Teletype and LPT for line printer survived into the operating systems of personal computers, long after both input/output devices fell from use.

83. Veit, *Stan Veit's History*.

84. Pugh et al., *IBM's 360 and Early 370 Systems* (Cambridge: MIT Press, 1991): 510–521.

85. Clifford Barney, "Award for Achievement [Alan F. Shugart], " *Electronics Week* (January 14, 1985) 40–44.

86. Jon Eklund, "Personal Computers," in Anthony Ralston and Edwin Reilley, eds., *Encyclopedia of Computer Science*, 3rd ed. (New York: van Nostrand Reinhold, 1993): 460–463.

87. Forrest Mims III, "The Tenth Anniversary of the Altair 8800," *Computers and Electronics* (January 1985): 62. Robert's account has been disputed by others and remains controversial.

88. Stephen Manes and Paul Andrews, *Gates: How Microsoft's Mogul Reinvented an Industry, and Made Himself the Richest Man in America* (New York: Doubleday, 1993): 63.

89. MITS Corporation, *Computer Notes* 1: 2 (July 1975): 6–7, National Museum of American History, Altair files.

90. Digital Equipment Corporation, "Introduction to Programming" (Maynard, MA, 1972): 9/4–9/5. Microsoft BASIC also broke with Dartmouth by allowing multiple statements on a line, by having "Let" and "End" optional, by recommending that a programmer "delete all REM [remark] statements … delete all unnecessary spaces from your program." (MITS Altair BASIC Reference Manual, 56; National Museum of American History, Altair Curatorial File.)

91. Manes and Andrews, *Gates*, chapters 2 and 3; for a discussion of John Norton, see Billy Goodman, "Practicing Safe Software," *Air & Space/Smithsonian* (September 1994): 60–67; also Paul Ceruzzi, *Beyond the Limits* (Cambridge: MIT Press, 1989), chapter 9.

92. This, too, is a matter of great dispute. Roberts insists that MITS had the rights to BASIC. In a letter to the newsletter "Computer Notes" on April 1976, Gates

stated, "I am not a MITS employee," but that was written after his rift with Roberts had grown deep. See also Stephen Manes and Paul Andrews, *Gates.*

93. MITS Corporation, "Computer Notes" (February 1976): 3. The open letter was distributed to many hobbyist publications and was widely read.

94. C. Gordon Bell, interview with the author, June 1992, Los Gatos, CA; Mark Bramhall, telephone interview with the author, 10 May 1997.

95. This term had been used, for example, with the IBM System/360 beginning in the late 1960s; see Pugh (1991), chapter 6.

96. C. Gordon Bell, interview with the author. Bell stated that he was the author of the PIP program, which found its way onto CP/M and in variations to MS-DOS; he says the name came from Edward Fredkin.

97. Pearson, *Digital at Work,* 64–65, 86; also C. Gordon Bell, interview with the author, June 1992.

98. Gary Kildall, "Microcomputer Software Design—a Checkpoint," *National Computer Conference* 44 (1975): 99–106; also Kildall, quoted in Susan Lammers, ed., *Programmers at Work* (Redmond, WA: Microsoft Press, 1989): 61.

99. Gary Kildall, "CP/M: A Family of 8- and 16-Bit Operating Systems," *Byte,* (June 1981): 216–229. Because of the differences between DEC minicomputers and the 8080 microprocessor, the actual code of CP/M was different and wholly original, even if the syntax and vocabulary were similar.

100. The above argument is based on PDP-10 and CP/M manuals in the author's possession, as well as conversations with Kip Crosby, to whom I am grateful for posting this question over an Internet discussion forum.

101. Jim C. Warren, "First Word on a Floppy-disc Operating System," *Dr. Dobb's Journal* (April 1976): 5.

102. Robert Slater, *Portraits in Silicon* (Cambridge: MIT Press, 1982), chapter 23.

103. Ibid.; also Stan Veit, *Stan Veit's History,* 64; and Digital Research, "An Introduction to C/M Features and Facilities" (1976), manual in the author's possession.

Chapter 8

1. C. Gordon Bell, interview with the author, 16 June 1992, Los Altos, California.

2. Dick Rubenstein, interview with the author; also Bell, interview.

3. Bell, interview.

4. C. Gordon Bell et al., *Computer Engineering: a DEC View of Hardware Systems Design* (Bedford, MA: Digital Press, 1978), graph on page 195.

5. Ibid., 13.

6. Perkin-Elmer later became known for its role in building the Hubble Space Telescope mirror; it was less well-known for building the critical optics assemblies needed to produce the photo masks used in chip-making.

7. Gould Electronics, "A Young Company with Deep Roots," undated brochure, ca. 1984; John Michels, "The Mega-mini Succeeds the Model T," *Datamation* (February, 1974): 71–74.

8. The word "virtual" later became popular as part of the term "virtual reality." It appears to have originated with IBM's marketing of System/370 and its memory-management architecture. The use of that word may have come from its use among Rennaissance artists, who spoke of a "virtual image" produced by a lens or a *camera obscura*.

9. Arthur Burks, Herman Goldstine, and John von Neumann, "Preliminary Discussion of the Logical Design of an Electronic Computing Instrument," 2nd ed., 2 September 1947, (Princeton, NJ: Institute for Advanced Study) 2, 4–7; Simon Lavington, "History of Manchester Computers," privately printed, Manchester, UK, 1975, 32–33; also T. Kilburn et al., "One-Level Storage System," *IRE Transactions on Electronic Computers*, EC-11 (1962): 223–235.

10. Lavington, "History of Manchester Computers," 34.

11. Franklin Fisher, *IBM and the US Data Processing Industry* (New York: Praeger, 1983): 343–344.

12. Bell, Mudge, and McNamara, *Computer Engineering: a DEC View of Hardware Systems Design* (Bedford, MA: Digital Press, 1978): 405–428.

13. Pearson, *Digital at Work*, 73.

14. Tracy Kidder, *The Soul of a New Machine* (Boston: Little, Brown, 1981). Kidder recounts how Data General resisted the use of a VAX-style "mode bit" to provide compatibility with its older line. One of the book's most dramatic episodes describes how Tom West, the engineer in charge of the new computer, surreptitiously opened a VAX at a customer site and disassembled it to see how it was designed (31–32).

15. I have been unable to verify this statement but have heard it from several sources. In light of DEC's weak support for UNIX, it suggests that Olsen did not care for the operating system; but others, more sympathetic, have said that he was only referring to a general trend (c.f. JAWS) that everyone wanted UNIX even though they did not know what to do with it.

16. One indication of this was the "Internet Worm," unleashed in 1988, which brought the Internet down. It was written by a student at Cornell and took advantage of some obscure flaws in VAX system software. A few years later such an attack would have been less damaging because the VAX no longer was the dominant machine.

17. Fisher et al., *IBM and the U.S. Data Processing Industry,* 442–444.

18. Bob O. Evans, "IBM System/360," Computer Museum Report (Summer 1984): 17.

19. D. C. Dykstra, "IBM's Albatross: A History Lesson for Micro Makers," *Computerworld* 18 (December 10, 1984): 134.

20. Partial copies are located at the Hagley Museum, Wilmington, Delaware, and at the Charles Babbage Institute, Minneapolis, Minnesota. The following summary of the trial is based on an examination of the transcripts at the Hagley. A synopsis of the trial, in agreement with its outcome, is found in Franklin Fisher's two books, cited above: Franklin Fisher et al., the *IBM and U.S. Data Processing Industry* (New York: Praeger, 1983); and Franklin Fisher, John J. McGowan, and Joel E. Greenwood, *Folded, Spindled, and Mutilated: Economic Analysis and U.S. vs. IBM* (Cambridge: MIT Press, 1983). A book that draws the opposite conclusion is Thomas Delamarter, *Big Blue: IBM's Use and Abuse of Power* (New York: Dodd, Mead, 1986).

21. *U.S. v. IBM,* testimony of F. Withington, 55989.

22. DeLamarter, *Big Blue,* xv.

23. Paul Carroll, *Big Blues: the Unmaking of IBM* (New York: Crown, 1994); the IBM that Carroll's book describes is one that apparently began with the introduction of the personal computer in 1981; also Charles Ferguson and Charles Morris, *Computer Wars: the Fall of IBM and the Future of Global Technology* (New York: Times Books, 1994).

24. The discussion of Palevsky's amassing a personal fortune of hundreds of millions of dollars in less than a decade was noted with some interest by the judge.

25. Fisher et al., *IBM,* 438; also Roy A. Bauer, Emilio Collar, and Victor Tang, *The Silverlake Project: Transformation at IBM* (New York: Oxford University Press, 1992).

26. Ivan T. Frisch and Howard Frank, "Computer Communications: How We Got Where We Are," *Proceedings NCC* 44 (1975): 109–117.

27. Lamont Wood, "The Man Who Invented the PC," *American Heritage of Invention & Technology* (Fall 1994): 64; also Pearson, *Digital At Work,* 90–92.

28. Pugh, *IBM's 360,* 606.

29. Ibid., 545–549.

30. Ibid., 550.

31. "AESOP: A General Purpose Approach to Real-Time, Direct Access Management Information Systems" (Bedford, MA: MITRE Corporation, June 1966), Report AD-634371, 7.

32. *Datamation* (October 1968): 17; also Robert Glass, *Computing Catastrophes* (Seattle: Computing Trends, 1983): 57–69; also W. David Gardner, "Route 128: Boston's Hotbed of Technology," *Datamation* (November 1981): 110–118.

33. Ibid.; also Viatron file, Box A30, Computer Museum, Boston, Historical Collections.

34. Letter, Daniel Whitney to Computer Museum, Ibid.

35. An Wang, *Lessons* (Reading, MA: Addison-Wesley, 1986).

36. Pearson, *Digital at Work*, 38; C. E. MacKenzie, *Coded Character Sets: History & Development* (Reading, MA: Addison-Wesley, 1980); Calvin Mooers, interview with Jon Eklund, Smithsonian Computer History Project, National Museum of American History.

37. Pugh, *IBM's 360*, 613.

38. Edwin McDowell, "'No Problem' Machine Poses a Presidential Problem," *New York Times* (March 24, 1981): C-7; see also Ibid., March 20, 26; March 16, 1, and March 27, 26. The *Times* editorial on March 20, with tongue in cheek, lamented that word processors would deprive future historians of the joy of uncovering a great figure's early thoughts, as recorded on rough drafts of manuscripts.

39. Charles Kenney, *Riding the Runaway Horse: the Rise and Decline of Wang Laboratories* (Boston: Little, Brown, 1992).

40. Kenney, *Riding the Runaway Horse*, 68–73; also Wang, *Lessons*, 182.

41. *Datamation* (June 1976): 48–61; also June 1, 1985, 50–51, 65; also Stephen T. McClellan, *The Coming Computer Industry Shakeout: Winners, Losers, and Survivors* (New York: Wiley, 1984), chapter 15.

42. The following account is based on a number of secondary sources, primarily Douglas Smith and Robert Alexander, *Fumbling the Future: How Xerox Invented, Then Ignored, the First Personal Computer* (New York: William Morrow, 1988), and George Pake, "Research at Xerox PARC: a Founder's Assessment," *IEEE Spectrum* (October 1985): 54–75.

43. Quoted in David Dickson, *The New Politics of Science* (New York: Pantheon Books, 1984): 122.

44. Arthur Norberg and Judy O'Neill, with Kerry Freedman, "A History of the Information Processing Techniques Office of the Defense Advanced Research Projects Agency" (Minneapolis, MN: Charles Babbage Institute, 1992).

45. Ibid.; also C. Gordon Bell and John E. McNamara, *High Tech Ventures: the Guide for Entrepreneurial Success* (Reading, MA: Addison-Wesley, 1991): 101; also Pake, "Research at Xerox PARC." Metcalfe was getting his Ph.D. from Harvard, but at the time he was recruited by PARC he had an ARPA-funded job at MIT.

46. J. C. R. Licklider, "Man-Computer Symbiosis," *IRE Transactions on Human Factors* 1 (March 1960): 4–11; Licklider and Taylor, "The Computer as a Communication Device," *Science and Technology* (April 1968).

47. Norberg and O'Neill, "A History of the Information Processing Techniques," 33–60.

48. Engelbart, in Adele Goldberg, ed., *A History of Personal Workstations* (Reading, MA: Addison-Wesley, 1988): 191.

49. William English, Douglas Engelbart, and Melvyn Berman, "Display-Selection Techniques for Text Manipulation," *IEEE Transactions on Human Factors in Electronics* 8 (March 1967): 5–15.

50. Douglas C. Engelbart and William English, "A Research Center for Augmenting Human Intellect," *Proceedings Fall JCC* 33-1 (1968): 395–410; also Goldberg, *History of Personal Workstations*, 202–206.

51. Douglas Smith and Robert Alexander, *Fumbling the Future: How Xerox Invented, Then Ignored, the First Personal Computer* (New York: William Morrow, 1988); Robert Metcalfe, "How Ethernet was Invented," *Annals of the History of Computing* 16: 4 (1994): 81–88; Tekla Perry and John Voelcker, "Of Mice and Menus: Designing the User-Friendly Interface," *IEEE Spectrum* (September 1989): 46–51.

52. Larry Press, "Before the Altair: the History of the Personal Computer," (1993): 27–33.

53. Goldberg, *History of Personal Workstations*, 265–289. Apparently Flip Wilson ad-libbed the phrase on an episode in 1969, while cross-dressed as his alter ego Geraldine Jones; see *Annals of the History of Computing* 17: 1 (1995), 5.

54. David Smith et al., "Designing the Star User Interface," *Byte* (April 1982): 242–282.

55. Phillip Ein-Dor, "Grosch's Law Re-revisited," *CACM* 28: 2 (1985): 142–151.

56. Peggy Kidwell and Paul Ceruzzi, *Landmarks in Digital Computing* (Washington, DC: Smithsonian Institution Press, 1994): 97.

57. Steven Manes and Paul Andrews, *Gates: How Microsoft's Mogul Reinvented an Industry, and Made Himself the Richest Man in America* (New York: Doubleday, 1993): 111. As this is being written (1997), Microsoft has agreed to invest a few hundred million dollars in Apple to rescue it.

58. Steven Wozniak, "The Apple II," *Byte* (May 1977); also interview of Wozniak by Gregg Williams and Rob Moore, "The Apple Story, Part 2," *Byte* (January 1985): 167–180.

59. Steven Wozniak, "The Making of an Engineer and a Computer," *Computer Museum Report* (Fall 1986): 3–8; also interview by Gregg Williams and Rob Moore, *Byte* (January 1985): 167–172.

60. Advertisement for Apple, *Byte* (July 1978): 14–15.

61. Steven Burke, "Visicalc Says Goodbye," *Infoworld* (June 24, 1985): 20–21; also Daniel Bricklin, "Visicalc and Software Arts: Genesis to Exodus," Computer Museum *Report* (Summer 1986): 8–10; also Susan Lammers, ed., *Programmers at Work* (Redmond, WA: Microsoft Press, 1989): 131–160.

62. Briklin, in Computer Museum *Report*, ibid., 9.

63. The IBM PC was not inherently restricted to addressing only 640 K of memory, but soon that became a de facto limit. It soon became the curse of the PC line of computers.

64. Jan Chposky, *Blue Magic: the People, Power and Politics Behind the IBM Personal Computer* (New York: Facts on File, 1988); also "Machine of the Year: the Computer Moves In," *Time* (January 3, 1983): cover, 14–37.

65. David Bradley, "The Creation of the IBM PC," *Byte* (September 1990): 414–420.

66. There are many variations of this story, including who chose the 8088 chip. In this brief summary I have relied on the account of Manes and Andrews in *Gates*, chapter 11.

67. There have been many charges that IBM appropriated the technology of small companies without giving what their creators felt was fair compensation. See, for example, An Wang's charge regarding his patent on core memory, in *Lessons* (Reading, MA: Addison-Wesley, 1986); and Erwin Tomash and Arnold Cohen's account of ERA's development of the drum memory, "The Birth of an ERA: Engineering Research Associates, Inc., 1946–1955," *Annals of the History of Computing* 1 (1979): 83–97.

68. Manes and Andrews, *Gates*, 160; also Tim Paterson, telephone interview with the author, 23 July 1996.

69. Interview with Paterson. A 1996 PBS television series, "Triumph of the Nerds," strongly insinuated that MD-DOS was "stolen" from CP/M, without offering any proof. See also G. Pascal Zachary, "Fatal Flaw," *Upside* (November 1994): 18–27.

70. For example, in CP/M the command PIP A:*.* B:*.* copied all the files on the second disk drive over to the first drive. To do that with MS-DOS one would write COPY B:*.* A:.

71. The above observations are based primarily on CP/M and MS-DOS manuals in the author's possession.

72. Quoted by Peggy Watt in *Infoworld* (Aug. 12, 1991): 48.

73. Tim Paterson, telephone interview with the author, 23 July 1996.

74. Bradley, "The Creation of the IBM PC," 420.

75. Chposky, *Blue Magic*, 180.

76. George Basalla, *The Evolution of Technology* (Cambridge, UK: Cambridge University Press, 1988).

77. Jef Raskin, letter to the editor, *Datamation* (August 1976): 8; also Raskin, "Holes in the Histories: Why the Annals of Apple have been Unreliable," MS privately circulated, 1994.

78. Raskin, interviewing Susan Lammers, ed., *Programmers at Work* (Redmond, WA: Microsoft Press, 1989): 227–245; also Ronald Baecker and William A. S. Buxton, *Readings in Human-Computer Interaction: a Multidisciplinary Approach* (Los Altos, CA: Morgan Kaufmann, 1987): 649–667.

79. See, for example, Steven Levy, *Insanely Great* (New York: Viking, 1994). Levy's book has so many factual errors that it cannot be relied upon, however; the most reliable account is Fred Guterl, "Design Case History: Apple's Macintosh," *IEEE Spectrum* (December 1984): 34–43.

80. Raskin, in Lammers, *Programmers at Work*, 230.

81. Indeed, one could not open the case without special tools. It was not long before third-party vendors began selling a "Mac Cracker" that combined the special "torx" screwdriver and prying tool need to open the case.

82. Tom Thompson, "The Macintosh at 10," *Byte* (February 1994): 47–54.

83. *Datamation* (June 1, 1985): 139–140.

84. This passage is based on a scanning of the issues of *Infoworld* during that period. The need to run Lotus 1-2-3 as a test of compatibility is said to have been the main reason that the 640 K memory barrier became so entrenched. 1-2-3 used memory addresses above 640 K for other functions, thus precluding that segment from ever being used for general storage.

Chapter 9

1. "Distributive Operating Multi-Access Interactive Network."

2. Mark Hall and John Barry, *Sunburst: the Ascent of Sun Microsystems* (Chicago: Contemporary Books, 1990): 60–61; also C. Gordon Bell and John E. McNamara, *High Tech Ventures: the Guide for Entrepreneurial Success* (Reading, MA: Addison-Wesley, 1991): 39–42, 323–325.

3. Stephen T. McClellan, *The Coming Computer Industry Shakeout: Winners, Losers, & Survivors* (New York: Wiley, 1984): 280–281.

4. Hall and Barry, *Sunburst*, chapter 1; Bell and McNamara, *High-Tech Ventures*, 325–326.

5. Peter Salus, *A Quarter Century of UNIX* (Reading, MA: Addison-Wesley, 1994); also D. M. Ritchie, "Unix Time-Sharing System: a Retrospective," *Bell System Technical Journal* 57 (1978): 1947–1969.

6. Salus, *A Quarter Century of UNIX*, 137–145.

7. See also "ed," "ln," "mv," and many others.

8. Donald A. Norman, "The Trouble with UNIX," *Datamation* (November 1981): 139–150.

9. Salus, *Quarter Century,* 137–142; 153–172.

10. Ibid., 153–172. Other accounts differ with Salus and state that Bolt Beranek and Newman, under an ARPA contract, was responsible for TCP/IP in UNIX.

11. Jamie Pearson, *Digital at Work* (Bedford, MA: Digital Press, 1992): 70–73; also C. Gordon Bell and John E. McNamara, *High Tech Ventures: the Guide for Entrepreneurial Success* (Reading, MA: Addison-Wesley, 1991): 37.

12. Glenn Rifkin and George Harrar, *The Ultimate Entrepreneur: the Story of Ken Olsen and Digital Equipment Corporation* (Chicago: Contemporary Books, 1988), chapters 25, 29, 30. The Rainbow was well-engineered and *almost* IBM compatible. But almost was not good enough—a fact that only a few realized at the outset, but which by 1982 was recognized by companies like Compaq as the only way to compete against IBM.

13. C. Gordon Bell, J. Craig Mudge, and John McNamara, *Computer Engineering: a DEC View of Hardware Systems Design* (Bedford, MA: Digital Press, 1978), chapter 17.

14. David A. Patterson, "Reduced Instruction Set Computers," *CACM* 28 (1985): 8–21.

15. John Markoff, "A Maverick Scientist Gets an I.B.M. Tribute," *New York Times,* 26 June 1990, D1; the "wild duck" memo is described by Herbert Grosch in *Computer: Bit Slices from a Life* (Novato, CA: Third Millenium Books, 1991): 258.

16. George Radin, "The 801 Minicomputer," *IBM J. Res. Dev.* 27 (May 1983): 237–246.

17. Patterson, "Reduced Instruction Set Computers," 16, 20; also John L. Hennessy and David Patterson, *Computer Architecture: a Quantitative Approach* (San Mateo, CA: Morgan Kaufmann, 1990).

18. R. Emmett Carlyle, "RISC-Y Business?" *Datamation* (February 15, 1985): 30–35.

19. John Hennessy and Norman Jouppi, "Computer Technology and Architecture: an Evolving Interaction," *IEEE Computer* 24 (1991): 18–29.

20. Hennessy and Patterson, *Computer Architecture,* 190; Hall and Barry, *Sunburst,* 163.

21. Hennessy and Patterson, *Computer Architecture*; Silicon Graphics Inc., Annual Reports for 1989–1993.

22. Grosch, *Computer,* 130–131.

23. Robert M. Metcalfe, "How Ethernet was Invented," *Annals of the History of Computing* 16: (1994): 81–88.

24. Metcalfe, "How Ethernet was Invented," 83.

25. R. Binder, N. Abramson, F. Kuo, A. Okinaka, and D. Wax, "ALOHA Packet Broadcasting: a Retrospect," in Siewiorek et al., *Computer Structures*, 416–428.

26. The term "ether" came from the "luminiferous aether" that physicists believed carried light, at least until Michaelson and Morley were unable to find evidence of its existence in their famous experiment of 1887. Physicists no longer believe in the existence of the ether, but computer scientists know it well.

27. Robert M. Metcalfe, and David R. Boggs, "Ethernet: Distributed Packet Switching for Local Computer Networks," in Siewiorek et al., *Computer Structures*, 429–438.

28. Metcalfe, "How Ethernet was Invented," 85.

29. C. Gordon Bell, in Adele Goldberg, ed., *A History of Personal Workstations* (New York: ACM Press, 1988): 19. The IBM network was a Token Ring system, a topology in which access to the channel was controlled by whichever computer held a "token," just as in the early days of U.S. railroads an engineer had to hold a unique token before he was allowed to take a train on a piece of unsignaled track, to prevent collisions.

30. One exception was Wall Street, where computer-savvy stock analysts developed sophisticated programs on SUN workstations to track price movements and recommend when to buy or sell a stock.

31. *Byte* (December 1984): 148.

32. As of this writing it appears that this equation may change. Linux, a free UNIX system that runs well on advanced Intel microprocessors, gives the owners of personal computers most of the power of an earlier generation of workstations. Likewise, Microsoft's operating system "Windows NT" is a direct competitor to UNIX-based workstations and also runs on personal computers. Perhaps the workstation may vanish as a class as a result.

33. The term "Packet Switching" probably originated with Donald Davies of the National Physical Laboratory in the U.K. See Martin Campbell-Kelly, "Data Communications at the National Physical Laboratory (1965–1975)," *Annals of the History of Computing* 9 (1988): 221–247. It may have been independently discovered by Paul Baran of the RAND Corporation at the same time. The RAND work was initially classified.

34. Janet Abbate, "From ARPANET to Internet: a History of ARPA-Sponsored Computer Networks, 1966–1988" (Ph.D. diss., University of Pennsylvania, 1994): 109.

35. For this section I am relying on the mainly unpublished work of Judy O'Neill, Janet Abbate, and Juan Rogers. I am grateful to them and to others who have shared their preliminary work with me. Some of the Internet's creators have

written "A Brief History of the Internet," which, not surprisingly, is available only on the Internet itself. They have published an abbreviated version in *CACM* (February 1997).

36. ARPANET was initially set up using a different protocol, NCP, but it was found to be ill-suited to connecting different networks to one another. ARPANET itself shifted to TCP/IP in January 1983.

37. Bob Metcalfe, "There Oughta be a Law," *New York Times*, 15 July 1996, C5.

38. Peter H. Salus, *Casting the Net: From ARPANET to INTERNET and Beyond* (Reading, MA: Addison-Wesley, 1995); chapters 5 and 9.

39. Most early modems worked at 300 Baud, which is not exactly the same as 300 bits per second but is in the same range.

40. Abbate, "From Arpanet to Internet"; also Ed Krol, *The Whole Internet Users' Guide and Catalog* (Sebastopol, CA: O'Reilly & Associates, 1992): 128–130.

41. Vannevar Bush, "As We May Think," *Atlantic Monthly*, 1945. The essay has been reprinted many times; for a discussion of its writing, publication, and early impact, see James Nyce and Paul Kahn, eds., *From Memex to Hypertext: Vannevar Bush and the Mind's Machine* (Boston: Academic Press, 1991).

42. Nelson, in *Dream Machines*, DM 44, 45.

43. Ibid., DM 19.

44. Engelbart's NLS (On-Line System) faded, but outliner programs later appeared for personal computers, where they have established a small but persistent niche. Examples include Thinktank, Lotus Agenda, and Ecco.

45. Academic work in Hypertext was summarized in a special issue of the *Communications of the ACM* 31 (July 1988).

46. By coincidence, one of the letters to the editor of the special issue of the *CACM* on hypertext, cited above, was by two program managers at ARPA, who discussed the impending dismantling of the ARPANET and the shifting of network activities elsewhere.

47. Tim Berners-Lee, "WWW: Past, Present, and Future," *IEEE Computer* 29 (October 1996): 69–77. Berners-Lee explicitly mentions Vannevar Bush, Doug Engelbart, and Ted Nelson as originators of the concepts that went into the Web.

48. Ibid., 70.

49. Ibid., 71.

50. Some of Andreesen's postings on the Internet have been preserved in an electronic archive. Since there is no way of knowing how long this material will be preserved, or whether it will remain accessible to scholars, I have not cited it here.

51. This chronology is mainly taken from an article in *Business Week*, July 15, 1996, 56–59.

52. IBM made profits even during the Great Depression, but it lost $2.8 billion in 1991, $5 billion in 1992, and $8 billion in 1993. It returned to profitability in 1995. DEC lost around $2 billion in 1994 and just barely started making money again in mid-1995. In early 1998, DEC was sold to Compaq.

53. Eric Weiss, "Eloge: AFIPS," *Annals of the History of Computing* 13: 1 (1991) 100.

Conclusion

1. Frederick I. Ordway, III, "*2001: A Space Odyssey* in Retrospect," in Eugene M. Emme, ed., *Science Fiction and Space Futures, Past and Present* (San Diego, CA: American Astronautical Association, 1982): 47–105. Ordway was a consultant to the film's director, Stanley Kubrick. The development of the character/computer HAL was the result of extensive consultations with IBM, Honeywell, RCA, General Electric, and other companies and technical experts. HAL seems to be physically much larger than on-board computers of the 1990s, but in its conversational user interface it is very close to what modern computer researchers hope to attain. For an assessment of how close we are to reproducing HAL, see David G. Stork, ed., *HAL's Legacy: 2001's Computer as Dream and Reality* (Cambridge: MIT Press, 1997).

2. E. J. Dijksterhuis, *The Mechanization of the World Picture* (Oxford: Clarendon Press, 1961).

3. Alan Turing, "On Computable Numbers, with an Application to the Entscheidungsproblem," *Proceedings London Mathematical Society, Series 2,* 42 (1936): 230–267.

4. I am indebted to Professor W. David Lewis of Auburn University for this concept.

5. *Electronics,* October 25, 1973; *Time,* January 3, 1983.

6. For example, that thesis is the basis for the Smithsonian's exhibition, "Information Age," which opened at the National Museum of American History in 1990.

7. See, for example, Clifford Stoll, *Silicon Snake Oil* (New York: Doubleday, 1995).

8. Bryan Pfaffenberger, "The Social Meaning of the Personal Computer, or Why the Personal Computer Revolution was no Revolution," *Anthropological Quarterly* 61 (January 1988): 39–47.

9. Thoreau's skepticism about technology was, of course, unusual. Recently I heard a historian assert that Thomas Jefferson would probably have been an enthusiastic proponent of modern computing and especially of the Internet (David K. Allison, "The Information Revolution in Jefferson's America," speech given at the University of Virginia for "Monticello Memoirs," May 30, 1996). The Library of Congress calls its Web site "Thomas" in Jefferson's honor.

Bibliography

Adams Associates. "Computer Characteristics Chart." *Datamation* (November/December 1960): 14–17.

Adams, Charles W., and J. H. Laning Jr. "The MIT Systems of Automatic Coding: Comprehensive, Summer Session, and Algebraic." *Symposium on Automatic Programming of Digital Computers* (1954): 40–68.

Aiken, Howard. "The Future of Automatic Computing Machinery." *NTF-4 Special Edition* (1956).

Allen, John. "The TX-0: Its Past and Present." *Computer Museum Report,* 8 (spring 1984): 2–11.

Allison, David K. "U.S. Navy Research and Development Since World War II." In *Military Enterprise and Technological Change,* edited by Merritt Roe Smith, 289–328. Cambridge: MIT Press, 1985.

Aspray, William. *John von Neumann and the Origins of Modern Computing.* Cambridge: MIT Press, 1990.

Aspray, William, ed. *Computing Before Computers.* Ames, Iowa: Iowa State University Press, 1990.

Aspray, William, ed. *Technological Competitiveness.* New York: IEEE Press, 1993.

Backus, John. "Programming in America in the 1950s—Some Personal Recollections." In *A History of Computing in the Twentieth Century,* 125–135. New York: Academic Press, 1980.

Baecker, Ronald, and William A. S. Buxton. *Readings in Human-Computer Interaction: A Multidisciplinary Approach.* Los Altos, CA: Morgan Kaufmann, 1987.

Barden, William Jr. *How to Buy and Use Minicomputers and Microcomputers.* Indianapolis, IN: Howard Sams, 1976.

Basalla, George. *"The Evolution of Technology."* Cambridge, UK: Cambridge University Press, 1988.

Bashe, Charles J., Lyle R. Johnson, John H. Palmer, and Emerson W. Pugh. *IBM's Early Computers.* Cambridge: MIT Press, 1986.

Bauer, F. L. "Between Zuse and Rutishauser: The Early Development of Digital Computing in Central Europe." In *A History of Computing in the Twentieth Century,* edited by J. Howlett, N. Metropolis, and Gian-Carlo Rota, 505–524. New York: Academic Press, 1980.

Bauer, Roy A., Emlio Collar, and Victor Tang. *The Silverlake Project: Transformation at IBM.* New York: Oxford University Press, 1992.

Baum, Claude. *The System Builders: The Story of SDC.* Santa Monica, CA: System Development Corporation, 1981.

Bell, C. Gordon. "The Mini and Micro Industries." *IEEE Computer* 17 (October 1984): 14–29.

Bell, C. Gordon, and Allen Newell. *Computer Structures: Readings and Examples.* New York: McGraw-Hill, 1971.

Bell, C. Gordon, J. Craig Mudge, and John McNamara. *Computer Engineering: A DEC View of Hardware Systems Design.* Bedford, MA: Digital Press, 1978.

Bell, C. Gordon, and John E. McNamara. *High Tech Ventures: The Guide for Entrepreneurial Success.* Reading, MA: Addison-Wesley, 1991.

Bell, Trudy E. "Patented 20 Years Later: The Microprocessor's True Father." *The Institute (IEEE)* 14 (November 1990): 1.

Beniger, James R. *The Control Revolution: Technological and Economic Origins of the Information Society.* Cambridge: Harvard University Press, 1986.

Bergstein, Harold. "An Interview with Eckert and Mauchly." *Datamation* (April 1962): 25–30.

Berkeley, Edmund. *Giant Brains, or Machines that Think.* New York: Wiley, 1949.

Bird, Peter. *LEO: the First Business Computer.* Wokingham, UK: Hasler Publishing, 1994.

Boehm, Barry. "Software Engineering." *IEEE Transactions on Computing* C25 (December 1976): 1226–1241.

Bolter, J. David. *Turing's Man: Western Culture in the Computer Age.* Chapel Hill: University of North Carolina Press, 1984.

Bradley, David. "The Creation of the IBM PC." *Byte* (September 1990): 414–420.

Brand, Stewart. "Spacewar: Fanatic Life and Symbolic Death Among the Computer Bums." *Rolling Stone* (7 December 1972), 50–58.

Braun, Ernest, and Stuart Macdonald. *Revolution in Miniature: The History and Impact of Semiconductor Electronics Re-explored in an Updated and Revised Second Edition.* 2nd ed. Cambridge, UK: Cambridge University Press, 1982.

Britcher, Robert N. "Cards, Couriers, and the Race to Correctness." *Journal of Systems and Software* 17 (1992): 281–284.

Brock, Gerald. *The U.S. Computer Industry: A Study in Market Power.* Cambridge: Ballinger, 1975.

Brooks, Frederick P. Jr. *The Mythical Man-Month: Essays on Software Engineering.* Reading, MA: Addison-Wesley, 1975.

Brooks, John. *The Go-Go Years.* New York: Dutton, 1984.

Brown, Kenneth A. *Inventors at Work; Interviews with 16 Notable American Inventors.* Redmond, WA: Tempus Books, 1988.

Brunetti, Cledo, and Roger W. Curtis. *Printed Circuit Techniques.* Circular #468. Washington, DC: National Bureau of Standards, 1947.

Buchholz, Werner. "Origin of the Word Byte." *Annals of the History of Computing* 3 (1981): 72.

Buchholz, Werner, ed. *Planning Computer System: Project Stretch.* New York: McGraw-Hill, 1962.

Burke, Steven. "Visicalc Says Goodbye." *Infoworld,* 24 June 1985, 20–21.

Burnham, David. *The Rise of the Computer State.* New York: Random House, 1983.

Caddes, Carolyn. *Portraits of Success: Impressions of Silicon Valley Pioneers.* Palo Alto, CA: Tioga, 1986.

Campbell-Kelly, Martin. "Programming the EDSAC: Early Programming Activity at the University of Cambridge." *Annals of the History of Computing* 2 (January 1980): 7–36.

Campbell-Kelly, Martin. *ICL: A Business and Technical History.* Oxford: Oxford University Press, 1989.

Carroll, Paul. *Big Blues: The Unmaking of IBM.* New York: Crown, 1994.

Ceruzzi, Paul. "Aspects of the History of European Computing." *Proceedings, 12th European Studies Conference,* Omaha, NB. 8–10 October 1987 (1987): 58–66.

Ceruzzi, Paul. *Beyond the Limits: Flight Enters the Computer Age.* Cambridge: MIT Press, 1989.

Chposky, Jan. *Blue Magic: The People, Power, and Politics Behind the IBM Personal Computer.* New York: Facts on File, 1988.

Cohen, Danny. "On Holy Wars and a Plea for Peace." *IEEE Computer* 14 (October 1981): 48–54.

Cohen, I. Bernard. *Revolution in Science.* Cambridge: Harvard University Press, Belknap Press, 1985.

Computers and Their Future. World Computer Pioneer Conference. Llandudno, UK: Richard Williams & Partners, 1970.

Cortada, James W. *Historical Dictionary of Data Processing—Technology.* Westport, CT: Greenwood Press, 1987.

Cress, Paul, Paul Dirkson, and J. Wesley Graham. *Fortran IV with WATFOR and WATFIV.* Englewood Cliffs, NJ: Prentice Hall, 1970.

Daly, R. P. "Integrated Data Processing with the UNIVAC File Computer." *Proceedings Western Joint Computer Conference* (1956): 95–98.

Darby, Edwin. *It All Adds Up: The Growth of Victor Comptometer Corporation.* [Chicago]: Victor Comptometer Corporation, 1968.

David, Paul. "Understanding the Economics of QWERTY: The Necessity of History." In *Economic History and the Modern Economist,* 30–49. Oxford: Blackwell, 1986.

Davis, E. M., W. E. Harding, R. S. Schwartz, and J. J. Corning. "Solid Logic Technology: Versatile, High Performance Microelectronics." *IBM Journal* 8 (April 1964): 102–114.

DeLamarter, Thomas. *Big Blue: IBM's Use and Abuse of Power.* New York: Dodd, Mead, 1986.

DeLanda, Manuel. *War in the Age of Intelligent Machines.* Cambridge: MIT Press, 1991.

Dennis, Michael A. *A Change of State: The Political Cultures of Technological Practice at the MIT Instrumentation Lab and the Johns Hopkins University Applied Physics Laboratory, 1930–1945.* Ph.D. diss., Johns Hopkins University, 1990.

Dickson, David. *The New Politics of Science.* Chicago: University of Chicago Press, 1984.

Digital Equipment Corporation. *PDP-11 Processor Handbook.* Maynard, MA: Digital Equipment Corporation, 1981.

Dijksterhuis, E. J. *The Mechanization of the World Picture.* Oxford: Clarendon Press, 1961.

Ditlea, Steve. *Digital Deli.* New York: Workman, 1984.

Dykstra, D. C. "IBM's Albatross: A History Lesson for Micro Makers." *Computerworld* 18 (10 December 1984): 134.

Eames, Charles and Ray, office of. Glen Fleck, ed. *A Computer Perspective: Background to the Computer Age.* Reprint, with an afterword by Brian Randell and foreword by I. Bernard Cohen. Cambridge: Harvard University Press, 1990.

Eckert, Wallace. *Punched Card Methods in Scientific Computation.* New York: Columbia University, The Thomas J. Watson Astronomical Computing Bureau, 1940.

Eckert, Wallace. "The IBM Pluggable Sequence Relay Calculator." *Mathematical Tables and Other Aids to Computation* 3 (1948): 149–161.

Ein-Dor, Phillip. "Grosch's Law Re-revisited." *CACM* 28 (February 1985): 142–151.

Eisler, Paul. *My Life with the Printed Circuit.* Edited with notes by Mari Williams. Bethlehem, PA: Lehigh University Press, 1989.

Engelbart, Douglas C. "A Research Center for Augmenting Human Intellect." *JCC* 33 (Fall 1968): 395–410.

Engineering Research Associates. *High Speed Computing Devices.* New York: McGraw-Hill, 1950.

English, William, Douglas Engelbart, and Melvyn Berman. "Display-Selection Techniques for Text Manipulation." *IEEE Transactions on Human Factors in Electronics* 8 (March 1967): 5–15.

Evans, Bob O. "SPREAD Report: The Origin of the IBM System/360 Project." *Annals of the History of Computing* 5 (January 1983): 4–44.

Evans, Bob O. "IBM System/360." *Computer Museum Report* (Summer 1984): 818.

Ferguson, Eugene S. *Engineering and the Mind's Eye.* Cambridge: MIT Press, 1992.

Fisher, Franklin, John J. McGowan, and Joen E. Greenwood. *Folded, Spindled, and Mutilated: Economic Analysis and U.S. v. IBM.* Cambridge: MIT Press, 1983.

Fisher, Franklin, James W. McKie, and Richard B. Mancke. *IBM and the U.S. Data Processing Industry.* New York: Praeger, 1983.

Fishman, Katherine Davis. *The Computer Establishment.* New York: Harper & Row, 1981.

Flamm, Kenneth. *Targeting the Computer: Government Support and International Competition.* Washington, DC: Brookings Institution, 1987.

Flamm, Kenneth. *Creating the Computer: Government, Industry, and High Technology.* Washington, D.C.: Brookings Institution, 1988.

Frank, Werner. "The History of Myth #1." *Datamation* (May 1983): 252–256.

Freiberger, Paul. *Fire in the Valley: The Making of the Personal Computer.* Berkeley, CA: McGraw-Hill, Osborne, 1984.

Friedel, Robert. *Zipper: An Exploration in Novelty.* New York: W.W. Norton, 1994.

Friedlander, Amy. *Natural Monopoly and Universal Service: Telephones and Telegraphs in the U.S. Communications Infrastructure, 1837–1940.* Reston, VA: Corporation for National Research Initiatives, 1995.

Frisch, Ivan T., and Howard Frank. "Computer Communications: How We Got Where We Are." *Proceedings NCC* 44 (1975): 109–117.

Gardner, W. David. "Route 128: Boston's Hotbed of Technology." *Datamation* (November 1981): 110–118.

Gardner, W. David. "Chip Off the Old Bloch." *Datamation* (June 1982): 241–242.

Gass, Saul I. "Project Mercury Real-time Computational and Data-flow System." *Proceedings Eastern JCC, Washington, DC* 20 (12–14 December 1961).

Geraghty, Jim. *CICS Concepts and Uses.* New York: McGraw-Hill, 1994.

Glass, Robert. *Computing Catastrophes.* Seattle: Computing Trends, 1983.

Goldberg, Adele, ed. *A History of Personal Workstations.* Reading, MA: Addison-Wesley, 1988.

Goldstine, Herman H. *The Computer from Pascal to von Neumann.* Princeton: Princeton University Press, 1972.

Goodman, Seymour E. "Soviet Computing and Technology Transfer: An Overview." *World Politics* 31 (July 1979): 539–570.

Gould, Inc. *A Young Company with Deep Roots: A History of Gould, Inc.* Rolling Meadows, IL: Gould [1984].

Grosch, Herb. *Computer: Bit Slices from a Life.* Novato, CA: Third Millenium Books, 1991.

Guterl, Fred. "Design Case History: Apple's Macintosh." *IEEE Spectrum* (December 1984): 34–43.

Hall, Eldon C. "The Apollo Guidance Computer: A Designer's View." *Computer Museum Report.* Boston: Computer Museum, 1982.

Hall, Eldon C. *Journey to the Moon: the History of the Apollo Guidance Computer.* Reston, VA: AIAA, 1996.

Hall, Mark, and John Barry. *Sunburst: The Ascent of Sun Microsystems.* Chicago: Contemporary Books, 1990.

Halstead, W. K., J. W. Leas, J. N. Marshall, and E. E. Minett. "Purpose and Application of the RCA BIZMAC System." *Proceedings Western Joint Computer Conference* (1956): 119–123.

Hanson, Dirk. *The New Alchemists: Silicon Valley and the Microelectronics Revolution.* Boston: Little, Brown, 1982.

Harvard University Computation Laboratory. *A Manual of Operation for the Automatic Sequence Controlled Calculator.* 1946, Reprint, with a foreword by I. Bernard Cohen and with an introduction by Paul E. Ceruzzi. Cambridge: MIT Press, Charles Babbage Institute Reprint series, vol. 8, 1985.

Harvard University Computation Laboratory. *Proceedings of a Symposium on Large-Scale Calculating Machinery.* Cambridge: Harvard University Press, 1948.

Hendrie, Gardner. "From the First 16-bit Mini to Fault-Tolerant Computers." *Computer Museum Report* 15 (spring 1986): 7.

Hennessy, John L., and David A. Patterson. *Computer Architecture: A Quantitative Approach.* San Mateo, CA: Morgan Kaufmann, 1990.

Hennessy, John L., and Norman Jouppi. "Computer Technology and Architecture: An Evolving Interaction." *IEEE Computer* 24 (September 1991): 18–29.

Heppenheimer, T. A. "How von Neumann Showed the Way." *American Heritage of Invention & Technology* (fall 1990): 8–16.

Hopper, Grace. "Log Book–ASCC." Cambridge, MA, 1944. Smithsonian Institution, National Museum of American History Archives, Grace Hopper Papers.

Hopper, Grace. *Compiling Routines.* Internal Memorandum. Philadelphia: Remington Rand, 1953. Smithsonian Institution, National Museum of American History Archives, Grace Hopper Papers.

Hopper, Grace. "Compiling Routines." *Computers and Automation* 2 (May 1953): 1–5.

Hord, Michael R. *The Illiac IV: The First Supercomputer.* Rockville, MD: Computer Science Press, 1982.

House, Robert W. "Reliability Experience on the OARAC." *Proceedings Eastern Joint Computer Conference* (1953): 43–44.

Hughes, Thomas Parke. *Networks of Power: Electrification in Western Society, 1880–1930.* Baltimore: Johns Hopkins University Press, 1983.

Hughes, Thomas Parke. *American Genesis: A Century of Invention and Technological Enthusiasm, 1870–1970.* New York: Penguin, 1989.

Hurd, Cuthbert C. "Early IBM Computers: Edited Testimony." *Annals of the History of Computing* 3 (April 1981): 163–182.

Johnson, L. R. *System Structure in Data, Programs, and Computers.* Englewood Cliffs, NJ: Prentice-Hall, 1970.

Johnson, Robert. "Interview (General Electric, ERMA)." *Annals of the History of Computing* 12: 2 (1990): 130–137.

Kemeny, John G. *Man and the Computer.* New York: Charles Scribner's Sons, 1972.

Kenney, Charles. *Riding the Runaway Horse: The Rise and Decline of Wang Laboratories.* Boston: Little, Brown, 1992.

Kidder, Tracy. *The Soul of a New Machine.* Boston: Little, Brown, 1981.

Kidwell, Peggy A., and Paul E. Ceruzzi. *Landmarks in Digital Computing: A Smithsonian Pictorial History.* Washington, DC: Smithsonian Press, 1994.

Kilburn, T., D. B. G. Edwards, M. J. Lanigan, and F. H. Sumner, "One-Level Storage System." *IRE Transactions on Electronic Computers*, EC-11 (April 1962): 223–235.

Kilby, Jack S. "Invention of the Integrated Circuit." *IEEE Transactions on Electronic Devices* 23 (July 1976): 648–654.

Kleiman, Herbert. "The Integrated Circuit: A Case Study of Product Innovation in the Electronics Industry." Ph.D. diss., George Washington University, 1966.

Knapp, Robert H., and and H. B. Goodrich. *The Origins of American Scientists*. 1952. Reprint, Chicago: University of Chicago Press, 1967.

Knuth, Donald. *The Art of Computer Programming*, 3 vols. Reading, MA: Addison-Wesley, 1968–1973.

Knuth, Donald. "Von Neumann's First Computer Program." *Computing Surveys* 2 (December 1970): 247–260.

Knuth, Donald. "The Early Development of Programming Languages." In *A History of Computing in the Twentieth Century, 197–273*, edited by N. Metropolis, J. Howlett, and Gian-Carlo Rota. New York: Academic Press, 1980.

Krol, Ed. *The Whole Internet Users' Guide and Catalog*. Sebastopol, CA: O'Reilly & Associates, 1982.

Kubie, Elmer. "Reflections of the First Software Company." *Annals of the History of Computing* 16: 2 (1994): 65–71.

Lammers, Susan, ed. *Programmers at Work*. Redmond, WA: Microsoft Press, 1986.

Lancaster, Don. *TTL Cookbook*. Indianapolis: Howard Sams, 1974.

Laning, J. Halcomb, and Neil Zierler. *A Program for Translation of Mathematical Equations for Whirlwind* I. Engineering Memorandum No. E-364. MIT. Cambridge, MA, 1954.

Lavington, S. H. *Early British Computers*. Bedford, MA: Digital Press, 1980.

Lavington, S. H. *History of Manchester Computers*. Manchester, UK: NCC Publications, 1975.

Levy, Steven. *Hackers: Heroes of the Computer Revolution*. Garden City, NY: Anchor Press/Doubleday, 1984.

Licklider, J. C. R. "Man-Computer Symbiosis." *IRE Transactions on Human Factors* 1 (March 1960): 4–11.

Licklider, J. C. R. "The Computer as a Communications Device." *Science and Technology* (April 1968).

Lukoff, Herman. *From Dits to Bits: A Personal History of the Electronic Computer*. Portland, OR: Robotics Press, 1979.

Lundstrom, David E. *A Few Good Men from UNIVAC.* Cambridge: MIT Press, 1987.

McClellan, Stephen T. *The Coming Computer Industry Shakeout: Winners, Losers, & Survivors.* New York: Wiley, 1984.

MacKenzie, C. E. *Coded Character Sets: History & Development.* Reading, MA: Addison-Wesley, 1980.

Mackenzie, Donald. *Inventing Accuracy: A Historical Sociology of Nuclear Missile Guidance.* Cambridge: MIT Press, 1990.

Mackenzie, Donald. "Negotiating Arithmetic, Constructing Proof: The Sociology of Mathematics and Information Technology." *Social Studies of Science* 23: 1 (1993): 37–65.

McMahon, A. Michal. "The Computer and the Complex: A Study of Technical Innovation in Postwar America." Manuscript, NASA History Office, Washington, DC, 1986.

"Machine of the Year: The Computer Moves In." *Time*, 3 January 1983, cover: 14–37.

Maddox, J. L., J. B. O'Toole, and S. Y. Wong. "The TRANSAC S-1000 Computer." *Proceedings Eastern Joint Computer Conference* (December 1956): 13–16.

Mahoney, Michael. "Software and the Assembly Line." (paper presented at workshop on technohistory of electrical information technology, Deutsches Museum, Munich, 1990).

Manes, Stephen, and Paul Andrews. *Gates: How Microsoft's Mogul Reinvented an Industry, and Made Himself the Richest Man in America.* New York: Doubleday, 1993.

Metropolis, N., J. Howlett, and Gian-Carlo Rota, eds. *A History of Computing in the Twentieth Century.* New York: Academic Press, 1980.

Mims, Forrest, III. "The Tenth Anniversary of the Altair 8800." *Computers and Electronics* (January 1985): 58–62, 81–82.

Misa, Thomas J. "Military Needs, Commercial Realities, and the Development of the Transistor, 1948–1958." In *Military Enterprise and Technological Change*, edited by Merritt Roe Smith, 253–287. Cambridge: MIT Press, 1985.

Monte, Tom, and Ilene Pritikin. *Pritikin: The Man Who Healed America's Heart.* Emmaus, PA: Rodale Press, 1988.

Moore, Gordon E. "Microprocessors and Integrated Electronics Technology." *Proceedings of the IEEE* 64 (June 1976): 837–841.

Moore School of Electrical Engineering. University of Pennsylvania. "Theory and Techniques for Design of Electronic Digital Computers; Lectures Given at the Moore School of Electrical Engineering, July 8–August 31, 1946." Philadelphia: University of Pennsylvania, 1947–1948; Reprint, Cambridge: MIT Press, 1985.

Morris, P. R. *A History of the World Semiconductor Industry.* IEE History of Technology Series 12. London: Peter Peregrinus, 1990.

Mounce, David C. *CICS: A Light Hearted Chronicle.* Hursley, UK: IBM, 1994.

Murray, Charles. *The Supermen: The Story of Seymour Cray and the Technical Wizards behind the Supercomputer.* New York: John Wiley, 1997.

Murray, Francis J. *Mathematical Machines.* Vol. 1, *Digital Computers.* New York: Columbia University Press 1961.

Myer, T. H., and I. E. Sutherland. "On the Design of Display Processors." *Communications of the ACM* 11 (June 1968): 410–414.

Naur, Peter, and Brian Randell. *Software Engineering; Report on a Conference Sponsored by the NATO Science Committee, 7–11 October, 1968.* Garmisch, Germany: NATO, 1969.

Nelson, Theodor. *Computer Lib.* South Bend, IN: Ted Nelson, 1974.

Noble, David F. *Forces of Production.* New York: Knopf, 1984.

Norberg, Arthur. "High Technology Calculation in the Early Twentieth Century: Punched Card Machinery in Business and Government." *Technology and Culture* 31 (October 1990): 753–779.

Norberg, Arthur, Judy O'Neill, and Kerry Freedman. *A History of the Information Processing Techniques Office of the Defense Advanced Research Projects Agency.* Privately printed typescript from CBI. Minneapolis, MN: Charles Babbage Institute, 1992.

Norman, Donald A. "The Trouble with UNIX." *Datamation* (November 1981): 139–150.

Noyce, Robert, and Marcian Hoff. "A History of Microprocessor Design at Intel." *IEEE Micro* 1 (February 1981): 8–22.

Noyes, T., and W. E. Dickenson. "Engineering Design of a Magnetic Disk Random Access Memory." *Proceedings, Western Joint Computer Conference* (7 February 1956): 42–44.

Nyce, James, and Paul Kahn, eds. *From Memex to Hypertext: Vannevar Bush and the Mind's Machine.* Boston: Academic Press, 1991.

O'Neill, Judy E. *The Evolution of Interactive Computing Through Time-Sharing and Networking.* Ph.D. diss., University of Minnesota, 1992.

Orchard-Hays, William. "The Evolution of Programming Systems." *IRE Proceedings* 49 (January 1961): 283–295.

Osborn, Roddy F. "GE and UNIVAC: Harnessing the High-Speed Computer." *Harvard Business Review* (July–August 1954): 99–107.

Padegs, A. "System/360 and Beyond." *IBM Journal of Research and Development* 25 (September 1981): 377–390.

Pake, George. "Research at XEROX PARC: A Founder's Assesment." *IEEE Spectrum* (October 1985): 54–75.

Patterson, David A. "Reduced Instruction Set Computers." *CACM* 28: 1 (1985): 8–21.

Pearson, Jamie. *Digital at Work.* Bedford, MA: Digital Press, 1992.

Petroski, Henry. *The Pencil: A History of Design and Circumstance.* New York: Knopf, 1990.

Pfaffenberger, Bryan. "The Social Meaning of the Personal Computer, or Why the Personal Computer Revolution Was No Revolution." *Anthropological Quarterly* 61 (January 1988): 39–47.

Phister, Montgomery Jr. "Quotron II: An Early Multiprogrammed Multiprocessor for the Communication of Stock Market Data." *Annals of the History of Computing* 11: 2 (1989): 109–126.

Pollack, S. *Studies in Computer Science.* Washington, DC: Mathematical Association of America, 1982.

Prasad, N. S. *IBM Mainframes: Architecture and Design.* New York: McGraw Hill, 1989.

Pugh, Emerson. *Memories that Shaped an Industry.* Cambridge: MIT Press, 1984.

Pugh, Emerson. *Building IBM: Shaping an Industry and Its Technology.* Cambridge: MIT Press, 1995.

Pugh, Emerson W., Lyle R. Johnson, and John H. Palmer. *IBM's 360 and Early 370 Systems.* Cambridge: MIT Press, 1991.

Randell, Brian, ed. *The Origins of Digital Computers: Selected Papers.* 2nd ed. Texts and Monographs in Computer Science. Berlin: Springer-Verlag, 1975.

Redmond, Kent C. and Thomas M. Smith, *Project Whirlwind: The History of a Pioneer Computer.* Bedford: Digital Press, 1980.

Reid, T. R. *The Chip: How Two Americans Invented the Microchip and Launched a Revolution.* New York: Simon & Schuster, 1985.

Rifkin, Glenn, and George Harrar. *The Ultimate Entrepreneur: The Story of Ken Olsen and Digital Equipment Corporation.* Chicago: Contemporary Books, 1988.

Ritchie, D. M. "Unix Time-sharing System: A Retrospective." *Bell System Technical Journal* 57 (July–August 1978): 1947–1969.

Roberts, H. Edward, and William Yates. "Exclusive! Altair 8800: the Most Powerful Minicomputer Project Ever Presented—Can be Built for Under $400." *Popular Electronics* (January 1975): cover, 33–38.

Robinson, L. P. "Model 30-201 Electronic Digital Computer." *Symposium on Commercially-available General-purpose Digital Computers of Moderate Price.* U.S. Navy, Office of Naval Research, Washington, DC (14 May 1952): 31–36.

Rodgers, William. *Think: A Biography of the Watsons and IBM.* New York: Stein & Day, 1969.

Rosen, Saul. "Electronic Computers: A Historical Survey." *Computing Surveys* 1 (March 1969): 7–36.

Rosen, Saul, ed., *Programming Systems and Languages.* New York: McGraw-Hill, 1967.

Rutishauser, Heinz. "Automatische Rechenplanfertigung bei Programmgesteuerten Rechenmaschinen." *ZAMP* 3 (1952): 312–313.

Rutishauser, Heinz. "Rechenplanfertigung bei programmgesteuerten Rechenmaschinen." *Mitteilungen aus dem Institut für Angewandte Mathematik der ETH* 3 (1952).

Salus, Peter. *A Quarter Century of UNIX.* Reading, MA: Addison-Wesley, 1994.

Salus, Peter. *Casting the Net: From ARPANET to INTERNET and Beyond...* Reading, MA: Addison-Wesley, 1995.

Sammet, Jean. *Programming Languages: History and Fundamentals.* Englewood Cliffs, NJ: Prentice-Hall, 1969.

Schneider, Ben Ross. *Travels in Computerland.* Reading, MA: Addison-Wesley, 1974.

Scott, Marilyn, and Robert Hoffman. "The Mercury Programming System." *Proceedings Eastern JCC* 20 (12–14 December 1961): 47–53.

Siewiorek, Daniel P., C. Gordon Bell, and Allen Newell. *Computer Structures: Principles and Examples.* New York: McGraw-Hill, 1982.

Simon, Herbert. *The Sciences of the Artificial.* 2nd ed. Cambridge: MIT Press, 1981.

Slater, Robert. *Portraits in Silicon.* Cambridge: MIT Press, 1992.

Smith, David C., Charles Irby, Ralph Kimball, and Bill Verplank. "Designing the Star User Interface." *Byte* (April 1982): 242–282.

Smith, Douglas, and Robert Alexander. *Fumbling the Future: How Xerox Invented, Then Ignored, the First Personal Computer.* New York: William Morrow, 1988.

Smith, Richard E. "A Historical Overview of Computer Architecture." *Annals of the History of Computing* 10: 4 (1989): 277–303.

Snively, George. "General Electric Enters the Computer Business." *Annals of the History of Computing* 10: 1 (1988): 74–78.

Snyder, Samuel S. "Influence of U.S. Cryptologic Organizations on the Digital Computer Industry." *Journal of Systems and Software* 1 (1979): 87–102.

Sobel, Robert. *IBM: Colossus in Transition.* New York: Bantam, 1981.

Stern, Nancy. *From ENIAC to UNIVAC: An Appraisal of the Eckert-Mauchly Computers.* Bedford, MA: Digital Press, 1981.

Stoll, Clifford. *Silicon Snake Oil.* New York: Doubleday, 1995.

Stockdale, Ian. "Vendors, International Standards Community Adopt Fortran 90." *NAS News (NASA Ames Research Center)* 7 (September 1992): 1–8.

Stork, David G., ed. *HAL's Legacy: 2001's Computer as Dream and Reality.* Cambridge: MIT Press, 1997.

Toffler, Alvin. *The Third Wave.* New York: Morrow, 1980.

Tomayko, James. *Computers in Spaceflight: The NASA Experience.* NASA Contractor Report 182505, Washington, DC: NASA, 1988.

Tomayko, James. "The NATO Software Engineering Conferences." *Annals of the History of Computing* 11: 2 (1989): 132–143.

Turing, Alan. "On Computable Numbers, with an Application to the Entscheidungsproblem." *Proceedings, London Mathematical Society* 2: 42 (1936): 230–267.

Turkle, Sherry. *The Second Self: Computers and the Human Spirit.* New York: Simon & Schuster, 1984.

"The TX-0: Its Past and Present." *Computer Museum Report* 8 (spring 1984): 2–11.

U.S. Air Force. *AESOP: A General Purpose Approach to Real-Time Direct Access Management Information Systems.* MTP-33; AD634371. Bedford, MA: MITRE Corporation, 1966.

U.S. Naval Mathematics Advisory Panel. *Symposium on Automatic Programming for Digital Computers.* Washington, DC: US Navy, 1954.

U.S. Office of Naval Research. *Symposium on Managerial Aspects of Digital Computer Installations.* Washington, DC, March 30, 1953.

Van Deusen, Edmund. "Electronics Goes Modern." *Fortune* (June 1955) 132–136, 148.

Veit, Stan. *Stan Veit's History of the Personal Computer.* Asheville, NC: WorldComm, 1993.

Wang, An. *Lessons.* Reading, MA: Addison-Wesley, 1986.

Warren, Jim. "Personal Computing: An Overview for Computer Professionals." *NCC Proceedings* 46 (1977): 493–498.

Watson, Thomas Jr. *Father, Son & Co.: My Life at IBM and Beyond.* New York: Bantam Books, 1990.

Weiss, Eric, ed. *Computer Usage: Fundamentals.* New York: Computer Usage Corporation. 1969.

Weizenbaum, Joseph. *Computer Power and Human Reason.* San Francisco: W. H. Freeman, 1976.

Wexelblat, Richard, ed. *History of Programming Languages.* New York: Academic Press, 1981.

Wilkes, Maurice. "The Best Way to Design an Automatic Calculating Machine" (paper presented at the Manchester Inaugural Conference, 1951). In *Computer Design Development: Principal Papers,* edited by Earl Swartzlander, Rochelle Park, NJ: Hayden Book Company, 1976): 266–270.

Wilkes, Maurice V. *Automatic Digital Computers.* London: Methuen, 1956.

Wilkes, Maurice V, David J. Wheeler, and Stanley Gill. *The Preparation of Programs for an Electronic Digital Computer.* Cambridge: Addison-Wesley, 1951.

Wolff, Michael F. "The Genesis of the IC." *IEEE Spectrum* (August 1976): 45–53.

Wolfe, Tom. "The Tinkerings of Robert Noyce." *Esquire* (December 1983): 346–374.

Wood, Lamont. "The Man Who Invented the PC." *American Heritage of Invention & Technology* (fall 1994): 64.

Woodbury, William. "The 603-405 Computer." *Proceedings of a Second Symposium on Calculating Machinery; Sept. 1949.* Cambridge: Harvard University Press, 1951.

Wozniak, Steve. "The Apple II." *Byte* (1977): 34–43.

Wozniak, Steve. "Interview." *Byte* (January 1985): 167–180.

Yates, JoAnne. *Control Through Communication: The Rise of System in American Management.* Baltimore: Johns Hopkins University Press, 1989.

Zuboff, Shoshanna. *In the Age of the Smart Machine.* New York: Basic Books, 1988.

Zuse, Konrad. *Planfertigungsgeräte.* Bonn: Gesellschaft für Mathematik und Datenverarbeitung. Zuse Archive, Folder #101/024, 1944.

Zuse, Konrad. "Der Programmator." *Zeitschrift für Angewandte Mathematik und Mechanik* 32 (1952): 246.

Index

URL (uniform resource locator), 302
Usenet, 298–299
User groups, 88, 215–216, 230, 264
U.S.S.R., computer developments in, 11
Utah, University of, 259, 285
Utopia, brought on through computers, 33, 237, 309–311

Vanguard, Project, 122
Veterans Administration, 137
VHSIC (very high speed integrated circuit), 172
Viatron, 252–254
  model 21, 253–254
Viet Nam War, 159, 258
Viola, software for Web, 303
Virtual memory, 245–246. *See also* DEC VAX; Ferranti Atlas; IBM System/ 370,
Virus, computer, 284
VisiCalc, 267–268, 276
VisiOn, graphical interface, 276
Visual Basic, 79. *See also* BASIC
VLSI. *See* Integrated circuits
VMS operating system, 246
von Braun, Wernher, 167
von Neumann architecture, 6, 23–24, 42, 178, 216. *See also* Architecture, computer
von Neumann, John, 21, 34, 44, 57, 59, 89

WAIS (Wide Area Information Service), 299–300
Wang Laboratories, 212, 254–257, 263
  LOCI calculator, 212
  Model 1200 word processor, 255
  Model 2200 computing calculator, 255
  WPS word processing system, 256–257
  Wang word processing, and personal computers, 293
Wang, An, 51, 212, 255
Waterloo University, computer science dept., 202–203
WATFOR, WATFIV, WATBOL, 203

Watson, Thomas, Jr., 44, 70, 288, 290
Watson, Thomas, Sr., 67
Wave-soldering, for assembling circuit boards, 193
Wayne State University, 43
West Coast computer design, 40
Western Electric, 65
Westinghouse, 181, 187
Wheeler Jump, 84
Wheeler, D. J., 84
Whirlwind, Project, 14, 140
  Whirlwind computer, 59, 86
Whole Earth Catalog, 207
WYSIWYG computer interface, 262
Whyte, William, 141
Wiedenhammer, James, 36
Wilkes, Maurice, 23, 148–149
Williams memory tubes, 34–35, 44–45
Williams, F. C., 34–35
WIMP interface, 261
Winchester disk storage, 200
Windows, graphical interface. *See* Microsoft
Wire-wrap, 72–73, 132, 179
Wooden Wheel, 43. *See also* IBM 650
Worcester *Telegram and Gazette*, computerized typesetting, 109
Word length, computer, 58, 125, 151. *See also* Architecture, computer
  minicomputer, 194
  PDP-10, 208
Word processing, for personal computers, 268
World Wide Web, 300–303
Wozniak, Steven, 264

X-Windows, 294
Xanadu, hypertext software, 301–302
Xerox Corporation, 167
  Alto, 261–262
  8010 Star Information System, 262–263
XDS 940 computer, 222. *See also* SDS 940
XDS division, 167
Xerox-PARC (Palo Alto Research Center), 257–261, 273–275, 291–292, 297